BIBLIOTECA DELL' «ARCHIVUM ROMANICUM»
Serie I: Storia, Letteratura, Paleografia

— 370 —

SUSANNA BARSELLA

IN THE LIGHT OF THE ANGELS

ANGELOLOGY AND COSMOLOGY IN DANTE'S *DIVINA COMMEDIA*

Leo S. Olschki Editore
MMX

Tutti i diritti riservati

Casa Editrice Leo S. Olschki
Viuzzo del Pozzetto, 8
50126 Firenze
www.olschki.it

Il presente volume è stato pubblicato
con il contributo della Fordham University Faculty of Arts and Sciences

ISBN 978 88 222 5974 5

TABLE OF CONTENTS

Acknowledgements	Pag.	VII
Introduction	»	IX
Chapter one – THE DEBATE ON ANGELS IN THE THIRTEENTH CENTURY: ISSUES RELEVANT TO DANTE'S *DIVINA COMMEDIA*	»	1
I. *Angelic Mediation and the Ideological Context of Dante's Angelology*	»	1
II. *The Debate on Angelology. Between Theology and Philosophy*	»	4
III. *Two Issues in the Angelology of the Thirteenth Century. Dante's Original approach*	»	9
a. The Creation of Angels and their Nature	»	9
b. The Contemplative and Active Functions of Angels	»	18
IV. Conclusions	»	25
Chapter two – DANTE AND CHRISTIAN ANGELOLOGY. LIMITS AND INFLUENCE OF THE PSEUDO-DIONYSIAN TRADITION	»	27
I. *The Interaction of Biblical and Pseudo-Dionysian Traditions in the Cosmological Context of the* Divina Commedia	»	28
a. Angelic Hierarchies and Cosmic Order. Dionysius, Aquinas, and Bonaventure	»	28
b. The Functional Interpretation of Angels. Saint Paul, Augustine, and Gregory the Great	»	32
II. *The Intellectual Light*	»	36
III. *The Angelic Hierarchy*	»	44
a. The Divine Taxis	»	46
b. The Angelic Episteme: Order and Science	»	51
c. Angelic *Energheia* and Imitative Theology	»	56
d. The Idea of Merit	»	60
e. The Divine Mirrors	»	64

TABLE OF CONTENTS

IV. *Conclusions*.	Pag.	68
Chapter three – MOVING THE STARS: ANGELS AND THE ORDER OF THE COSMOS	»	71
I. *Cosmology and Angelology*	»	72
a. The 'Book' of Stars	»	72
b. The Blessed Movers: Angels and Intelligences	»	77
c. Dante's Eschatological Cosmology: the Order of the Holy Spirit	»	84
d. Love, Knowledge, and Dante's Epic Journey. *Paradiso* I	»	91
e. The Desire of Motion. *Paradiso* I	»	97
f. Order and Ontology. *Paradiso* II	»	99
g. The Angels and the Mechanics of the Spheres	»	106
II. *Astrological Determinism and Free Will*	»	108
III. *Conclusions*	»	122
Chapter fur – OTHER ANGELIC OPERATIONS: ANGELS IN *PURGATORIO*, THE HEAVENLY MESSENGER, AND BEATRICE	»	125
I. *The Angels and the Blessed. A Premise*	»	126
II. *The Role of Angels in* Purgatorio. *Beatitudes and the Gifts of the Holy Spirit*	»	133
III. *The Heavenly Messenger* (Inferno, IX, *79-103*)	»	144
a. The Messo's Veil	»	148
b. The Poetic Impasse	»	150
c. The Virtuous Descent	»	154
d. The Angelic Mercury: Eloquence and Wands	»	157
IV. *Beatrice's Angelic Light*	»	163
a. The Exceptional Virtue of the *Gentilissima* Beatrice	»	165
b. Beatrice's Angelic Operations	»	170
c. The Analogical Model	»	176
V. *Conclusions*	»	181
Concluding Remarks	»	183
Works cited	»	187
Index	»	201

ACKNOWLEDGMENTS

My first thought of gratitude goes to my teachers at Johns Hopkins University, Salvatore Camporeale and Pier Massimo Forni, who taught me that reading a text is an act of philological and historical interpretation. The passionate conversations on angels, Aquinas, and Dante with 'Campo' many years ago were the first stimuli to this project. His memory lives in part in this book. Amilcare Jannucci, Guglielmo Gorni, and Robert Hollander introduced me to the study of Dante while still a graduate student and commented on early versions of specific parts of the book while still in its very first stage. I wish to add with gratitude Zygmunt Baranski to the list of these early commentators. A special thank goes to Giuseppe Mazzotta, for his invaluable mentorship in guiding me to become a Dante scholar, and to Marcello Ciccuto for his enlightening advise in the many discussions engaged over the years. I am grateful to the friends and colleagues who have contributed valuable criticism and suggestions at various stages on parts or on the entire manuscript: Giuseppe Mazzotta, Pier Massimo Forni, Marcello Ciccuto, Bruno Nacci, Angela Capodivacca, Susanne Hafner, Viviane Mahieux, Carey Kasten, Maria Enrico, who helped to make my English sound less Italian, and Marco Cangiano, patient husband and inflexible editor. I am grateful to the Dante Society of America for giving me the opportunity to present part of this book publicly and to Fordham University for granting me the fellowship that allowed me to complete this book. I dedicate it to my family.

Part of the section on «The Heavenly Messenger» appeared in an earlier version as *The Mercurial Integumentum of the Heavenly Messenger (Inferno IX, 79-103)*, «Letteratura Italiana Antica», IV, 2003, pp. 371-395. Unless otherwise indicated, translations are mine.

INTRODUCTION

This book is the first organic study of angelic operations in the *Divina Commedia*. By setting Dante's angelology within the context of the medieval theological and philosophical debate on angelic nature and functions, it analyses the structural and poetic role of angelic operations in Dante's cosmo-ethical system. More than a syncretism of classical and theological elements, the angelology of the *Commedia* represents a conscious synthesis of different traditions, coherently reflecting Dante's vision of the cosmos in which all dimensions – material and spiritual, philosophical and theological – reconcile in the providential design of salvation. Situated at the limits of material and immaterial worlds, angels are the nexus that ties together human and divine, thus assuring the equilibrium of nature and history, time and eternity, necessity and freedom.

Critics of the *Commedia* tend to see angels as part of the structural order of the poem, but to leave in the margins the effects of their operations when interpreting the poem and its major figures. My analysis shows that angels are the fulcrum on which the relation between creation and cosmology hinges, and that Dante's angelology is an essential element to the interpretation of the structure and the poetics of the *Commedia*. The importance Dante attributed to spiritual creatures in the order of the cosmos coherently reflects the intellectual historical reality in which the *Commedia* was conceived. Angels were so prominent in the theological, philosophical, and scientific debates of the twelfth and the thirteenth centuries that Thomas Aquinas and Bonaventure of Bagnoregio were known respectively as *Doctor angelicus* and *Doctor seraficus* for their fundamental contributions to medieval angelology. However, even without considering the historical context, the pervasive presence of angels in the poem provides evidence of the special place they occupy in Dante's vision of the ultramundane space. Fallen angels oversee divine justice in *Inferno*; anthropomorphic angels are guardians and guides of Purgatorial terraces; incorporeal Intelligences move the heavens in *Paradiso* and like the divine servants of Old Testament actively intervene in influencing the sublunar sphere.¹ All these figures are expressions of the poetic angelology of the poem.

Yet, angels in the *Commedia* have received limited critical attention. Most studies have examined the classical and theological sources of angelic figures, but none has investigated exhaustively the importance of angelic operations in the structure of poem.² In particular, no study has analyzed the innovative character of Dante's doctrine of angelic operations, investigated how the connection he established between contemplative and cosmological functions affects the relation between creation and cosmology, and argued for their relevance in the interpretation of key figures such as the heavenly messenger, the angels of *Purgatorio*, and Beatrice. This book takes on the challenge.

The adoption of a holistic approach that takes into account the interplay of science, philosophy, and theology, has allowed me for reading the poem as expression of its historical and ideological contexts. In this perspective, the

---

[1] The Fathers of the Church generally shared the notion that angels could be sent on mission to human beings. See ATTILIO MELLONE, *Angelo*, in *Enciclopedia Dantesca (ED)*, 6 vols., U. Bosco and G. Petrocchi eds., Roma, Istituto della Enciclopedia Italiana, 1970-78, pp. 268-271.

[2] Specifically dedicated to the angels in the *Divina Commedia* and their scriptural sources are CARLO ZANINI, *Gli angeli nella Divina Commedia. In relazione ad alcune fonti sacre*, Milano, Cogliati, 1908, and ROMANO GUARDINI, *L'Angelo nella Divina Commedia*, in his *Studi su Dante*, Brescia, Morcelliana, 1967, pp. 13-130. Guardini presents a brief excursus on angelic presences in the *Commedia*. See also VINCENZO DI GIOVANNI, *Gli angeli nella Divina Commedia*, in *Dante e il suo secolo*, Firenze, 1865, pp. 317-341. For a study of the influence of the Aristotelian theory of intelligences on Dante, see STEPHEN BEMROSE, *Dante's Angelic Intelligences*, Roma, Edizioni di Storia e Letteratura, 1983. Among the numerous readings on the cantos XXVIII and XXIX of the *Divina Commedia*, see among others ZYGMUNT BARANSKI, *Dante tra dei pagani e angeli cristiani*, «Filologia e critica», IX, 1984, pp. 298-299, ATTILIO MELLONE, *Il canto XXIX del Paradiso. Una lezione di angelologia*, «Nuove Letture Dantesche», VII, 1974, pp. 198-200. A study of Dante's angelological sources is in VIRGINIA JEWISS, *On Men and of Angels: The Poetic of Angelology in the Works of Dante Alighieri*, Ph.D Dissertation, Yale University, 1995. See also, ATTILIO MELLONE, *Angelo*, in *Enciclopedia Dantesca*, 6 vols., Roma, Istituto della Enciclopedia Italiana, 1970-1976, I, pp. 268-273. For an extensive treatment of the theme of angels in medieval thought see G. BAREILLE – J. MISKGIAN – J. PARISSOT – L. PETIT – A. VACANT, *Ange*, in *Dictionnaire de Théologie Catholique*, A. Vacant, E. Mangenot, E. Amann eds., Paris, Letouzey et Ané, 1899-1972, pp. 1189-1248; and L. PETIT – J. PARISOT – J. MISKGIAN – A.VACANT, *Angélologie*, in *Dictionnaire de Théologie Catholique*, pp. 1248-1271. MARCO BUSSAGLI, *Storia degli angeli*, Milano, Rusconi, 1995 presents a history of angels from antiquity to modern times illustrating the syncretism of mythological and Biblical traditions that characterizes angelologic tradition. For the relevance of angels in medieval philosophical thought see TIZIANA SUAREZ-NANI, *Les anges et la philosophie. Subjectivité et function cosmologique des substances séparées à la fin du XIIIe siècle*, Paris, Vrin («Études de Philosophie Médiévale», LXXXII), 2002; and of the same author, *Connaissance et language des anges selon Thomas d'Aquin et Gilles de Rome*, Paris, de Vrin, 2002 («Études de Philosophie Médiévale», LXXXV); and *Angels in Medieval Philosophical Inquiry. Their Function and Significance*, I. Iribarren and M. Lenz eds., Hampshire, Ashgate, 2008 («Ashgate Studies in Medieval Philosophy»). For the influence of Franciscan spirituality on medieval angelology see DAVID KECK, *Angels and Angelology in the Middle Ages*, Oxford, Oxford University Press, 1998. For an iconographic history of the representation of angelic choirs in the Middle Ages, see BARBARA BRUDERER EICHBERG, *Les neuf choeurs angéliques: origine et évolution du thème dans l'art du Moyen Age*, Poitiers, Centre d'Etudes Supérieures de Civilisation Médiévale, 1998.

INTRODUCTION

Commedia appears as a poetic *summa* that reflects the complexity of the issues related to the angelological debate of Dante's times, particularly those related to the definition of the nature and functions of the angels in the divine project of creation. One of my conclusions is that by defining angelic nature as fully actualized (as referred to the existence of angels) and by making angelic operations both philosophically and poetically necessary to the harmonization of the ontological and ethical orders of the *Commedia*, Dante found innovative solutions to these issues.

When considered from the point of view of angels, Dante's poem appears as a 'system', an elaborated artistic mechanism that articulates and reproduces in a poetic microcosm the blueprints of creation and its providential destiny. In this system, angels operate as illuminating instruments of divine Providence that regulate physical and spiritual movements in the cosmos. By illuminating and moving the heavenly spheres and assisting human beings in their path to God that winds through the search of wisdom and virtue, angels influence the universal dynamics of individual and historical events.

Two additional observations to clarify further the methodology followed in this book. First, while reconstructing Dante's complex network of sources, and examining the ones that could have been more likely accessible to the author, this book privileges the investigation of the intellectual issues the author faced in elaborating an angelic doctrine that would fit the eschatological cosmology of the *Commedia*. By reading the poem in this light, this work shows the complexity of Dante's intellectual and poetic challenges, and the originality of his solutions. Furthermore, and second, by privileging the point of view of the 'maker' the book looks at Dante's poetry as 'poiesis', relying on the assumption that poetry cannot be separated from the philosophical and theological elements that support and form the angelic structure of the *Commedia*. Rather, it maintains that the structural elements of the poem form a 'realistic' map of the world beyond the senses that obeys to supreme poetic necessity. Form and content, signified and signifier, are intimately connected and reciprocally illuminating. Structure allows for articulating expression in a coherent and systematic discourse, whereas poetic expression transforms the inexpressible reality of the divine into narrative experience.

Prior to describing the structure of the book, it is necessary to illustrate why angelology is relevant to the interpretation of the *Commedia*, and what is the insight we gain from reading it from this point of view.

Angelology provides us with a privileged perspective to look at the *Commedia* from the point of view of the author. It makes possible to perceive Dante's intent to reconcile not just the spiritual and material sides of universe, but also the traditions of thought that stem from Scriptural and classical wisdoms.

To this end, Dante needed to design a mechanism that connected the Empyrean to the material part of the universe. Angelic operations provided such link, but to shape angels as intermediaries between the spiritual universe of theology and the concrete world of physics involved a series of philosophical issues that Dante needed to address. The influence of mainstream medieval angelology does not fully explain the originality and complexity of Dante's soteriological vision of angelic operations. His adaptation and combination of Neo-Platonic and Aristotelian doctrines of angelic Intelligences need to be considered as Dante's attempt to shape angelic structure to suit his ambitious design of a cosmological angelology. To substantiate this reading, the book shows that the philosophical foundations of Dante's angelology, and the interlacing of Patristic, Dionysian, and Aristotelian traditions that characterize its composite nature serve the purpose of placing the cosmological and illuminative functions of angels at the core of the conceptual harmonization of creation and cosmology of the *Commedia*.

Angels are not only central to the cosmological design of the *Commedia*, but are also crucial to the poetic harmony of the three canticles. They constitute the exemplars upon which Dante modeled Beatrice's operations, and those of other angelic figures that intervene in the poem, ultimately casting light on Dante's own writing. By extending to the poem the metaphor of mirrors as applied to the angelic mode of illumination, one can say that Dante's poetry is similar to an angelic language of revelation that manifests – as do the angels and Beatrice – what is otherwise obscure. The *Commedia* thus emerges as an illuminated and illuminating 'angelic' operation intended to elicit in its readers the desire to undertake a cognitive voyage beyond the pillars of reason to pursue virtue in view of salvation.

Different aspects of medieval science and thought enrich the microcosm of the *Commedia*. Scholarship has shown the direct and indirect influence of Dante's contemporary thought on his vision of the cosmos, where many disciplines converge to form a coherent design. Most studies on Dante's angels have focused on their theological (Romano Guardini, Diego Sbacchi) or classical sources (Stephen Bemrose). These studies have identified the Pseudo-Dionysius the Areopagite and Aristotle as the principal sources of Dante's angelology. A few scholars, such as Carlo Zanini and Edmund G. Gardner, have tried to establish – but only with partial success – a structural correspondence between the hierarchical ordering of the angels and that of the heavens in *Paradiso*. Recent research has shown the relevance of metaphysics (from the seminal works by Bruno Nardi to the recent contribution by Christian Moevs), the strategic importance of astrology (Alison Cornish, Richard Kay), and the complexity of Dante's cosmology (Giorgio Stabile, Cesare Vasoli, Giuseppe Mazzotta).[3] These fields are all relevant to Dante's angelology. Over the years the contributions of eminent scholars have improved and deepened our perception of specific aspects of the cosmology illustrated in the *Commedia*. These studies constitute the necessary backdrop against which this book proposes its reading of angelic figures.

Dante's philosophical design of angelic functions in his eschatological cosmology is one of the most innovative aspects of his angelology. Not only it allows for the spatial connection between spiritual and material sides of the universe, intellectually and materially operated by the angels, but it also harmonizes the temporal dimensions implied in the *Commedia*. As chronocrators, angels govern both historical and sacred time, for they supervise the human voyage from the physical to the spiritual sphere as part of the universal journey that leads the dispersed multiplicity of creation back to union with the One. Angel's instrumental role as providential mediators, allows for embracing in a unitary vision both these temporal dimensions.

By regulating the dynamics of celestial movement, angels become instrumental causes of all processes of generation and corruption, and of all individual and historical events moving towards the final resolution of time into eternity at the Final Judgment. Dante's individual voyage reflects, as in an 'angelic' mirror, the universal return of creatures to God and exemplarily embodies the dialectics between eternal and contingent that marks the ideological architecture of the *Commedia*, which revolves around the intellectual poles of Christian eschatology and Aristotelian causality. This architecture relies on the place angels occupy in the liminal space where material and spiritual worlds, sacred and natural history, converge and find expression in a language where the beauty of the contingent embeds the content of eternity.

The numerous passages dedicated to the exposition of different aspects of Dante's angelic doctrine evidence his effort to provide philosophical ground to angelic operations in the *Commedia*. As wise teacher and initiator of Dante to the divine truths of creation, Beatrice explains in *Paradiso* VII, 130-132 and XXIX, 31-45 the doctrine of the nature and creation of the angels; in *Paradiso* XXVIII, 98-126 the ordering of angelic hierarchies; and in *Paradiso* I, 103-126 and II, 112-141, she illustrates the link between the angelic transmission of light and cosmological order. To complete Beatrice's theological expositions, Aquinas presents a synthesis of the doctrine of angels as mirrors of divine goodness in *Paradiso* XIII, 58-60, and Carlo Martello (Charles Martel) illustrates in *Paradiso* VIII, 97-105 and 123-135 the relation between angelic hierarchies, cosmol-

---

[3] Specific bibliography is given in the footnotes of the chapters.

ogy, and astrological influence within the context of the co-existence of free will and predestination. These themes constitute the topics discussed in the four chapters of this book.

Following this introduction, the first chapter sets Dante's angelology in its historical context. It analyzes the philosophical and theological issues that emerged in the twelfth and thirteen-century debates on angels and that are relevant to the design of the *Commedia*. In particular, this chapter focuses on Dante's original response to two highly debated themes concerning the definition of angelic nature and the notion of angelic operations. In this chapter, I argue that, far from assuming an Averroist 'divinization' of the Intelligences, Dante's vision of angelic nature was in reality compatible with Aquinas's distinction between act and potency in the angels. Related to the definition of angelic nature was that of angelic functions. Unlike mainstream angelology, Dante's relied on the strict relation he established between speculative and active operations. This chapter shows that this relation is the philosophical keystone of Dante's angelology, for it is on this basis that he could assign to angels their role as instruments of providential order and eliminate the hiatus between physical and metaphysical domains.

Chapter II addresses the issue of Dante's reinterpretation of the *De coelesti hierarchia* (*Celestial Hierarchy*), the most influential text on angelology authored by the Pseudo-Dionysius the Areopagite in the fifth century A.D. While most studies have concentrated on the elements Dante derived from Pseudo-Dionysian angelology, in this chapter I discuss the limits of its influence and show that, while providing a viable model of cosmic harmonization, the Dionysian system did not offer Dante a complete theory of angelic operations that defined the modes in which angels influence the actions of human beings. Thus, while relying on a theurgical structure derived from the Pseudo-Dionysius, Dante adapted it in the light of the angelological doctrines of his time, particularly Aquinas and Bonaventure's, to reconcile the theological and philosophical issues posed by the cosmological function he attributed to angels. Equally relevant to the structure of Dante's angelology was the accent the Dionysian interpreters, and particularly Gregory the Great, placed on angelic ministerial functions, which had its roots in Biblical exegesis, from Saint Paul to Augustine. The chapter illustrates that these various streams of Dionysian interpretations merged and interacted in Dante's definition of hierarchies, and formed a synergetic synthesis with the cosmological attributes of Aristotelian Intelligences.

Against the conceptual backdrop explored in the first two chapters, chapter III discusses the angels' function as heavenly movers and investigates the relation between cosmology and creation from an angelological point of view.

INTRODUCTION

The chapter sets Dante's knowledge of contemporary scientific sources within the context of medieval rediscovery of the Arab and Greek texts on astronomy and astrology. Starting from the eleventh century, these texts changed forever the Western medieval vision of cosmos and dramatically influenced theological thought. Directly and indirectly known by Dante, these texts constitute the essential background to analyze the connections he established among angels, cosmology, and astrology. Based on the analysis of the notions of order, desire, and inclinations, the chapter shows the existence of three hierarchically connected principles of cosmological order: the Holy Spirit, as inspiring divine love; Providence, as its governing principle, and angels, as instruments of Providence who actualize divine dispositions. Moved by their desire to become as much as possible in God's likeness, angels move the celestial spheres by reflecting onto them the divine light they contemplate, thus infusing their virtues in stars and planets and indirectly influencing human inclinations. In performing their operations as movers, angels are thus necessary links in the order of creation and assure equilibrium between free will and predestination.

The final chapter analyses other angelic operations that do not involve the motion of the heavenly spheres, but are connected to the governing functions of spiritual substances. It shows that by guiding human beings in their journey to beatitude through acquisition of virtue and cooperation with grace, angels fulfill their task as divine ministries of cosmic order. After exploring the analogy Dante established between the operations of angelic hierarchies and those of the Blessed in section I, section II investigates the presence of angels and angelic figures in *Purgatorio*. It argues that angels on the purgatorial terraces transmit to the atoning souls (and to the poet-pilgrim) the knowledge of the seven gifts of the Holy Spirit necessary to complete their purification process. The third section shows Dante's intentional syncretism of pagan and Christian motives in designing the figure of the heavenly messenger (*messo celeste*) that intervenes to resolve the textual and hermeneutical impasse before the gates of Dis in *Inferno* IX. The fourth section proposes a new interpretation of Beatrice as analogical model of the angel. In the light of the analogy Dante established among the angels, Mary and Christ, and between angels and blessed souls, he attributed to Beatrice the power to perform angelic operations. Her illuminating action in guiding the pilgrim's transhumanization through the heavens, suggests a substantial 'angelization' of Beatrice, in accordance with the general design of Dante's angelology and the ethical purpose of his poetic invention.

To conclude, the figures who assist the pilgrim's itinerary in the otherworld exemplify the different modes in which Dante articulates the necessity of supernatural intervention in the universal journey of the soul to God. They are examples of the necessity of angelic ministries not only in the order of the cosmos but also in the poetic order of the *Commedia* and testify to the new insight angelology casts on the interpretation of the poem.

CHAPTER ONE

THE DEBATE ON ANGELS IN THE THIRTEENTH CENTURY: ISSUES RELEVANT TO DANTE'S *DIVINA COMMEDIA*

This chapter presents Dante's angelology within the context of the philosophical debates on angelic nature and functions as related to the controversial question of the angels' mediation between divine and human worlds. The first two sections introduce the main sources, themes, and issues of the debate on angelology in the twelfth and thirteenth centuries, and concentrate on the aspects relevant to Dante's theory of angelic Intelligences. In the third section, the first paragraph discusses the creation and nature of angels as presented in *Paradiso* XXIX. The second paragraph addresses Dante's original elaboration of the doctrine of angelic operations.

I. ANGELIC MEDIATION AND THE IDEOLOGICAL CONTEXT OF DANTE'S ANGELOLOGY

Dante's angelic doctrine developed within a complex philosophical and theological context, in which different visions of the role and nature of spiritual substances existed. Common to theologians of different schools, from the Victorines to the Scholastics, was however the importance they attributed to angels in order to make philosophical truths compatible with the Sacred Scriptures. By occupying an intermediate position between divine and human spheres, angels represented a necessary link in the chain of beings, and were essential to explain the connection between the two. How their being, existence, knowledge, language, and functions should be defined was object of debate. Dante's interest in these topics emerges in various works, from the *De vulgari eloquentia* (*On Vernacular Language*), to the *Convivio* (*The Banquet*) and the *Monarchia* (*The Monarchy*). It is however in the *Commedia* that angels assume a structural role in his vision of the Christian cosmos as instrumental intermediaries between spiritual and material worlds.

CHAPTER ONE

Various texts and traditions concurred to form its poetic design. Among the most important were Aristotle's Intelligences; the medieval Christianized mythology; Scriptural angelology; the Neo-Platonic Greek tradition; the angelic doctrine of Gregory the Great (*Moralia in Job* XXXII, 48, and *Homily* XXXIV), Thomas Aquinas (particularly *Summa Theologiae* I^a qq. 50-64; 106-114), Bonaventure (especially *Collationes in Hexaëmeron*, XXI-XXIII); the mystic tradition of Bernard of Clairvaux (*De considerazione* V); and, among the literary sources, Brunetto Latini (*Li Livres dou Tresor*, I, xii, 5).[1] All these medieval authors referred – with variations- to the most authoritative source of medieval angelology, the Pseudo – Dionysius the Areopagite, and his treatise on angels, *De coelesti hierarchia* (*On Celestial Hierarchy*).[2] Crucial to a full understanding of Dante's knowledge and adaptation of these sources to his poetic needs was the mediation of the great commentaries on Dionysius's *De coelesti hierarchia*, such as those written by Hugh of Saint Victor, and by Aquinas's teacher, Albert the Great.

Against the backdrop of medieval angelological doctrines, Dante created a system in which the angels' hierarchical communication of divine science and their cosmological role as movers of the celestial spheres can be seen as a response to the theological and philosophical issue of the mediation between divine and sensible spheres. As this and the following chapters will show, Dante's angelic Intelligences, from Fortuna in *Inferno* VII to the heavenly choirs incessantly moving the material spheres of heaven in the Empyrean,

---

[1] The ordering of the angelic hierarchies Dante presented in *Convivio* II, vi, was the same Brunetto Latini and Isidore of Seville adopted, which was in turn based on Gregory the Great's *Moralia in Job*. For the main doctrines on the ordering of the hierarchies different or derived from the Pseudo-Dionysius the Areopagite, see C.A. PATRIDES, *Renaissance Thought on the Celestial Hierarchy: The Decline of a Tradition*, «Journal of the History of Ideas», XX, vol. 2, 1959, pp. 159-162. Hereafter now on I will refer to the Pseudo-Dionysius the Areopagite as 'Dionysius' or 'Areopagite'.

[2] Until the Middle Ages, the Pseudo-Dionysius the Areopagite was identified with the judge of the Areopagus that Saint Paul converted to Christianity in Athens (*Acts*, XVII: 34). According to Dionysius of Corinth (Eusebius, *Historia Ecclesiastica*, III, iv) Dionysius would have been the first Bishop of Athens. The references to late ecclesiastical documents, the language, and the evident influence of Plotinus and Proclus' Neo-Platonism, led scholars at the end of the nineteenth century to suggest that he was probably a disciple of Proclus, possibly writing in the second half of the fifth century. Although many hypotheses have been proposed, the real identity of the author of the works attributed to the Areopagite (*Corpus Dionysiacum*) remains unknown. In the Middle Ages, another version of Dionysius's life presented him as a martyr in Paris. In the ninth century, Hilduin, the first translator of the *Corpus Dionysiacum*, authored the *Incipit passio sancto Dionysii*, in which he connected the two legends and made Dionysius the first bishop of Paris and founder of his abbey at Saint Denis. The legend of the apostolic authority of the Areopagite spread in the Middle Ages and was commonly accepted as true. Dante endorsed the medieval tradition, and affirmed that Saint Paul revealed to Dionysius the ordering of the separate substances in *Par.* XXVIII, 138-139: «ché chi 'l vide, qua su liel discoperse / con altro assai del ver di questi giri», [«it was disclosed / to him by one who saw it here above – / both that and other truths about these circles»].

were the spiritual links connecting divine and material worlds, analogously but subordinately participating in Christ's work of mediation.

In structuring his visionary architecture, not only Dante combined the prevalent Scholastic angelologies of the thirteen-century, but he also assimilated the approach of the Fathers, who considered the mediation of intellectual substances crucial to the enactment of Christ's works of mediation and redemption. Indeed, Incarnation and Christological doctrines strongly mark Dante's poem. It is crucial to set the angelology of the *Commedia* in this perspective, for by conveying divine enlightenment angels act as providential instruments within a salvific perspective. By making the invisible visible, they inspire in humankind the desire to imitate Christ, and this imitation is the source and the way to salvation. The angels and their operations – visible signs of their governing functions – are thus necessary tesserae in the organic mosaic of a universal Trinitarian order, and as such are essential to the poetic structure of the poem.

The attribution of cosmological governing functions to angelic hierarchies raised a delicate issue, related to the more general problem of the role of angels as active mediators between celestial and human worlds. The harmonization of the Old Testament tradition, in which angels were essentially divine intermediaries, with Christ's unique power of mediation was a crucial and controversial topic underlying the theological and philosophical debates on angelology since the time of the Fathers of the Church. Aquinas was among the theologians who firmly warned about the risk of interpreting the angels as mediators of grace. He was aware that the assimilation of Dionysian hierarchies into a systematic vision of the cosmos could conflict with the assumption that Christ was the unique mediator of divine grace. As we will see in the following chapter, the Areopagite's doctrine of angels subliminally maintained the ancient role of mediation they held in the Old Testament. Most medieval theologians who adopted his hierarchical structure, such as Aquinas and Bonaventure, tried to 'correct' his system by stressing the angels' subordinate role in Christ's redemptive work.

Facing the same risk of attributing too much autonomy to angels as mediators, Dante needed to reformulate their functions as purely instrumental to Christ's work of mediation in which they participated as executors of divine will. Similar to Bernard of Clairvaux's theory of Marian intercession, the angelic structure of the *Commedia* functions like an 'acquaeductus' that conveys the flow of light descending from the Empyrean down to earth through the hierarchies. The angels' mirroring function as divine *specula* (mirrors) is inscribed in Dante within a theological framework of analogical mediation of grace and divine wisdom, so that angelic doctrine results shaped by and centered on Incarnation. The mediation angels operate can be illustrated in terms of attributive analogy between the angels and the enlightening power of the Word. Chapter IV of this book will explore the terms of this analogy. The following sections discuss the issues related to the definition of angelic nature and operations, and show that the solutions Dante adopted were aimed at providing a philosophical basis for the role of instrumental mediation he attributed to angels in the *Commedia*.

II. The Debate on Angelology. Between Theology and Philosophy

The relevance of angels in the *Commedia* reflects the preeminent role of angelology in the theology of the twelfth and the thirteenth centuries. Desiring to reconcile the exegesis of creation with logical and metaphysical premises, theologians stressed the role of angels as necessary linchpin in the order of creation. In their views, angelic ministries were instrumental to the communication between celestial and sublunary worlds, thus ensuring the harmony and perfection of creation.

The *De coelesti hierarchia* was the first systematic treatise on angels, authored in the fifth century A.D. by a possible disciple of Proclus known as the Pseudo-Dionysius the Areopagite. The author claimed to be a converted disciple of Saint Paul and to have received revelation on angelic hierarchies directly from the apostle.[3] His claim remained unchallenged until the fifteenth century, when Lorenzo Valla disputed his apostolic authority, and dated his works to the fifth century A.D.[4] Prior to Dionysius's systematic

---

[3] See Marta Cristiani, *Dionigi Areopagita (Pseudo)*, in *ED* II, pp. 460-462. For the influence of the Areopagite on Western thought, see Hyacinthe François Dondaine et al., *Influence du Pseudo-Denis en Occidént*, in *Dictionnaire de Spiritualité*, Paris, Beauchesne, 1957, III, pp. 318-378. See also Edmund G. Gardner, *Dante and the Mystics; a Study of the Mystical Aspect of the Divina Commedia and its Relations with Some of its Medieval Sources*, New York, Octagon, 1968 (1913), pp. 77-110, Id., *Dante's Ten Heavens, A Study of the Paradiso*, New York, Haskell, 1970 (1898), pp. 195-214; and C.A. Patrides, *Renaissance Thought*, pp. 155-166. For an account of angelological doctrines and the revival of Dionysian studies in early scholastic authors, see Marcia L. Colish, *Early Scholastic Angelology*, «Recherches de Théologie ancienne et médiévale», LXII, 1995, pp. 80-109. For the possible influence on Dante of Dionysius and of the mystic tradition derived from Dionysius, particularly through the medium of Albert the Great's commentary on the *De coelesti hierarchia*, see Diego Sbacchi, *La presenza di Dionigi l'Aeropagita nel «Paradiso» di Dante*, Firenze, Olschki («Biblioteca di "Lettere Italiane". Studi e Testi», LXVI), 2006, pp. 1-20; Marco Ariani, *"E sí come di lei bevve la gronda / de le palpebre mie" (Par I. XXX. 8): Dante e lo Pseudo-Dionigi Areopagita*, in *Leggere Dante*, Lucia Battaglia Ricci ed., Ravenna, Longo, 2003, pp. 131-152. See also Molly Morrison, *Looking at God: Imagery for the Divinity in Dante's Paradiso*, «Forum Italicum», XXXV, 2, 2001, pp. 307-317 and in particular for the figure of Beatrice, see Antonio Rossini, *Il Dante sapienziale. Dionigi e la bellezza di Beatrice*, Pisa, Serra, 2009, pp. 59-73.

[4] The first one to defend the debated authenticity of Dionysius's works was John of Scythopolis

treatment of the *De coelesti hierarchia*, the writings of the Early Fathers – who were more interested in the modes in which specific angels interacted with human beings – focused on allegorical interpretations of angelic presences in Biblical and Neo-Testamentary sources. A brief excursus on the transmission and interpretations of the Dionysian *Corpus* is a necessary premise to understand fully its influence on Dante's angelology.

The Dionysian *Corpus* of writings, composed by four surviving treatises on *De divinis nominibus* (*The Divine Names*), *De mystica theologia* (*The Mystical Theology*), *De coelesti hierarchia, De ecclesiastica hierarchia* (*On the Ecclesiastical Hierarchy*) and ten *Letters*, arrived in the West relatively late, as a gift of the Byzantine Emperor Michael the Stammerer to King Louis the Pious at Compiegne, in the first half of the ninth century.[5] Before this date, Eastern commentators such as John of Scythopolis and Maximus the Confessor engaged in an effort to clear the Dionysian writings from all possible traces of the heresies of Monophysitism and Donatism; a major concern of the orthodox readers of the sixth and the seventh centuries.[6] Their purging efforts facilitated

---

(Scholia, ca. 540), supported by Maximus the Confessor. The Lateran Council eventually accepted the authenticity as demonstrated in 649. The identification of the Neo-Platonic author of the *Corpus Dionysiacum* with Dionysius was first disputed by Lorenzo Valla and Nicholas of Cusa, followed by Erasmus and Tommaso de Vio (Cajetan). Lorenzo Valla challenged the authenticity of the *Dionysian Corpus* in his *Collatio Novi Testament*. See LORENZO VALLA, *Opera*, E. Garin ed., Torino, Bottega d'Erasmo, 1962, vol. 1, p. 852 b., republished and commented by SALVATORE CAMPOREALE in his *Lorenzo Valla. Umanesimo e Teologia*, Firenze, Istituto Nazionale di Studi sul Rinascimento, 1972, pp. 428-430.

[5] In the ninth century, John Scotus Eriugena's translation into Latin (a few years after Hilduin of Saint Denis's questioned translation) begun the series of the numerous and authoritative commentaries that testify to the vast echo Dionysius's works had in medieval thought. Among these, of particular relevance to estimate the presence of Dionysian elements in the *Commedia* are the commentaries by Hugh of Saint Victor, Peter Lombard, Robert Grosseteste, Vincent of Beauvais, Bonaventure, Thomas Aquinas, and the most extensive of all by Albert the Great. For a detailed account of these translations and commentaries, see C.A. PATRIDES, *Renaissance Thought*, pp. 156-161. For a short history of the transmission of the text see JEAN LECLERCQ, *Influence and Noninfluence of Dionysius in the Western Middle Ages*, in *Pseudo-Dionysius. The Complete Works*, Colm Luibheid, Paul Rorem, Rene Roques eds., Introductions by Jaroslav Pelikan, Jean Leclercq, Karlfried Froehlich, New York, Paulist Press, 1987, pp. 25-32. For the medieval knowledge and transmission of the *Corpus Dionysiacum*, see HYACINTHE F. DONDAINE, *Le Corpus Dionysien de l'Université de Paris au XIIIe siècle*, Roma, Edizioni Storia e Letteratura, 1953. See also MARCO ARIANI, "*E sí come di lei bevve la gronda*", pp. 132-136.

[6] Monophysitism was a doctrine that maintained that Incarnated Christ had only one nature, namely human, which eventually evolved in divine. Its opposite, Nestorianism, argued that Christ had only divine nature. Both were condemned by the Third Council of Costantinople in 681. The Donatists, named after the bishop of Carthage Donatus Magnus, where a schismatic body of the Northern African Church. They maintained the invalidity of the sacraments administered by the bishops that had surrendered the Scriptures under Diocletian's persecution. This implied that the validity of sacraments depended on the dignity of those who administered them. In 409, Emperor Honorius decreed the Donatists heretical. For the reception of the Pseudo-Dionysius in the East, see JAROSLAV PELIKAN, *The Odyssey of Dionysian Spirituality*, in *The Complete Works* (pp. 11-24).

and prepared the reception of the *Corpus* in the catholic West. In the second half of the ninth century, Hilduin, abbot of Saint Denis (Paris), completed the first translation of the complete works of the Areopagite. Judging his translation difficult and obscure, in 862 Charles the Bald commissioned a new translation to John Scotus Eriugena. This new translation, accompanied by a commentary, was revised by the papal librarian Anastasius in 875, and became the main channel of transmission of the Dionysian texts in the Middle Ages.

Since the beginning of the diffusion of the Dionysian *Corpus* in the West, two main lines of interpretation emerged and later converged. On the one side, theologians incorporated and developed the mystical elements of the Dionysian system, focusing on the anagogical process of return to unity with God; the dialectical process of affirmative/negative theology; and the symbolic nature of reality arranged in hierarchical theophanies. On the other side, the systematic approach displayed in Dionysius's writings stimulated the philosophical speculation of Scholastic theologians. The *Commedia* presents a synthesis of these two interpretations. It created a poetic bridge across mystical and philosophical components to structure an angelological system in which Dionysian and Aristotelian, theological and philosophical, sources combined.

Although present in the libraries of many monasteries, Dionysian works exerted little influence on monastic tradition.[7] The symbolic theology of Dionysius, however, appealed to the mystic theologians of Cathedral Schools. Anselm of Laon first introduced passages from Dionysius's works in his *Glossa Interlinearis*, a medieval authoritative work, thus giving impulse to the diffusion of these texts. Based on the translations and commentaries of Eriugena (known as the 'old translation') and of Sarrazin (known as the 'new translation'), Hugh and Richard of Saint Victor continued the tradition of Dionysian commentaries revealing their interest not only in the moral aspects of angelology, but also in its metaphysical implications. Their filter is of particular relevance, as they contributed to place Dionysius within the context of Augustine's Neo-Platonic philosophy.[8]

In the twelfth century, early scholastic authors were mainly concerned with harmonizing the Dionysian hierarchies with the account of creation contained in the book of *Genesis*. These authors focused on themes related to the

---

[7] On the Dionysian influence on monasticism, see JEAN LECLERCQ, *Influence and Noninfluence*, and LOUIS BOUYER, *La spiritualità dei Padri (III-VI secolo) Monachesimo antico e Padri*, in *Storia della spiritualità*, vol. 3/B, Bologna, Edizioni Dehoniane, 1968 (pp. 129-156).

[8] HUGH OF SAINT VICTOR, *Commentary on the Celestial Hierarchy of Saint Dionysius the Aeropagite*, PL 175, 925-1154. See also RICHARD OF SAINT VICTOR, *Beniamin Major*, PL 196, 63-202. For the influence of Pseudo-Dionysius on Hugh and Richard of Saint Victor, see MARCIA L. COLISH, *Medieval Foundations of the Western Intellectual Tradition 400-1400*, New Have, Yale University Press, 1997, pp. 229-233.

creation of angels, and on ethical issues related to angelic will, grace, and merit.⁹ Among these scholars there was a general implicit agreement on the angelic arrangement in triadic hierarchies. Only a few, however, concurred on whether the angels could have both contemplative and active functions, parallel to the two types of life. Contemplative and active ministries corresponded on the one side to the incessant contemplative movement of angels in the Empyrean, and, on the other side, to their operations as guardian angels, messengers, defenders of realms, and other tasks they performed in accordance with the Biblical tradition.¹⁰ Whether the angels could actually perform both functions was later object of debate, and in Dante's times, the Dionysian model of purely speculative ministries tended to prevail.

Later in the twelfth century, as more Aristotelian texts started to be reintroduced in the West accompanied by the commentaries of Arab philosophers, the interest in systematic and 'scientific' theology grew. Theologians of this period recovered Dionysian angelology by assimilating the elements that would contribute to the design of a coherent theological system in which angels played a major role as divine instruments in the economy of salvation. The doctrine of angels gradually moved away from metaphysical issues and questions about the nature of angels, the definition of hierarchy, and the connection between angels and cosmology entered the core of theological speculation.¹¹ In this intellectual context, the Dionysian systematic scheme appealed to the new philosophical sensitivity prompted by the reintroduction of Aristotelian texts to the West, for Dionysian texts brought about a new perspective on the doctrines of the One and of the Good, two themes central to

---

⁹ For an account of the development of the doctrine of angels in the twelfth century, and the main issues treated in this period, see MARCIA L. COLISH, *Early Scholastic Angelology*, pp. 80-109.

¹⁰ See for example Alexander of Hales, who quotes BERNARD OF CLAIRVAUX's *On Consideration* V, in his *Glossa*: «Sic in anima est vis practica et speculativa, ita in angelo ministrativa et contemplative: hac fruuntur, illa merentur». [«As the soul has practical and speculative capacities, so angels have administrative and contemplative functions: the former gives them merit, the latter gives them delight»]. ALEXANDER OF HALES, *Glossa in quatuor libros sententiarum Petri Lombardi*, Florence, Quaracchi, 1952, Book II, d. X, 2. As G. Bareille observed: «The Fathers place under angelic operations and supervision the world, animated and inorganic matter, the stars, the earth, the elements, the meteorological phenomena, the plants, the animals, the nations, the peoples, the human being [...]», G. BAREILLE, *Ange*, in *Dictionnaire de Theologie Catholique*, I, p. 1214. For a comparison of different functions attributed to angelic hierarchies and orders based on the Dionysian angelology, see DAVID E. LUSCOMBE, *Angels as Exemplar of World Order: the Hierarchies in the Writings of Alain de Lille, William of Auvergne, and St. Bonaventure*, in *Angels in Medieval Philosophical Inquiry*, pp. 15-28.

¹¹ Starting in the eleventh century, translations of Aristotelian texts that included *Metaphysics* and *Ethics to Nicomachus* begun to circulate in the West giving impulse to a renovated interest in speculative and practical philosophy. See EDWARD CRANZ, *The Reorientation of Western Thought c. 1100 A.D. The Break with the Ancient Tradition and its Consequences for Renaissance and Reformation*, paper delivered at the Duke University Center for Medieval Studies, March 24, 1982, pp. 1-24.

the theology of Neo-Platonic orientation. These elements contributed to consolidate the Areopagite's authority. His angelic hierarchical system provided a general framework that allowed for explaining the communication of divine science to human beings, and a rational access to unintelligible reality through an anagogical process guaranteed by the hierarchical arrangement of beings and knowledge. These elements particularly influenced the Victorine masters, who attributed both moral and cognitive values to the mystical process of return to God.[12] The influence of Dionysius through the Victorine filter is evident in the general conception of Dante's journey, as it reveals a similar ethic-cognitive character.

As Marie-Dominique Chenu observed, the Dionysian angelology provided a «total hypothesis» that conquered philosophers, theologians, and those poets who, like Dante, felt the urge to express a systematic and encyclopedic view of the universe:

> Such was the grand vision that intoxicated Pseudo-Dionysius and provided the master plan within which the universe and man, God and Christ, the sacraments and contemplation, body and soul, light and shadows, symbols and negations all found a sublime explanation.[13]

As the interests of theologians gradually shifted toward philosophical issues, their concern about moral aspects of angelology became secondary, but did not vanish. Authors such as Richard and Hugh of Saint Victor, William of Auxerre, and Alexander of Hales combined ethical and metaphysical elements in their commentaries on the Dionysian works.[14] They brought the main features of the Dionysian symbolic-anagogic angelology into the two main angelological systems elaborated in the thirteenth century by Bonaventure and Aquinas.

Chapter II discusses the Dionysian definition of hierarchy, its nature, and function, and shows how Dante departed from his system to adapt it to the

---

[12] See in particular the two works by RICHARD OF SAINT VICTOR, *De Arca Noe morali libri IV*, in *Patrologiae cursus completus. Series Latina* (PL), Jacques-Paul Migne ed., Paris, Garnier fratres, 1844-1903, PL 176, 617-680 D; and *De Arca mystica* PL, 176, 681-704 A.

[13] MARIE-DOMINIQUE CHENU, *Nature, Man, and Society in the Twelfth Century. Essays on New Theological Perspectives in the Latin West*, J. Taylor and L.K. Little eds. and transl., Toronto, University of Toronto Press, 1997 (1957), p. 82.

[14] William of Auxerre died in 1231. He was Archdeacon of Beauvais and discussed angelic hierarchies in his main work, the *Summa aurea*. Alexander of Hales (also known as *Doctor Irrefragabilis* and *Theologorum Monarcha*), was educated in a monastery at Hales but studied and became teacher at the University of Paris. He entered the Franciscan order in 1222 and died in 1245. In his *Glossa* he commented angelic hierarchies following Saint Bernard. See ALEXANDER OF HALES, *Glossa in quatuor libros sententiarum*, Book II, and BERNARD OF CLAIRVAUX, *Five Books on Consideration. Advice to a Pope*, Kalamazoo, Cistercian Publications, 1976, Book V.

cosmological and poetic design of the *Commedia*. Prior to doing so, however, it is necessary to investigate the philosophical issues of angelic nature and operations, and the solutions Dante sought in defining the structural setting of his poem. This is done in the next section.

III. Two Issues in the Angelology of the Thirteenth Century. Dante's Original approach

The two main issues, central to the twelfth and thirteenth-century debate on angelology are of particular relevance to the *Commedia*: the definition of angelic nature, and the doctrine of angelic operations. As we will see, Dante approached these issues in view of the instrumental function as intermediaries between intelligible and unintelligible worlds he assigned to angels.

a. *The Creation of Angels and their Nature*

The introduction of a philosophical perspective in the elaboration of a doctrine of angels led to a systematic inquiry about angelic nature. The definition of the nature of separate substances implied finding a criterion to distinguish divine, angelic, and human natures. Without such a criterion, it was impossible to conceptualize a continuum of creatures as separate beings from their creator in philosophical terms, and to elaborate a cosmological structure free of the influences of Neo-Platonic emanatism.

Most theologians of Dante's time adhered to universal hylomorphism, and applied the Aristotelian definition of substance as 'synolon' composed of matter and form to angels.[15] Scholastic terminology, interpreting form as the actualizing principle of matter, generally identified form with act and matter with potency. The sustainers of universal hylomorphism affirmed that God alone was pure act and pure form. All other creatures, including angels, presented in their essence a combination of both matter and form. This application of Aristotelian metaphysical principles led however to a double contradiction. Matter would make angels' nature corruptible (in contrast with the shared view that they were created but immortal), thus preventing any distinction between angels and human beings. If angels were pure form, however, it

---

[15] Aristotle considered the Intelligences as purely spiritual substances, for matter would make them subject to change. According to Aquinas (*S. th.* I, q. 50, a. 2), Avicebron in his *Fons Vitae* (beginning of eleventh century) was the first to apply Aristotelian hylomorphism universally. On the debate on angelic nature in the thirteenth century, see David Keck, *Angels and Angelology*, pp. 93-114.

would be impossible to distinguish them from God. In trying to solve this conundrum, some theologians, and Bonaventure among them, introduced untenable distinctions of different kinds of matter.[16]

The debate took a turn with Aquinas' distinction among being (*ens*), essence (*essentia*), and existence (*esse*, or *actus essendi*).[17] Aquinas denied the composite character of angelic essences, and affirmed their pure spirituality. He postulated, however, a difference between essence and existence, and argued that they coincided only in God. Angels received their 'act of being', or existence, from God's act of creation, and since they derived their existence from an external cause, essence and existence were in them in the same relation as potency and act. Therefore, for the *Doctor Angelicus*, angels were pure form, and could not be pure acts. In composite substances, there were two different ways of considering potency and act, with respect to essence and with respect to existence:

Unde in rebus compositis est considerare duplicem actum, et duplicem potentiam. Nam primo quidem materia est ut potentia respectu formae, et forma est actus eius; et iterum natura constituta ex materia et forma, est ut potentia respectu ipsius esse, in quantum est susceptiva eius. Remoto igitur fundamento materiae, si remaneat aliqua forma determinatae naturae per se subsistens, non in materia, adhuc comparabitur ad suum esse ut potentia ad actum: non dico autem ut potentiam separabilem ab actu, sed quam semper suus actus comitetur. Et hoc modo natura spiritualis substantiae, quae non est composita ex materia et forma, est ut potentia respectu sui esse; et sic in substantia spirituali est compositio potentiae et actus, et per consequens formae et materiae; si tamen omnis potentia nominetur materia et omnis actus nomine-

---

[16] Bonaventure eventually concluded that matter must be of the same kind in angels as in other composite natures. See *Coll.* IV, 6-13, St. BONAVENTURE, *Collationes in Hexaëmeron*, in *Opera Omnia*, 10 vols., Quaracchi ed., Firenze, Collegium S. Bonaventura, 1892-1902. English translation, *Collations on the Six Days*, New Jersey, St. Anthony Guild Press, 1970. On the medieval debate on angelic hylomorphism, see DAVID KECK, *Angels and Angelology*, pp. 93-99. For the issue of angelic subjectivity and individuation in Thomas Aquinas, and the condemnation in 1277 of Aquinas' propositions on angelic nature, see TIZIANA SUAREZ-NANI, *Les anges et la philosophie*, pp. 25-53 and 75-85.

[17] Aquinas presented his doctrine on angelic nature and his objections to universal hylomorphism in his *Summa Theologiae* I, q. 50, a. 2. The distinctions among being, essence, and existence discussed in the *Summa* appear in other works such as *Summa contra gentiles* II, q. 52, 2nd arg. (c. 1261), in the *Quaestiones disputatae de spiritualibus creaturis* (1267-8); and mainly in his *De ente et essentia (On Being and Essence)* (1252-56). See JOSEPH BOBIK, *Aquinas On Being and Essence*, Notre Dame, Notre Dame University Press, 1970. For Aquinas' criticism of universal hylomorphism, see also JAMES D. COLLINS, *The Thomistic Philosophy of the Angels*, Washington DC, The Catholic University of America (Philosophical Series), LXXXIX, 1947, pp. 50-74. Latin texts of Aquinas's works are from the Leonine edition of Aquinas's complete works, SANCTI THOMAE AQUINATIS DOCTORIS ANGELICI, *Opera omnia*, iussu Leonis XIII P.M. edita, cura et studio fratrum ordinis praedicatorum, Romae, 1882, sqq. English translations are from *The Summa Theologica of Saint Thomas Aquinas*, edited by the Fathers of the English Dominican Province, 5 vols., Christian Classics, 1981.

tur forma. Sed tamen hoc non est proprie dictum secundum communem usum no minum. (*De Spiritualibus creaturis*, a. 1r).¹⁸

Hence in composite objects there are two kinds of act and two kinds of potency to consider. For first of all, matter is as potency with reference to form, and the form is its act. And secondly, if the nature is constituted of matter and form, the matter is as potency with reference to existence itself, insofar as it is able to receive this. Accordingly, when the foundation of matter is removed, if any form of a determinate nature remains which subsists of itself but not in matter, it will still be related to its own existence as potency is to act. But I do not say, as that potency which is separable from its act, but as a potency which is always accompanied by its act. And in this way the nature of a spiritual substance, which is not composed of matter and form, is a potency with reference to its own existence; and thus there is in a spiritual substance a composition of potency and act, and, consequently, of form and matter, provided only that every potency be called matter, and every act be called form; but yet this is not properly said according to the common use of the terms.

Therefore, for the *Doctor Angelicus*, as spiritual substances angels were not composed of matter and form, but, with respect to their existence, they were however composed of act and potency.¹⁹

Aquinas' position influenced the debate on essence and existence as it developed in the thirteenth century among theologians such as Giles of Rome, William of Ockham, and Duns Scotus, and it also surfaces in the expository *Epistle* XIII to Cangrande attributed to Dante. It is not surprising to find a reflection of this crucial debate in Dante's doctrine of creation in *Paradiso* XXIX, 22-30, where he illustrated the creation of form, matter, and their simultaneous composition (in terms of nature and not in terms of time):

> Forma e matera, congiunte e purette,
> usciro ad esser che non avía fallo,
> come d'arco tricordo tre saette.

---

¹⁸ AQUINAS, *De spiritualibus creaturis*, critical ed., L. Keeler ed., Roma, Universitas Gregoriana, 1937. Translation by M.C. Fitzpatrick and J.J. Wellmuth, Milwaukee, Marquette University Press, 1949.

¹⁹ «In hoc ergo differt essentia substantiae compositae et substantiae simplicis quod essentia substantiae compositae non est tantum forma, sed complectitur formam et materiam, essentia autem substantiae simplicis est forma tantum. [...] Huiusmodi ergo substantiae quamvis sint formae tantum sine materia, non tamen in eis est omnimoda simplicitas nec sunt actus purus, sed habent permixtionem potentiae», [«Therefore, the essence of a composite substance and that of a simple substance differ in that the essence of a composite substance is not form alone but embraces both form and matter, while the essence of a simple substance is form alone [...]Although substances of this kind are form alone and are without matter, they are nevertheless not in every way simple, and they are not pure act; rather, they have an admixture of potency [...]»]. AQUINAS, *De ente et essentia*, P. Porro ed. and transl., Milano, Bompiani, 2002 IV. English translation by Robert T. Miller, 1997, Medieval Sourcebook, Centre for Medieval Studies, Fordham University, http://www.fordham.edu/halsall/basis/aquinas-esse.html.

> E come in vetro, in ambra od in cristallo
> raggio resplende sí, che dal venire
> all'esser tutto non è intervallo,
> > così il triforme effetto del suo sire
> > nell'esser suo raggiò insieme tutto
> > sanza distinzione in essordire (*Par.* XXIX, 22-30).[20]
>
> Then form and matter, either separately
> or in mixed state, emerged as flawless being,
> as from a three-stringed bow, three arrows spring
> > and as a ray shines into amber, crystal,
> > or glass, so that there is no interval
> > between its coming and its lightning all,
> > > so did the three – form, matter, and their union –
> > > flash into being from the Lord with no
> > > distinction in beginning: all at once.

Divine Love produces form, matter, and their combination in a simultaneous act. Dante's expression in line 30, «sanza distinzione in essordire» («with no distinction in the beginning»), by contrast alludes to the distinction present in the final effect of creation in time. Although simultaneous creation (creation *simul*) from nothing (*ex-nihilo*) was logically intended as being before time, it was however understood as perfected in time. Dante's insistence on the doctrine of creation *simul*, which theologians widely discussed presenting a range of different positions, is of particular relevance for his angelology.[21] Indeed, the doctrine of simultaneous creation is a necessary premise to Dante's identification of angels with the Aristotelian celestial movers. In *Paradiso* XXIX, 43-45 we find the correlation of these two elements of his angelology:

> e anche la ragione il vede alquanto,
> che non concederebbe che ' motori
> sanza sua perfezion fosser cotanto (*Par.* XXIX, 43-45).
>
> and reason, too, can see in part this truth,
> for it would not admit that those who move
> the heavens could, for so long, be without their perfect[ing] task.

---

[20] The Italian text of the *Divina Commedia* is from GIORGIO PETROCCHI's critical edition *La Commedia secondo l'antica vulgata di Dante Alighieri*, Milano, Mondadori, 1966-67, republished by Einaudi, Torino, 1977. English translations are from ALLEN MANDELBAUM, *The Divine Comedy of Dante Alighieri*, Berkeley, University of California Press, 3 vols., 1980-1982.

[21] The time of creation of angels was a chief theme in the philosophical debates of the first half of the twelfth century. The main issue was to reconcile *Genesis*, where the creation of angels is not explicitly mentioned, with the consolidated tradition of the existence and importance of angels in the order of creation. For the main trends in this debate, see MARCIA COLISH, *Early Scholastic Angelology*, pp. 80-86.

According to Aristotle, every nature has an intrinsic end that is realized in the act that is proper to that nature. Just as a seed realizes its perfection by becoming a plant, so angels became perfect in accomplishing their function as movers. In Dante the motion of the spheres is a consequence of their act of contemplation. The proximity of the hierarchies to the center of creation grants them the perfect vision of the source of life. This vision in turn enriches the angels with a desire that drives them to circle around God, and this movement causes the motions of the celestial spheres. The tercet quoted above stresses the importance of the angelic motion of the heavens, for it originates the planetary influence on the human world. As Chapter III of this book will show, this influence is vital to the realization of the grand design of salvation, upon which the ultimate meaning of angelic operations relies. Within the eschatological perspective of the *Commedia*, the heavens constitute the «perfezion» of the angels. Through the heavens they exercise their most important governing function, which extends beyond the stars to reach the dominion of human experience. Since angelic Intelligences are the actualizing principles of the motion and virtues of the spheres, their creation must have been simultaneous with that of the material spheres, for there cannot be act without potency ready to receive it, and the celestial bodies could not produce their effects in the sublunar world without the cause of their movement. The harmony that ties all creatures together through angelic influence reflects Dante's purpose to establish continuity between physical and spiritual spheres in an essentially Christological perspective of mediation. The same harmony emerges in the poetic dance of language and images of cantos XXVIII and XXIX, where structure bends to poetry and poetry manifests structure in what Hans Urs von Balthasar defined as Dante's theological aesthetics.[22]

Theologians had different positions on the doctrine of simultaneous creation. Aquinas neither rejected nor accepted it, and his arguments varied in his various works.[23] He recognized that it was rational to suppose that angels and heavens were created at the same time, but he also affirmed that angels could

---

[22] «This cosmos, permeated with divine energies, is understood in a Christian way. Its entire aesthetic power of expression serves to support a Christian theological aesthetics». HANS URS VON BALTHASAR, «Dante», in *The Glory of the Lord. A Theological Aesthetics*, vol. III (*Studies in Theological Style: Lay Style*), San Francisco, Ignatius Press, 1986 (1962), pp. 9-104, 69.

[23] In his *Summa Theologiae* I, q. 61, a. 3 Aquinas argued that angels must be considered part of the created universe, and not a separate reality on their own. Contrary to the opinions of Early Fathers such as Jerome, Basil, and Nazianzen, he admitted simultaneous creation as probable. In his *Quaestiones disputatae de potentia* I, q. III, a. 18, Aquinas still defended creation *simul* as possible, but he rejected the 'perfection' argument as contrary to the logic of the hierarchical arrangement of creatures.

have been created before the visible world, for the perfection of a superior being cannot depend on the existence of an inferior one, just as a cause cannot depend on its effects. Possibly, as Bruno Nardi suggested, the idea Dante displayed in *Paradiso* XXIX, 43-45 reflects the influence of Averroist philosophy. In his commentary to *Metaphysics*, Averroes (Abū l-Walīd Muhammad ibn Ahmad Muhammad ibn Rushd) (following Aristotle) identified the natural end of Intelligences in the motion of celestial bodies, thus supporting the doctrine of simultaneous creation of spiritual and visible worlds.[24] Dante followed the Greek-Arab theory, against the authority of Aquinas, for it was necessary to and coherent with his poetical cosmogony, where his eschatological vision of creation relied upon angelic intervention in human destiny through astrological influence.[25] This notwithstanding, he did not accept Aristotle's idea that the number of Intelligences coincided with the number of movements of celestial bodies. This non-identification, however, involved critical philosophical issues that Dante tried to solve by reformulating the notion of angelic operations, a topic discussed in the next section of this chapter.

Having established the contemporaneous creation of material and spiritual worlds, Dante turned to the hierarchical structure of creation in *Paradiso* XXIX. In his exposition, the visible cosmos and the laws that govern it from the inaccessible regions of invisible reality emerge as the product of a unique act of gratuitous love. The simultaneous unfolding of form, matter, and their combination («costrutto»), coincides with the immediate establishment of order («ordine»):

> Concreato fu ordine e costrutto
> a le sustanze; e quelle furon cima
> nel mondo in che puro atto fu produtto;
>   pura potenza tenne la parte ima
> nel mezzo strinse potenza con atto
> tal vime, che già mai non si divima (*Par.* XXIX, 31-36).

> Created with the substances were order
> and pattern; at the summit of the world

---

[24] BRUNO NARDI, *Tutto il frutto ricolto del girar di queste spere*, in ID., *Dante e la cultura medievale*, Roma-Bari, Laterza, 1983 (1942), pp. 245-264.

[25] Aquinas neither accepted nor rejected this opinion. His cautious position may be explained by the fact that if angelic perfection required celestial bodies, it would be possible to claim that angelic nature needed some sort of matter. See *S. Th.* I, q. 61, a. 3.

were those in whom pure act had been produced;
   and pure *potency* possessed
the lowest part; and in the middle, act
so joined *potency* that they
   never disjoin.[26]

The tercets indicate that the creation of angels coincided with their disposition in a hierarchical order, which reflected their different degrees of vicinity to and participation in the divine. The angels' relative hierarchical ranks correspond to their different degrees of contemplative penetration of divine science, which in turn determines their various ministries.

All commentators have interpreted «pure act» in line 33 as referring to angels, as the spiritual creatures at the top of the ladder of beings. The line and its interpretations are however problematic. To define angelic substances as «pure acts» would be technically incorrect, for all theologians agreed that only God is pure act. For this reason most of them considered the angels as composed of form and matter, although a special, superior kind of matter. The hypothesis that Dante wrote «act» to mean «form» does not seem entirely convincing. It is contradictory to assume that Dante in line 33 used scholastic terminology loosely, when he employed the same terminology quite appropriately throughout both cantos XXVIII and XXIX, where he 'technically' exposed his angelological design.[27] Bruno Nardi suggested that the definition of angels as pure acts in line 33 indicates the presence of concepts ascribable

---

[26] Italics indicate the changes I have made to Allen Mandelbaum's translation. I have substituted in lines 34-35 «potentiality» with the philosophical term «potency» to stress Dante's philological accuracy in using technical philosophical language.

[27] In his thorough study on Dante's metaphysics of matter and form, Christian Moevs admits that Dante's definition of angels as pure acts is problematic and follows the hypothesis of Dante's incorrect use of scholastic terminology, as it was common among theologians especially before the debate originated by AQUINAS's *De ente et essentia*. CHRISTIAN MOEVS, *The Metaphysics of Dante's Comedy*, Oxford, Oxford University Press, 2005, pp. 43-44. Before Moevs, Bruno Nardi expressed a different opinion and observed that it seems unlikely that Dante would use inappropriately a terminology he had employed correctly on other occasions: «Ancora "una imprecisione verbale"? Mi pare che queste imprecisioni verbali siano un po' troppo frequenti, in un uomo che conosceva così bene, secondo certi commentatori, la somma *Contra gentiles*! E perché non pensare piuttosto a residuo di concetto averroistico, che un tempo avesse esercitato su Dante qualche seduzione»? [«Another "inaccurate word"? It seems to me that these inaccurate words occur a bit too frequently for a man who knew so well (according to certain commentators) the *Contra gentiles*! What if it was instead a residue of some Averroist concept that could have seduced Dante in the past?»]. BRUNO NARDI, «Dal *Convivio* alla *Commedia*», in *Dal Convivio alla Commedia. Sei saggi danteschi*, Roma, Istituto Storico Italiano per il Medio Evo, 1992 (1960), p. 47. I believe that Dante was aware of the correct terminology, but, relying of the commonly accepted interchange of 'act' and 'form', he spoke of 'pure act' to stress the importance of the continuous activity of all angels.

to Averroist philosophy of possible appeal to Dante, even though he had probably already abandoned them in the *Commedia*. In his commentary to *Metaphysics* Averroes, following Aristotle, considered Intelligences as pure forms, perfectly actualized in their operation as movers. To define Intelligences as pure acts, and therefore as divinities, was not an issue in pagan polytheist culture, but it was a major obstacle in monotheistic Christianity.[28] As a possible solution to an issue that was too obvious to be ignored, one should look at the general context of the angelic doctrine that emerges in the *Commedia*, which suggests that Dante did not consider angels to be pure acts in absolute terms (as they were created beings), but only with respect to their ministerial functions.

If the definition of angels as pure acts was in contrast with the thesis of the sustainers of universal hylomorphism, it could be however reconciled with Aquinas' doctrine of «ens» and «essentia». As we have seen, while maintaining the Aristotelian distinction between form and matter on the one side, and act and potency on the other, the *Doctor Angelicus* argued that while angels were incorporeal and immaterial, it was possible to distinguish both act and potency in them. They possess potency with respect to their coming into being, since they are the product of God's creation, but they can be considered pure acts with respect to their essence, for no residual potency remained unexploited in them after their creation. As purely intellectual creatures angels are eternally in act with respect to their ministerial functions directed to carry out and administer providential decrees. A passage in the *Monarchia* seems to support this view. In comparing angelic and possible intellects, Dante observed that:

Nam, etsi alie sunt essentie intellectum participantes, non tamen intellectus earum est possibilis ut hominis, quia essentie tales speties quedam sunt intellectuales et non aliud, et earum esse nichil est aliud quam intelligere quod est quod sunt; quod est sine interpolatione, aliter sempiterne non essent (*Monarchia* I, iii, 7).[29]

For while there are indeed other beings who like us are endowed with intellect, nonetheless their intellect is not "potential" in the way that man's is, since such

---

[28] In Aristotle's cosmology, the Intelligences were immaterial separate substances that could be considered pure acts and assimilated to divinities. The same was not however possible in Christian monotheistic religion. In his commentary to Aristotle's *Metaphysics*, Averroes interpreted the Intelligences in this way. For B. Nardi's discussion of the possible influence of Averroes on Dante's angelic doctrine, see B. NARDI, *Dal Convivio alla Commedia*, pp. 37-47.

[29] Latin text is from Bruno Nardi's edition of *Monarchia*, in DANTE ALIGHIERI, *Opere minori*, Milano-Napoli, Ricciardi, 1979, pp. 241-503. English translation from *Dante's Monarchia*, R. Kay transl., Toronto, Pontifical Institute of Medieval Studies, 1998 («Studies and Texts», 131).

beings exist only as Intelligences and nothing else, and their very being is simply the act of understanding that their own nature exists; and they are engaged in this ceaselessly, otherwise they would not be eternal.

The being of Intelligences is their very act of understanding what they are and consequently what they are for («quod est quod sunt»), and in this understanding their nature is perpetually and fully in act. Unlike human beings, the Intelligences know immediately both universals and particulars, without the mediation of concepts or any form of sense perception («sine interpolazione»).[30] Their act of understanding («intelligere») of the divine is at the same time their being and their operation. Whatever the angels' role in the design of the cosmos, it derives from this immediate relation between understanding and operation. The «pure act» of line 33 thus seems to find a logical explanation if «act» is interpreted as the natural end of Intelligences, created to operate upon the other creatures, and not to be acted upon. Pietro di Dante, Benvenuto da Imola, Francesco da Buti, and Bernardino Daniello, among others, seem to support this interpretation, as they viewed Dante's controversial «pure act» as an expressionist use of Scholastic language to stress the importance of angelic operations in his poem.

Another possible hypothesis is that the «pure act» of line 33 may refer to God, and not to angels. If this hypothesis were correct, the line would read, «Angels occupied the highest place in the order of creation, in which pure act produces itself (God)». Like in Latin language, the passive form «fu produtto» could be interpreted as an impersonal form, leaving an intentional ambiguity to the line. The ambiguity would allow for a definition of angels as pure forms, still at the top of the hierarchy of substances, but without incurring in the theological error of defining them as pure acts.

Whatever the interpretation, the crucial point seems that Dante wanted to stress the lack of any form of materiality in the nature of separate substances, and their condition of perpetual activity. There may be another reason behind Dante's attribution of so much importance to the pure spirituality of angels –

---

[30] The same concept reappears in *Paradiso* XXIX, 76-81: «Queste sustanze, poi che fur gioconde / de la faccia di Dio, non volser viso / da essa, da cui nulla si nasconde: / però non hanno vedere interciso / da novo obietto, e però non bisogna / rememorar per concetto diviso», [«These beings, since they first were gladdened by / the face of God, from which no thing is hidden, / have never turned their vision from that face, / so that their sight is never intercepted / by a new object, and they have no need / to recollect an interrupted concept»]. As B. Nardi notices: «L'espressione "sine interpolazione" risponde esattamente a "vedere interciso da novo obietto", checché vi abbiano almanaccato su traduttori e commentatori» [«Whatever the fantasizing speculations of translators and commentators, the expression "sine interpolazione" corresponds exactly to "vedere interciso da novo obietto"»]. BRUNO NARDI, *Dal Convivio alla Commedia*, p. 44.

besides the Averroist hypothesis, and besides his need for perfectly active separate substances. It is in Aquinas' opinion that any suspicion on the presence of matter in angelic nature would limit the importance of angelic intellectual operations.[31] If angels were not pure form, they would not be able, from a philosophical point of view, to apprehend and transmit knowledge in a simultaneous intellectual act through hierarchies, and therefore would not be able to perform their governing operations. This may perhaps be the main reason the discussion on angelic nature is so relevant in *Paradiso*.

b. *The Contemplative and Active Functions of Angels*

In *Convivio* II, iv, Dante discussed the doctrine of angelic operations and placed it within the context of philosophical tradition. His purpose was to reconcile the Christian assumption of an indeterminate number of angels with Aristotle's of a number of Intelligences limited to the number of celestial movements. Aristotle had introduced the Intelligences with the purpose to explain the motion of the celestial bodies and identified them with this operation. He denied the existence of other Intelligences on the ground that if deprived of their proper finality, they would be «idle», and therefore unnecessary.[32] Although interested in a universal reconciliation of theological and philosophical cosmologies, Dante in the *Convivio* encountered difficulties in overcoming the contradictions that emerged from the combination of the two systems. As Nardi argued, at this stage of Dante's inquiry on angels,

---

[31] See *S. Th.* I, q. 50, a. 2 and *De spiritualibus creaturis*, I, 1. Thus Etienne Gilson explains Aquinas's argument: «Things are ready to fall into the grasp of intelligence in the measure in which they are free from matter. Forms which are inserted into matter, for example, are individual forms. They cannot [...] be apprehended as such by the intellect. Pure intelligence, whose object is the immaterial as such, must therefore also be free from all matter. The total immateriality of angels is therefore demanded by the very place they occupy in the order of creation». ETIENNE GILSON, *The Christian Philosophy of St. Thomas Aquinas*, Notre Dame, University of Notre Dame Press, 1994 (1956), p. 164.

[32] «Furono certi filosofi, de' quali pare essere Aristotile nella sua Metafisica (avegna che nel primo di Cielo incidentemente paia sentire altrimenti), [che] credettero solamente essere tante queste, quante circulazioni fossero nelli cieli, e non più; dicendo che l'altre sarebbero state etternalmente indarno, sanza operazione: ch'era impossibile, con ciò sia cosa che loro essere sia loro operazione». [«Some philosophers, including, it seems, Aristotle, believed that there were only as many Intelligences as there were revolutions in the heavens. Their reasoning was that any other Intelligences would live an eternally purposeless existence, since they would lack any activity, and this would be an impossibility, because activity constitutes the very essence of an Intelligence's being»]. See CESARE VASOLI's commentary on *Convivio* II, iv, 3, where he discusses Bruno Nardi's opinions about the influence of Averroes on this aspect of Dante's angelic doctrine. Quotations are from DANTE ALIGHIERI, *Convivio*, in *Opere minori*, tome I, part II, C. Vasoli and D. De Robertis eds., Milano-Napoli, Ricciardi, 1988 («La letteratura Italiana. Storia e testi», 5). English translation from *Dante's "Il Convivio" (The Banquet)*, R.H. Lansing ed. and transl., New York-London, Garland, 1990 («Garland Library of Medieval Literature Series B», 65).

the influence of Aristotle and Averroes still seems evident. Speculation on the nature and kind of angelic operations represented a necessary step toward harmonizing philosophical and theological premises of Dante's angelology.

In seeking to justify philosophically his assumption of an indeterminate number of angels, in chapter iv of the second book of *Convivio*, Dante reviews different theories on Intelligences and shows the existence of opinions other than Aristotle's among the philosophers who both preceded and followed him. Unlike his old disciple, Plato had posited the existence of as many Intelligences as the number of forms things may take (e.g., universals and particular determinations of things), thus identifying Intelligences and ideas.[33] Pagan religion attributed divine value to the ideas and identified them with the gods. The ancients, however, did not possess the science and the enlightened reason of Christians, which – Dante says – makes it evident that angels are in far greater number than the movements of the stars:

ché pur per ragione vedere si può in molto maggiore numero essere le creature sopra dette, che non sono li effetti che [per] li uomini si possono intendere (*Convivio* II, iv, 8).

for even by reason alone it can be perceived that the creatures mentioned above are of far greater number than are the effects which men can apprehend.

Here, as later in *Paradiso*, Dante does not argue that the number of angels is infinite, but only that it is much greater than the number of celestial motions calculated by the astronomers.[34] This, he says in *Convivio*, is a truth that

---

[33] «Altri furono, sì come Plato, uomo eccellentissimo, che puosero non solamente tante Intelligenze quanti sono li movimenti del cielo, ma eziandio quante sono le spezie delle cose (cioè le maniere delle cose): sì come è una spezie tutti li uomini, e un'altra tutto l'oro, e un'altra tutte le larghezze, e così di tutte. E volsero che, sì come le Intelligenze delli cieli sono generatrici di quelli, ciascuna del suo, così queste fossero generatrici dell'altre cose ed essempli, ciascuna della sua spezie; e chiamale Plato "idee", che tanto è a dire quanto forme e nature universali». [«Others, such as Plato, a man of quite outstanding ability, held that there were not simply as many Intelligences as there were movements of the heavens, but that there were in addition as many as there were particular species of things. They maintained that just as the Intelligences who move the heavens bring those heavens into existence, each being the source of its particular heaven, so these other Intelligences bring into existence all other things and are their exemplars, each with respect to its particular species. Plato calls them "Ideas", signifying that they are forms or universal natures»]. *Convivio* II, iv, 4-5.

[34] Dante's opinion was based on the Biblical authority of the Book of Daniel: «*Milia milium ministrabant ei et decies milies centena milia assitebant ei* (*Dan.* VII, 10)» [«The number of angels is not infinite, but so great that the human mind cannot count it»]. This interpretation was shared by many commentators, as for example Johannis de Serravalle (1416-17) and Benvenuto da Imola (1375-80), who stated that: «non quia angeli sint infiniti, sed quia non sunt comprehensi sub numero certo», [«the number of angels is not infinite, but indeterminate»]. The difference between assuming an infinite or an indefinite number of angels was crucial, because to assume a limited number of an-

only probable and not demonstrative arguments can support, for it is part of a mystery not accessible to the human mind.[35] Based on the authority of the Scriptures, for Dante the number of angels probably corresponds to the immensely vast but yet determined number of species or forms actualized in the created world. As chapter III of this book will show, this idea takes its definite form in the *Commedia*, where a connection between angels and created forms takes place through the informing action of angels on the celestial bodies, which in turn influence the determinations of material reality.

Dante's argument in support of the thesis that there are more Intelligences than celestial motions reflects his intent to complement Aristotelian and Christian angelologies, and more generally (and in this he is close to Aquinas) to reconcile philosophic and theological rationales. Within this perspective, the number of angels depended on what we cannot see and we do not know, and must correspond to the functions the separate substances were called to perform. The implication was that angelic functions include but do not coincide with their moving tasks. Motion is an essential part of the broader set of operations required by their role in the order of creation.

Within this context, an investigation on the nature and end of angelic operations acquired crucial importance. In the *Convivio*, however, before Dante exploited the potentials of the comprehensive Dionysian scheme, he still faced obstacles in reconciling Aristotelian and theological doctrines. In imagining the cosmos as a sort of universal kingdom that angels govern and administer, Dante explicitly assigned them a purely active life. In contrast, the Dionysian doctrine predicated purely speculative angels. Furthermore, he knew from Aristotle's *Ethics to Nicomachus* that contemplative life was superior to active life, and that speculative intellect was superior to practical intellect. How could then the highest and most perfect creatures (in terms of similitude to God) not enjoy the superior state of pure contemplation? A possible answer

---

gels links them to the innumerable but finite determinations of material reality. Biblical quotations in Latin are from *Biblia Sacra Vulgatae Editionis*, Milano, San Paolo, 1995. English translations are from the Douay-Rheims version of the Bible available online at www.drbo.org.

[35] «Né si maravigli alcuno se queste e altre ragioni che di ciò avere potemo, non sono del tutto dimostrat[iv]e; ché però medesimamente dovemo amirare loro eccellenza – la quale soverchia li occhi della mente umana, sì come dice lo Filosofo nel secondo della Metafisica –, e afferma[r] loro essere». [«It should cause no surprise if the reasons given, and others which we could adduce on this point, do not constitute exhaustive demonstrations, for this is itself cause for admiring the excellence of these creatures – which, as the Philosopher says, transcends what the human mind can clearly perceive – and for affirming their existence»]. *Convivio* II, iv, 16. In commenting on this passage, Cesare Vasoli observes that Dante at this stage was aware that Aristotelian and theological teachings on Intelligences were in conflict, and he tried (unsuccessfully) to solve it. By affirming that the question on the number of angels was an objet of probable and not deterministic demonstration, Dante left this question open to interpretation.

was to create a logical nexus between contemplative and active functions. In the *Convivio*, Dante took a first step toward the solution he later adopted in the *Commedia* by arguing that separate substances performed both contemplative and active functions:

> Onde, con ciò sia cosa che quella che è qui l'umana natura non pur una beatitudine abbia, ma due, sì com'è quella della vita civile e quella della contemplativa, inrazionale sarebbe se noi vedemo quelle avere [la] beatitudine della vita attiva, cioè civile, nel governare del mondo, e non avessero quella della contemplativa, la quale è più eccellente e più divina. E con ciò sia cosa che quella che ha la beatitudine del governare non possa l'altra avere, perché lo 'ntelletto loro è uno e perpetuo, conviene essere altre fuori di questo ministerio, che solamente vivano speculando (*Convivio* II, iv, 10-11).

> Consequently, since human nature as it exists here has not only one blessedness but two, namely that of the civil life and that of the contemplative life, it would be illogical for us to find that these beings have the blessedness of the active (that is, of the civil) life, in governing the world, and not that of the contemplative life, which is more excellent and more divine. And since the one that has the blessedness of governing cannot have the other because their intellect is one and perpetual, there must be others outside this ministry who live by contemplation alone.

This opinion, common among the Early Fathers and in general among not-rigorously Aristotelian theologians, was however problematic. At this stage of his elaboration, and probably still under the influence of Averroism, Dante believed that the two functions could not co-exists in angels. Attributing to angels both speculative and active functions would be equivalent to endowing them with a certain degree of potency.[36] Purely intellectual substances always in act could not pass from one function to another, for that would imply a passage from potency to act. A possible solution seemed to postulate the existence of two different categories of angels: those with active roles such as

---

[36] On this point, in the *Convivio*, Dante follows Aristotle and Averroes, for, as Nardi noticed, Aquinas admitted instead a «certain succession» («quaedam intelligentiarum succession») in the angelic cognitive process. This «succession» cannot be interpreted in temporal terms (angels' understanding is immediate and eternal), but in logical terms, as indicating the passage of angels from an act to another act («quum non succedat actus potentiae, sed actus actui»). In the *Commedia*, however, Dante seems to make a synthesis of both positions by making contemplation and action temporally and logically coincide. He thus implicitly accepted Aquinas's argument and designed his hierarchies as simultaneously in contemplation and action, defining both functions as 'operations'. In the 'monarchic' vision of the cosmos present in both Aquinas and Dante angels represented the administering power of Providence, and it was essential to consider them as eternally operative. In this sense, Dante shared Aquinas's opinion that operation was the essence of angels («sua actio est sua essentia»). See BRUNO NARDI, *Dal Convivio alla Commedia*, pp. 44-45; 56. Quotations are from Aquinas's *Contra Gentiles* II, c. 101 as quoted in B. NARDI.

the celestial movers, and those exclusively dedicated to contemplation. Before Dante, Giles of Rome had opted for this hypothesis, which originated from Avicenna (Abū Alī al-Ḥusayn ibn 'Abd Allāh ibn Sīnā).[37] The distinction of angels in two categories was however difficult to maintain, as it required a rational explanation of the relation between speculative and active Intelligences. Indeed Dante is not explicit on this point. He only says, rather vaguely, that the motion of the heavens is a consequence of the speculative activity of «some» («certe») Intelligences:

> E non è contra quello che pare dire Aristotile nel decimo dell'Etica, che alle sustanze separate convegna pure la speculativa vita. Come pure la speculativa convegna loro, pure alla speculazione di certe segue la circulazione del cielo, che è del mondo governo; lo quale è quasi una ordinata civilitade, intesa nella speculazione delli motori (*Convivio* II, iv, 13).

> This does not run counter to what Aristotle seems to say in the tenth book of the Ethics, namely that the contemplative life alone befits separate substances. Although the contemplative life alone befits them, to the contemplative life of just a certain number of them falls the circular movement of the heaven, which is the governing of the world, which is a kind of civil order conceived within the contemplation of its movers.

The causal relation implied by the term «segue» employed in the passage above does not explicitly identify how heavenly motion comes into being. Either Dante assumed that only some Intelligences were movers, and they moved as a consequence of their contemplation (which would contradict what he had stated only a few paragraphs earlier), or that there were subordinate Intelligences (the active ones) whose action depended on the superior, contemplative, angels.

These issues remained unsolved in the *Convivio*. Dante indirectly returned to the question of the relation between speculative and active functions in the first book of the *Monarchia*, where he discussed the nature of human beings and the goal they should pursue to fulfill their nature. In setting once again his discourse within the framework of Aristotelian ethics, Dante affirms that all things have been created with a purpose:

> Propter quod sciendum primo quod Deus et natura nil otiosum facit, sed quicquid prodit in esse est ad aliquam operationem. Non enim essentia ulla creata ultimus

---

[37] For a more detailed discussion on Avicenna's hypothesis, see chapter III. Giles of Rome, was an Augustinian theologian who authored four series of *Questions* on angels in the second half of the thirteenth century. Giles of Rome, unlike Aquinas and other theologians, followed the Arab philosopher and assumed the existence of inferior and superior Intelligences. On Giles of Rome's doctrine on angelic knowledge, see TIZIANA SUAREZ-NANI, *Connaissance et language des anges*, pp. 77-163.

finis est in intentione creantis, in quantum creans, sed propria essentie operatio: unde est quod non operatio propria propter essentiam, sed hec propter illam habet ut sit (*Monarchia* I, iii, 3).

Consequently the first point to bear in mind is that God and nature do nothing in vain; on the contrary whatever they bring into being is designed for a purpose. For in the intention of its creator *qua* creator the essential nature of any created being is not an ultimate end in itself; the end is rather the activity which is proper to that nature; and so it is that the activity does not exist for the sake of the essential nature, but the essential nature for the sake of that activity.

Everything that exists, thus, exists in view of a «certain operation». The operation proper to the nature of intellectual substances consists in their act of understanding, namely in their contemplation of the divine being. This proposition, conflicts with what Dante had affirmed in the *Convivio*, when he said that Intelligences were purely active substances. In another passage, however, he seems to indicate a possible way out of this conundrum. Speaking again of the possible intellect, specific to human beings, Dante affirms that speculative intellect becomes «by extension» practical, «its goal being doing and making»:

Et huic sententie concordat Averrois in comento super hiis que *De anima*. Potentia etiam intellectiva, de qua loquor, non solum est ad formas universales aut speties, sed etiam per quandam extensionem ad particulares: unde solet dici quod intellectus speculativus extensione fit practicus, cuius finis est agere atque facere (*Monarchia* I, iii 9).

And Averroes is in agreement with this opinion in his commentary on the *De anima*. Now the intellectual potentiality of which I am speaking is not only concerned with universal ideas or classes, but also (by extension as it were) with particulars; and so it is often said that the theoretical intellect by extension becomes practical, its goal then being doing and making.

Such 'extension' («quondam extensionem») derives from the capacity of speculative intellect to deal with 'particulars' or species. Although the above passage deals with human intellect, the observation could be extended to angelic intellect, for angels knew simultaneously universals and particulars.[38] This is the direction Dante seems to have taken in designing the functions of angelic Intelligences in the *Commedia*. The notion that speculation extends and transforms into practical action is evident at all levels of the poem, from the angels

---

[38] According to Aquinas, angels know objects in their singularity 'a priori', without sensible mediation. See AQUINAS, *S. Th.* I, q. 57, a. 2, and Tiziana Suarez-Nani discussion on the philosophical implications of the angelic knowledge of individual realities, essential to perform their role as instruments of divine Providence. TIZIANA SUAREZ-NANI, *Connaissance et langage des anges*, pp. 52-54.

to Dante himself. The references to the author's 'mission' as 'scribe' convey the idea that the poem was a transcription of the knowledge his ultra-mundane vision had conveyed to him.

The idea Dante enunciated in the *Monarchia* finds application in the *Commedia*. In the heaven of the Primum Mobile, when Beatrice shows to Dante the vision of the angelic choirs, there is no longer any distinction between contemplative and active functions. The implicit corresponding division between inferior and superior Intelligences disappears. Angels contemplate and incessantly circle around the luminous point at the centre of the universe. Their desire to reunite with God makes them move, and by moving they set the entire cosmological machine in motion. Their understanding is thus immediately transformed into the operations specific to their nature. The most important of these operations, the one that attracts Dante's curiosity in canto XXVIII, is the motion of the heavenly spheres, which he assigns to all orders, and will be discussed in Chapter III.

In the *Commedia*, the concept of operation is not only philosophically consistent, but – as act inspired by the Holy Spirit – is also consistent with the Christological foundation of the poem. The central importance of operations in a Christian perspective emerges in *Paradiso* XXV where Saint James examines Dante on hope.

In naming the sources that had inspired hope in him, Dante recalls James' epistle (line 77); a fundamental text for the significance of works in the doctrine of faith and justification:

> Tu mi stillasti, con lo stillar suo,
> ne la pistola poi; sì ch'io son pieno,
> e in altrui vostra pioggia repluo (*Par.* XXV, 76-78).
>
> And just as he instilled, you then instilled
> with your Epistle, so that I am full
> and rain again your rain on other souls.

A verset from David's *Psalm* IX («"sperino in te" [...] "color che sanno il nome tuo"», *Par.* XXV, 73-74) introduces Dante's answer to James' question on hope.[39] The connection is not a casual one: Dante explains that David's

---

[39] It is David's *Psalm* IX, 11, «Sperent in te qui noverunt nomen tuum», which underlines the relation between faith and hope. For the importance of the Sapiential tradition for the ethical-educative mission of the *Commedia*, and the bibliography on this topic see ANTONIO ROSSINI, *Il Dante sapienziale*. See also LINO PERTILE, *La puttana e il gigante. Dal Cantico dei Cantici al Paradiso Terrestre di Dante*, Ravenna, Longo, 1998, and PAOLA NASTI, *Favole d'amore e "saver profondo". La tradizione salomonica in Dante*, Ravenna, Longo, 2007.

wisdom and James' epistle are the two sources that have enlightened him on the meaning of this theological virtue. The epistle, however, does not explicitly treat the theme of hope. Rather, it is one of the most relevant documents for the doctrine of works. In the epistle, the Apostle states that «Sicut enim corpus sine spiritu mortuum est, ita et fides sine operibus mortua est» [«For even as the body without the spirit is dead; so also faith without works is dead». St. James, 2:26]. Faith, works, and hope are intimately bound together in this passage, establishing a relation that the Davidic chant emphasizes, and which finds its climax in the conclusion of line 78, where Dante returns on the theme of his own poetic mission: «e in altrui vostra pioggia repluo». The linking of faith, works, and hope is visible and strengthened by the indirect reference to Dante's individual experience, which opens the canto and provides a subliminal hermeneutic key for the interpretation of the meaning of the subsequent tercets. Canto XXV opens with Dante's reflection on his own work, which he hopes may gain him his return from exile.[40] The theme of hope is thus introduced by, and extended to, Dante personal experience. He wishes that his poetic work, the fruit of his painful expatriation from his origins, may reopen the doors of Florence to him, and give him hope of a possible existential 'salvation'. In Dante's individual experience, work and hope are connected, thus reflecting the message contained in James' epistle. Dante's choice of Saint James as the pilgrim's examiner in *Paradiso* XXV could be seen in the light of this connection. Angelic operations and human works are but two different although analogous aspects of the participative theology of the *Commedia*. Both are inscribed in the providential order governing creation, and both testify to the tropological dimension of the poem.

CONCLUSIONS

In this chapter, we have seen how the debates on angelic hylomorphism and the functions of angels in the hierarchical structure of the universe were

---

[40] «Se mai continga che 'l poema sacro / al quale ha posto mano e cielo e terra, / sì che m'ha fatto per molti anni macro, / vinca la crudeltà che fuor mi serra / del bello ovile ov'io dormi' agnello, / nimico ai lupi che li danno guerra; / con altra voce omai, con altro vello / ritornerò poeta, e in sul fonte / del mio battesmo prenderò 'l cappello». [«If it should happen ... If this sacred poem- / this work so shared by heaven and by earth / that it has made me lean through these long years- / can ever overcome the cruelty / that bars me from the fair fold where I slept, / a lamb opposed to wolves that war on it, / by then with other voice, with other fleece, / I shall return as poet and put on, / at my baptismal font, the laurel crown»]. *Par.* XXV, 1-9.

central to define the role angels played in their mediating function between divine and human spheres. Theologians presented a variety of different positions, rarely definitive and often difficult to reconcile. The debates on these questions constituted the intellectual backdrop against which Dante structured his angelology and made it a key element in the poetics of the *Commedia*. We have seen that Dante approached the philosophical issues concerning angelic nature and operations in view of making angels material instruments in the communication between intelligible and sensible worlds. By finding original solutions to the much-debated questions on the angels' role as mediators between supernatural and natural spheres, their nature, their functions, Dante weaved together theology and philosophy, and superseded both in a superior poetic expressive synthesis. Crucial to this end, was his theory of angelic operations. The link Dante established between contemplation (of divine science) and action (to direct creatures in their process of reunification with God) is indissoluble and crucial to define the role of angels as governing instruments of creation. It provided Dante with the philosophical ground necessary to design his Intelligences as mirrors conveying divine illumination that connected supernatural and natural worlds. As seen, a critical change intervened between the *Convivio* – where a philosophical (Aristotelian) approach predominated – and the *Commedia* – where Dante adapted the Dionysian hierarchical structure to the cosmological functions of angels. In combining Greek-Arab Intelligences and Christianized Neo-Platonic angels, he reconciled the Aristotelian metaphysical foundation of his cosmology with the theological doctrine of creation, and found answers to questions that a purely philosophical approach had raised but left unsolved. How Dante adapted Dionysian angelology to serve his poetic purposes is the subject of the next chapter.

CHAPTER TWO

DANTE AND CHRISTIAN ANGELOLOGY. LIMITS AND INFLUENCE OF THE PSEUDO-DIONYSIAN TRADITION

In the Middle Ages, the Areopagite was the acknowledged authority in angelology. In the twelfth and thirteenth centuries, an unprecedented revival of Dionysian studies followed centuries of partial oblivion. During this period, Scholastic thought gradually assimilated the Areopagite's theology. His tripartite hierarchical ordering became the cornerstone of medieval angelology and ecclesiology, merging with and prevailing over a tradition less systematic and more focused on angelic functions, whose main representative was Gregory the Great. The universality of Dionysius's system – essentially a Christian transposition of the metaphysics of participation of Neo-Platonic influence – appealed to both mystics, such as the Victorines, and Scholastic theologians, such as Aquinas and Bonaventure, who partly mediated between Dionysian and Gregorian approaches to angelology. This chapter analyzes Dionysius's doctrine of angelic hierarchies and focuses on the modifications Dante introduced to adapt it to the structural needs of the *Commedia*. The first section investigates the interaction of Gregory the Great and Dionysius's angelic doctrines through the filter of Aquinas and Bonaventure's Scholastic theology. It presents the assimilation of these traditions as a necessary premise to Dante's design of a cosmological role of angels that harmonized Christian Trinitarian theology and Aristotelian philosophy. Core Dionysian concepts such as light, hierarchy, imitation, merit, and angelic mirroring function therefore needed reinterpretation. Section II analyses the notion of light in the *Commedia* within the context of medieval theories of light. Section III discusses the notions of hierarchy, imitation, merit, and mirroring function in Dante.

CHAPTER TWO

I. THE INTERACTION OF BIBLICAL AND PSEUDO-DIONYSIAN TRADITIONS IN THE COSMOLOGICAL CONTEXT OF THE *DIVINA COMMEDIA*

Dante's references to Dionysius in *Paradiso* (X, 115-17; XXVIII, 130-32) and in the *Epistles* (IX, 16; XIII, 60) suggest that to some extent he knew at least the *De coelesti hierarchia*, *De divinis nominibus*, and some of the authoritative commentaries on these works, such as Albert the Great's.[1] Particularly relevant for their influence on Dante were Aquinas and Bonaventure's doctrines on angels, and particularly their *Summa Theologiae* (I, a. 50-64; 106-114) and *Collationes in Hexaëmeron* (*Conf.* XXI-XXIII) respectively.[2] Indicative of the importance of the cosmological context for Dante's angelology, and of the influence of these Doctors of the Church in mediating Dionysian thought in the *Commedia*, is the fact that Bonaventure and Aquinas lead the dance of the two crowns of wise spirits in the heaven of the Sun (*Paradiso* X-XIII), where Dionysius appears among the Dominican splendors. The following paragraph analyzes the presence of Aquinas and Bonaventure's themes in the relation between angels and cosmic order in *Paradiso* X.

---

[1] For Dante's knowledge of Albert the Great's commentary on Dionysius's *De colesti hierarchia*, see DIEGO SBACCHI, *La presenza di Dionigi Areopagita*, pp. 1-58. Diego Sbacchi shows that only Albert the Great's commentary provides an explanation for the correspondence of the angelic order of the Thrones with the contemplative souls and of the order of Dominations with that of the souls of justice. Even in this case, Dante seems to have paid special attention to the functional aspect of hierarchical arrangement. On the influence of Dionysius's treatise on the *Divine Names* see PIERO SCAZZOSO, *I nomi di Dio nella "Divina Commedia" e il "De divinis nominibus" dello Pseudo-Dionigi*, «La Scuola Cattolica», LXXXVI, 1958, pp. 198-213. For a description of the *Corpus Dionysiacum* and bibliographical references on the influence of the Areopagite on Dante see chapter I of this book, pp. 1-9.

[2] Aquinas wrote on angels in numerous works. Of particular relevance, beyond the *Summa Theologiae*, are his *Summa contra Gentile*, and the opuscule *On spiritual substances*. Another important text for Bonaventure's angelology is the *Praenotata* to *Distinctio IX*, in his *Lectura* to Book II of Peter Lombard's *Sentences*. As Barbara Faes de Mottoni has shown, this text presents differences with the angelology of the *Collationes in Hexaëmeron*, where a mystical reading of Dionysius's hierarchies seems to prevail, and reveals Bonaventure's early intention to mediate between Dionysian and the Gregorian traditions. See BARBARA FAES DE MOTTONI, *Bonaventura e le gerarchie angeliche*, «Freiburger Zeitschrift für Philosophie und Theologie», 3, 1993, pp. 312-358, and, of the same author, *San Bonaventura e la scala di Giacobbe. Letture di angelologia*. Napoli, Bibliopolis, 1995, particularly pp. 39-103. See also the exposition of Bonaventure's angelology in ÉTIENNE GILSON, *La philosophie de Saint Bonaventure*, Paris, Vrin, 1953, pp. 192-216. For Bonaventure's influence on medieval angelology, see also DAVID KECK, *Angels and Angelology in the Middle Ages*, and J.P. BOUGEROL, *Saint Bonaventure et la hiérarchie dionysienne*, «Archives d'Histoire Doctrinale et Littéraire du Moyen Âge», XXXVI, 1969, pp. 131-167. For Aquinas' angelology see JAMES D. COLLINS, *The Thomistic Philosophy of the Angels*, and ETIENNE GILSON, *The Christian Philosophy of St. Thomas Aquinas*.

a. *Angelic Hierarchies and Cosmic Order. Dionysius, Aquinas, and Bonaventure*

In the opening lines of canto X, Dante exhorts his readers to contemplate and admire the order of creation and establishes a visible link between cosmological order and the moving function of the angelic Intelligences:

> Guardando nel suo Figlio con l'Amore
> che l'uno e l'altro etternalmente spira,
> lo primo ed ineffabile Valore
>   quanto per mente e per loco si gira
> con tant'ordine fé, ch'essere non puote
> sanza gustar di lui chi ciò rimira (*Par.* X, 1-6).
>
> Gazing upon His Son with that Love which
> one and the Other breathe eternally,
> the Power-first and inexpressible-
>   made everything that wheels through mind and space
> so orderly that one who contemplates
> that harmony cannot but taste of Him.

The order regulating material and immaterial regions («quanto per loco e per mente si gira») is the result of the divine art of creation, where «girare» (to move) hints at the physical («per loco») and intellectual («per mente») moving operations of the angels. The contemplation of the harmony of creation becomes in Dante's poetic vision a «taste» («gustar») of God, where the lemma «taste» indicates the natural perception of divinity attainable through contemplation, conceived as a progressive ascent from sensible to spiritual worlds. The idea that it was possible to achieve knowledge of God (limited to what is accessible to human mind) by contemplating His creation and to arrive at the immaterial through the material – *per visibilia ad invisibilia* – reveals the latent presence of a symbolic theology of Dionysian derivation. These elements are however cast within the Trinitarian doctrine of creation, which surfaces in the image of the Father (the implicit subject of the line) gazing «with Love» on His Son (line 1) and suggests an influence of Franciscan, and particularly Bonaventure's, thought.[3]

---

[3] Giuseppe Mazzotta notices that Bonaventure's Trinitarian theology surfaces in the opening lines of canto X: «The Trinitarian theology at the opening of *Paradiso* 10 differs markedly from Saint Augustine's *De Trinitate* as well as from Boethius' *De Trinitate* and Aquinas commentary on it. Dante's view of God as generous source or inexaustible *fons* appropriates Saint Bonaventure's doctrine in the *Collationes*, where in creation as well as man emerge *ex nihilo*». GIUSEPPE MAZZOTTA, *The Heaven of the Sun: Dante between Aquinas and Bonaventure*, in *Dante for the New Millennium*, T. Barolini and H. Wayne Storey eds., New York, Fordham University Press, 2003, pp. 152-168. For the influence of Franciscan theology on the *Divine Commedia*, and particularly on *Paradise* see *Dante and the Franciscans*, S. Casciani ed., Leiden, Brill, 2006 (The Medieval Franciscans, 3), and NICK HAVELY, *Dante and the Franciscans: Poverty and the Papacy in the "Commedia"*, Cambridge, Cambridge University Press, 2004 (Cambridge Studies in Medieval Literature).

The *incipit* of canto X presents several of the threads Dante wove into the cantos of the Sun, and implicitly hints at Bonaventure and Aquinas as the main sources behind his adaptation of Dionysian angelology to a theological perspective centered on Christ's redemptive work, to which angels actively co-operate. The nexus between angels and cosmic order implicitly established anticipates an essential characteristic of the theological cosmogony of the *Commedia*, where separate substances are instrumental causes for the realization of providential order in the cosmos.[4]

If the accent on Trinitarian theology subliminally introduces the figure of Bonaventure, the vision of the order of creation in *Paradiso* X, 1-6 constitutes the backdrop against which Dante presented Aquinas as 'mediator' of Dionysius's angelology. Almost at the end of the canto, in *Paradiso* X, 115-117, Aquinas shows Dionysius among the contemplative souls of the Dominican circle of lights.[5] The theologian anticipates Beatrice's revelation of *Paradiso* XXVIII, by indicating Dionysius as the unique source of true angelic vision:[6]

> Appresso vedi il lume di quel cero
> che giù in carne più addentro vide
> l'angelica natura e 'l ministero (*Par.* X, 115-117).
>
> Next you can see the radiance of that candle
> which, in the flesh, below, beheld most deeply
> the angels' nature and their ministry.

The dittology «nature and ministry» qualifies the two aspects of angelology that receive preeminent attention in the poem, and that ultimately play a decisive role in its structure. The word «ministero» (ministry), placed in a privileged position at end of line 117, indicates the significance Dante, unlike Dionysius, attributed to angelic operations.

The tercet subliminally links the Neo-Platonic Dionysian system of the *De coelesti hierarchia* to Aquinas' systematic interpretation of his thought, both cast within the context set at the beginning of the canto. While Dionysius

---

[4] The relation between angelic orders and divine Providence will be further explored in Chapter III.

[5] This could be seen as an indirect reference to the major Scholastic commentaries to his works: the commentary on *De divinis nominibus* by Aquinas's, and the commentary on *De coelesti hierarchia* by Albert the Great.

[6] In *Paradiso* XXVIII, 130-132, Beatrice attributes to Dionysius the correct angelic ordering, and in *Paradiso* XXVIII, 136-139 she restates the commonly accepted idea that Saint Paul had revealed to Dionysius the true angelic ordering: «E se tanto secreto ver proferse / mortale in terra, non voglio ch'ammiri: / ché chi 'l vide qua sù gliel discoperse /con altro assai del ver di questi giri». [«You need not wonder if a mortal told / such secret truth on earth: it was disclosed / to him by one who saw it here above- / both that and other truths about these circles»].

had provided Dante with a general scheme of universal relations linking creatures to their transcendent origin, Aquinas had attempted a systematic reconciliation of Aristotelian philosophy with Christian theology. Most importantly, he had openly addressed the question of the possible cosmological functions of angels. Although he had not devised a rational structure to make Dionysian and Aristotelian separate substances truly compatible with Christian cosmology, he had nonetheless admitted that angels were responsible for celestial motion. The same desire of harmonizing Dionysian and Aristotelian systems characterizes the ideological perspective of the *Commedia*, where the theology of angelic hierarchies relies on the Aristotelian metaphysics of the celestial movers to connect material and immaterial sides of the universe.

Dante's interest in angelic operations descended from his vision of the cosmos as ordered theocracy, and reflected the functional interpretation that the theologians influenced by Biblical and Patristic traditions attributed to the governing operations of angels. Beyond maintaining a visible link with the Biblical tradition of performing angels, the term «ministero» in *Paradiso* X, 117 evokes ecclesiastical and political offices, and leads the reader to parallel celestial and worldly functions in ecclesiastical and secular institutions. Dionysius's *De ecclesiastica hierarchia* had explicitly illustrated the political nuances implicit in the idea of hierarchy, as it established that ecclesiastical and secular orders should mirror the nine-tiered celestial hierarchical arrangement.[7] The idea that things below should mirror things above greatly appealed to twelfth and thirteenth – century theologians such as Alain de Lille (*Hierarchia*) and William of Auvergne (*De universo*), who influenced both Aquinas and Bonaventure and contributed to form the angelological background of Dante's poem.[8] It is not surprising to find echoes of this aspect of the philosophical speculation on angelic substances resounding in Dante's carefully chosen word «ministero».

In this respect, the influence of Bonaventure seems particularly relevant. In his *Collationes in Hexaëmeron*, the Franciscan theologian related the three Dionysian hierarchies to the Trinity according to three principles: eternal exemplarity, integrity, and origin of monarchy. Each distinction corresponded to an interpretation of the hierarchical orders that focused on their functions as models for human behavior. According to integrity, for example, he de-

---

[7] The relation between worldly and heavenly orders was an aspect of Dionysian angelology that appealed to most theologians of the twelfth and the thirteenth centuries. See JOHN MARENBON and DAVID E. LUSCOMBE, *Two Medieval Ideas: Eternity and Hierarchy*, in *The Cambridge Companion to Medieval Philosophy*, A.S. McGrade ed., Cambridge, Cambridge University Press, 2003, pp. 51-72.

[8] For the different arrangements of the hierarchies in the writings of Alan de Lille, William of Auvergne, and Bonaventure, see DAVID E. LUSCOMBE, *The Hierarchies in the Writings of Alan de Lille, William of Auvergne and Bonaventure*, in *Angels in Medieval Philosophical Inquiry*, pp. 15-28.

fined hierarchies in terms of sacred power, sacred knowledge, and sacred action, which corresponded to the three kinds of life in heaven and on earth: contemplative, mixed, and active. Accordingly, knowledge pertained to the highest hierarchy, power (or order) to the intermediate, and action to the lowest, which consisted of the angels sent to assist human beings. Whether the lowest hierarchy only or all angels could be sent on mission to actively intervene in the human sphere was object of debate, and opinions vary even in different works of the same author. Unlike most theologians of his time, Dante seems to attribute the power to intervene and act in the human sphere to all hierarchies and not only to the lowest order. In Bonaventure's distinctions, the role of angels in directing action is evident, as is his strong interest in the implications of the Dionysian ordering in terms of governing functions. The *Commedia* develops these implications within a context centered on the physical and metaphysical meaning of angels. In Dante's design of angelic operations, both Aquinas's mediation of the Aristotelian doctrine of the Intelligences and Bonaventure's synthesis of Dionysian and Gregorian angelologies play a relevant role. The interaction of these traditions is determinant to understand fully how Dante expanded angelic functions to include that of celestial movers.

b. *The Functional Interpretation of Angels. Saint Paul, Augustine, and Gregory the Great*

Speculation on angelic «ministeri» as operations directed to assure divine intervention in the human world was not prevalent in Dionysius.[9] Rather, it belonged to the angelological tradition that originated with the Early Fathers and stemmed from Saint Paul, who in his epistle to the Hebrews wrote of the angels in terms of their role in the work of salvation:

Nonne omne sunt *administratorii* spiritus, in *ministerium* missi propter eos, qui haereditatem capient salutis? (*Hebr.* 1:14).[10]

Are they not all *ministering* spirits, sent to *minister* for them, who shall receive the inheritance of salvation?

---

[9] Dionysius interpreted angelic ministries in terms of their enlightening function. In *De coelesti hierarchia* he stated that these creatures have been called 'angels', namely messengers, because they manifest to us divine revelation: «Propterea et ultra omnia cognominatione anelica selectim dignae factae sunt, eo quod primo in seipasas edunt divinam illuminationem et per se in nos deferent, quae supra nos sunt, manifestationes». [«That is why they have a preeminent right to the title of angels or messenger, since it is they who first are granted the divine enlightenment and it they who pass on to us these revelations which are so far beyond us»]. *CH* IV, 2, 180B.

[10] My italics.

In Saint Paul angels are «missi» (sent) from heaven (like the divine messenger of *Inferno* IX) to operate by 'definition' for and among humankind.[11] A similar focus on angelic ministries was present in Augustine, for whom the noun «angel» (etymologically meaning *messenger*) did not indicate the nature, but the offices of spiritual substances:[12]

Spiritus autem Angeli sunt; et cum spiritus sunt, non sunt angeli; cum mittuntur, fiunt angeli. Angelus enim officii nomen est, non naturae. Quaeris nomen huius naturae, spiritus est; quaeris officium, angelus est: ex eo quod est, spiritus est; ex eo quod agit, angelus est. Vide illud in homine. Nomen naturae homo, officii miles: nomen naturae vir, officii praeco; homo enim fit praeco, id est, qui homo erat fit praeco; non qui erat praeco fit homo. Sic ergo qui erant iam spiritus conditi a creatore Deo, facit eos angelos, mittendo eos nuntiare quod iusserit; et ignem flagrantem facit ministros suos. Legimus apparuisse ignem in rubo, legimus etiam missum ignem desuper, et implesse quod praeceptum est. Ministravit ergo, cum impleret: cum esset, in natura sua erat; cum egit quod iussum est, ministerium implevit. Sic secundum litteram in creatura (*Enarrationes in Psalmos*, 103, I: 15).

The angels are spirits, and when they are spirits they are not angels; when they are sent they become angels. For angel is a name of office, not of nature. Thou askest the name of this nature, it is spirit; thou askest the office, it is angel; by that it is, it is spirit; by that it does, it is angel. See the same in man. Man (homo) is the name of nature, soldier of office: man (vir) is the name of nature, herald of office. For a man becometh a herald; that is, he that was before a man becomes a herald, not he that was before a herald a man. thus, therefore, those who were already made spirits by God the Creator, them He maketh angels, by sending them to declare what He had commanded, and He maketh a flaming fire His ministries. We read that a fire appeared in the bush. We read also that fire was sent from above, and fulfilled what was commanded, it fulfilled a ministry. Thus after the letter in the creature.

As «soldier» or «lawyer», Augustine says, the term «angel» designates the ministries the spiritual substances perform, according to the literal meaning of their names. Following in Paul and Augustine's steps, Dante designed his an-

---

[11] The essential function of angelic substances in both New and Old Testament was that of revealing and announcing the divine to humankind. The word angel derives from the Hebraic word *mal'akh*, which means 'messenger'. The term was translated in Greek as *anghelos* and in Latin as *angelus*. Spiritual entities having the function of messengers and revealing hidden divinity are present in many ancient traditions and in the Scriptures. For a history of angels and the confluence of the function of messengers typical of the ancient pagan tradition into the Biblical and Christian angelology, see MARCO BUSSAGLI, *Storia degli Angeli*, Milano, Rusconi, 1995, pp. 13-80.

[12] The Latin text of the *Enarrationes in Psalmos* is from AUGUSTINE, *Esposizioni sui Salmi*, («Opere di Sant'Agostino», III), Roma, Città Nuova, 1982 (1967). English translation is from Augustine's *Expositions on the Psalms*, vol. V, Oxford, J.H. Parker ed., 1853, pp. 83-84.

gels as spiritual officers who perform a variety of tasks by holding the cosmos in balance as instruments of grace, servants of Christ, and enlightening guides of souls on their path to salvation. Their operations thus assume a soteriological value, evident in the structural role Dante assigned them in the *Commedia* where not only they aid, guard, and illuminate the atoning souls and Dante the wayfarer, but also preside over the functioning of the material cosmos within a providential perspective. The angels' necessary presence in moments of impasse, as in *Inferno* IX, before the gates of Dis, or in *Purgatorio*, leads the reader to perceive these radiant beings as officers attending to their governing functions, equally essential to the order of universe as to that of the *Commedia*.

Gregory the Great was the major interpreter of the 'functional' approach to angels we find in Saint Paul and Augustine, which he developed within a Dionysian hierarchical setting by stressing the practical more than the anagogical aspect of angelic ministries.[13] His works represented the most important point of reference for those who, like Dante, were interested in a soteriological angelology. Most likely, this was among the reasons Dante followed Gregory's ordering in *Convivio*, where his interest in reconciling ethics and theology was already evident.[14] In the *Commedia*, however, he chose to adopt the comprehensive Dionysian system as a more appropriate springboard to solve some of the philosophical issues left unresolved in the *Convivio*, as seen in chapter I.

The Gregorian angelological tradition started to merge significantly with the one stemmed from the Dionysian *Corpus* in the twelfth century, and developed consistently during the thirteenth century. The result was a system in which both contemplative and active angelic functions had relevance,

---

[13] In the arrangement adopted by Gregory the Great in his *XL Homiliarum in Evangelia*, II, xxxiv, 7, PL, 76, 1249 D - 1250 C and which also Bernard of Clairvaux followed in his *De considaratione* V, the positions of the Virtues and the Principalities are reversed with respect to the order proposed by Dionysius. In his *Moralium libri, sive expositio in librum B. Job* XXXII, xxiii, 48 Gregory the Great proposed a completely different order, adopted in parallel also by his contemporary Isidore of Seville in *Etymologiarum sive Originum libri* XX, VII, v. The ordering is the following: 1) Seraphim, Cherubim, Powers; 2) Principalities, Virtues, Dominations; 3) Thrones, Archangels, Angels. See ATTILIO MELLONE, *Gerarchia angelica*, in *ED*, pp. 122-124, and C.A. PATRIDES, *Renaissance Thought*, p. 160. Gregory the Great classified the hierarchies according to their different functions, assuming a relation of analogy between angels and human beings. Angels were archetypes of human behavior: as angels imitated God, so men imitated angels. G. BAREILLE, *Angeologie d'apres les Pères (Ange)*, *Dictionnaire de Théologie Catholique*, pp. 1192-1211.

[14] The centrality of the illuminating function in Dionysius's angels is likely another reason he became his major reference in the *Commedia*, and rejected Gregory's the Great's in *Paradiso* XXVIII, 130-135. The stress Dionysius placed on angels as 'mirrors' of God, and therefore mediators of light, underlines the harmony of human and divine wills in the vision of salvation that characterizes the *Commedia*. See ATTILIO MELLONE *Angelo*, and MARCO BUSSAGLI, *Storia degli angeli*, pp. 193-196.

although they were not yet systematically coordinated from a philosophical point of view. The synergetic combination of Neo-Platonic and Patristic doctrines prepared the terrain for the reconciliation of Christian angelology with the Aristotelian doctrine of the moving Intelligences we find in Dante. Aimed at assuring communication between heaven and earth, Dante's angelology implied a role of philosophical and theological mediation to angels, subordinately and in cooperation with Christ's mediation. His reinterpretation of hierarchies as enlightening celestial movers transformed the mystic participative angelology of Dionysius into one essentially serving the practical purpose of defining divine intervention in the cosmos and the human world, in accordance with the eschatological perspective of the *Commedia*. While the Dionysian contemplative-illuminative structure provided the theological ground for the role Dante intended to attribute to the angels as vectors of divine knowledge interfacing immaterial and material spheres, the Gregorian tradition contributed to define the ethical impact of angelic operations through their influence on the sublunar world. Rather than the result of an unconscious syncretism, Dante's combination of these two traditions through the filter of Scholastic theology appears as an intentional choice aimed at establishing a harmonious interplay of cosmic and ethical orders. In so doing, reconciled Christian and Aristotelian perspectives within a Trinitarian theology centered on Incarnation as absolute mediation between God and humankind. Although interlacing Gregorian and Dionysian themes, Dante's angelology thus conceptually differed from both.

The combination of these two streams of thought strengthened the ontological relation between contemplative and active functions that marks the angelic invention of the *Commedia*, and constitutes the philosophical premise to the angels' role as chronocrators. In Dante's angelology, contemplation leads to action, and cognitive and active moments are neither logically nor temporally separable. Understanding and communicating, knowing and operating, are reciprocally necessary moments in the eschatological perspective of the poem. Cognition has an eminently practical end, for angels understand what is necessary to carry out their ministries. In the same way, Dante's vision of God in the Empyrean is a cognitive vision that produces the action of writing, an action initiated by Mary, and assisted by the enlightening power of Dante's guides. Writing is after all *techne* (art), deeply rooted in intellectual understanding, granted by merit and grace, and therefore intrinsically ethical.

CHAPTER TWO

II. THE INTELLECTUAL LIGHT

Light was a key notion in the Dionysian hierarchical system, instrumental to the participative character of his angelology. Although a thorough investigation of light in terms of metaphysics goes beyond the scope of the present work, to analyze the nature and function of light within the angelic setting of the *Commedia* is a necessary premise to understand Dante's approach to angelology. Scholars, with diverging opinions, have suggested a direct or indirect influence on Dante of medieval theories of light (grouped under the loose definition of 'metaphysics of light').[15] Dante's use of light in the *Commedia* seems however to combine Neo-Platonic thought with a scientific tradition of Aristotelian-Arab derivation.[16]

---

[15] Among the scholars who argued in favor of the Neo-Platonic influence of 'metaphysics of light' are Bruno Nardi, Joseph Mazzeo, P. Egidio Guidubaldi, and Maria Corti. Of a different orientation are instead Attilio Mellone and Simon Gilson. See in particular BRUNO NARDI, *La dottrina dell'Empireo nella sua genesi storica e nel pensiero dantesco*, in *Saggi di filosofia dantesca*, Firenze, La Nuova Italia, 1967², pp. 167-214; JOSEPH MAZZEO, *Light, Love, and Beauty in the Paradiso*, «Romance Philology», XI, 1957-58, pp. 1-18; ID., *Cultural Tradition in Dante's "Commedia"*, Ithaca, Cornell University Press, 1960, in particular chapters II and III; EGIDIO GUIDUBALDI, *Dal "De luce" di R. Grossatesta all'isalmico "Libro della scala". Il problema delle fonti arabe una volta accettata la mediazione oxfordiana*, Firenze, Olschki, 1978; MARIA CORTI, *Metafisica della luce come poesia*, in *Percorsi dell'invenzione. Il linguaggio poetico e Dante*, Torino, Einaudi, 1993, pp. 147-163; ATTILIO MELLONE, *Luce*, in *ED*, pp. 712-713; SIMON GILSON, *Light Reflection, Mirror Metaphors, and Optical Framing in Dante's Commedia: Precedents and Transformations*, «Neophilologus», LXXXIII, 1999, pp. 241-252. For a detailed bibliography on works on the 'metaphysics of light' and a discussion of the various positions of scholars supporting the influence of Neo-Platonic texts, see SIMON GILSON, *Medieval Optics and Theories of Light in the Works of Dante*, Lewinston, Edwin Mellen Press, 2000, particularly chap. V on *Dante and the "Metaphysics of Light": A Reassessment*, pp. 151-169. Gilson critically revises the notion of light metaphysics, first proposed by Clemes Baeumker in 1908. This term, now widely used, induces the false belief that a systematic and commonly shared doctrine of light existed in the Middle Ages: «The term "light metaphysics" is, however, often used uncritically and, given the diversity of contexts and differing strands of thought subsumed within this heading, it is misleading to regard medieval ideas about light as forming a unified, coherent, or even exclusively Neoplatonic body of doctrine». SIMON GILSON, *Medieval Optics*, p. 168.

[16] Of particular influence on the medieval theory of optics were the translations of Euclid's treatise on Geometry, *Elements* (first translated into Latin by Adelard of Bath in the twelfth century and quoted by Dante in *Inferno* IV, 142), and the treatise *Optica* attributed to Euclid; the *Book of Optics* by Alhazen (Abū 'Alī al-Ḥasan ibn al-Ḥasan ibn al-Haytham), translated into Latin at the end of the twelfth century; Al Kindi's (Abū Yusuf Ya'qūb ibn Isḥaq al-Kindī, ninth c.) treatises *De aspectibus* and *De radiis stellatis*, on stellar radiations; Roger Bacon's *Perspettiva*, a part of his *Opus maius*. The literature on medieval perspective is vast. I limit myself to refer to the already cited work by SIMON GILSON, *Medieval Optics*, and, in particular for the analysis of Islamic science, to GRAZIELLA FEDERICI VESCOVINI, *Studi sulla prospettiva medievale*, Torino, Giappichelli, 1965, and ID., *Le teorie della luce e della visione ottica dal IX al XV secolo. Studi sulla prospettiva medievale e altri saggi*, Perugia, Morlacchi, 2003. See also GIUSEPPE BOFFITO, *La teoria della visione in Dante e in Cecco d'Ascoli*, «Rivista di fisica, matematica e scienze naturali», s. 2, IV, 1929-30, pp. 67-73; ID., *Quali esperienze e leggi dell'ottica furono note a Dante e Cecco d'Ascoli?*, ivi, pp. 225-228; ALESSANDRO PARRONCHI, *La perspettiva Dantesca*, «Studi danteschi», XXXVI, 1959, pp. 5-103; ID., *Perspettiva*, in *ED*, pp. 438-439. The

In the Neo-Platonic tradition, light was the visible sign of divine emanation. Whether descending from God, or identified with Him, it penetrated the universe through a hierarchy of luminous beings. Early Christian adaptations of Neo-Platonism, such as Dionysius's, developed a language based on the metaphorical expansion of light as revelatory sign of the divine presence that penetrated the universe through luminous streams.[17] The inflow of scientific Greek and Arab texts after the eleventh century influenced the development of medieval optics (*perspettiva*) by introducing a nexus between theories of light and causality.[18] Among these texts, particularly influential were the anonymous *Liber de causis* (twelfth c.), which Dante repeatedly quoted in the *Convivio*, and the *De intelligentiis* (thirteenth c.) attributed to the Pseudo-Vitellione (or Adam of Belladonna). These Neo-Platonic treatises provided the ground for later developments of theories of light within a Christian perspective, and merged with the Augustinian doctrine of illumination. Light was further associated to gnosiological causality, leading to view illumination as a necessary condition to the intellectual contemplation of the divine.

---

motive of light is particularly relevant in the figuration of Islamic Paradise, and particularly in the Kitab al-Mi'raj (*Book of the Ladder*) attributed to Abu'l-Qasim). The book was first translated into Latin as (*Liber Scale Machometii*) in 1264 for Alphonso X of Castillia el Sabio. For a possible Dante's knowledge of this text see MIGUEL ASÍN PALACIOS, *La Escatologia musulmana en la "Divina Comedia"*, Madrid, Instituto Hispano-Árabe de Cultura, 1961³, which contains the history of the debate that Palacios's discovery generated among Dante scholars. Maria Corti suggested that in this period Dante's mentor Brunetto Latini was at the court of Alphonso X and could have brought a copy of the Kitab al-Mi'raj to Dante. See MARIA CORTI, *La Commedia di Dante e l'oltretomba Islamico*, in *Scritti su Cavalcanti e Dante. La Felicità mentale. Percorsi dell'invenzione e altri saggi*, Torino, Einaudi, 2003, pp. 365-379. See also ID., *Metafisica della luce come poesia*, p. 126 and p. 160, n. 11.

[17] For the sinestesia in metaphors of light and water, typical of these texts and recurrent in Dante, see MARCO ARIANI, *'Metafore assolute': emanazionismo e sinestesie della luce fluente*, in *La metafora in Dante*, M. Ariani ed., Firenze, Olschki, 2009, pp. 193-219.

[18] Dante refers to the *perspettiva*, or medieval optics, in *Convivio* II, iii, 6, and II, xiii, 27. In the first passage Dante speaks about the number of the heavens according to Tolomeus, and places the *perspettiva* among the disciplines that «determine and manifest» the position of the stars: «Sì che secondo lui [Tolomeus], secondo quello che si tiene in astrologia ed in filosofia poi che quelli movimenti furono veduti, sono nove li cieli mobili; lo sito delli quali è manifesto e determinato, secondo che per un'arte che si chiama perspettiva, e [per] arismetrica e geometria sensibilemente e ragionevolemente è veduto, e per altre esperienze sensibili». [«So that according to him and according to the received opinion in astrology and in philosophy since the time those movements were first perceived, there are nine moving heavens; and their position is manifest and determined by the art called optics, and by arithmetic and geometry, as is perceived by the senses and by reason, and by other demonstrations to the senses»]. The second passage classifies *perspettiva* as auxiliary discipline to geometry, associated with the heaven of Jupiter: «la Geometria è bianchissima, in quanto è sanza macula d'errore e certissima per sé e per la sua ancella, che si chiama Perspettiva». [«Geometry is furthermore most white insofar as it is without taint of error and most certain both in itself and in its handmaid, which is called Optics»]. For the importance of geometry as form of knowledge and discipline auxiliary to *perspettiva* in *Paradiso*, see GIUSEPPE MAZZOTTA, *Spettacolo e geometria della giustizia (Paradiso XVIII-XX): L'Europa e l'universalità di Roma*, in *Dante e l'Europa*, Ravenna, Centro Dantesco dei Frati Minori Conventuali, 2004, pp. 59-77.

Both of the above quoted texts affirmed that light was a substantial form (*forma substantialis*), and while in the *Liber de causis* God was only the source of divine illumination, the *De intelligentiis* even identified Him with light. The assumption of formal substantiality of light was a fundamental common trait among the Christian Neo-Platonic authors generally included in the category of 'metaphysics of light' such as Robert Grosseteste, Roger Bacon, and Bartholomew of Bologna.[19] Many scholars have suggested a possible influence of Batholomew of Bologna's treatise *De luce* particularly on *Convivio* III, xiv, where Dante describes the distinctions among light («luce»), illumination («splendor»), and ray («raggio»), which play an important role in Dante's engineering of light transmission in the *Commedia*.[20] These notions were however common knowledge among the scholars of optics, and Bartholomew of Bologna seems more relevant to Dante because of his Christological approach to light. A commentary on the evangelical identification of Christ as «lux», the *De luce* reoriented the approach to the doctrine of light from a creational to a soteriological perspective.

---

[19] In his opuscule *De luce*, Grosseteste posited light as the principle, first cause, and substantial form of matter. For Grosseteste light was corporeity. It allowed matter to assume dimensionality, and its expansion and contraction produced the universe. Evident in Grosseteste's writings is the influence of Augustinian Neo-Platonism, the *Liber de causis*, and Arab authors such as, possibly, Avicebron's *Fons vitae*, in which the philosopher introduced the idea of luminous rays emanating from substances. In his introduction to Grosseteste's opuscola, Pietro Rossi shares J. McEvoy's opinion that the original trait of Grosseteste's works on light was his attempt to base his synthesis of *Genesis* and Aristotelian cosmologies on a mathematical structure of reality. See *Roberto Grossatesta, De luce seu inchoatione formarum*, in *Metafisica della luce*, P. Rossi ed., Milano, Rusconi, 1986, pp. 109-123. See also JAMES MCEVOY, *Medieval Cosmology and Modern Science*, in *Philosophy and Totality*, J. McEvoy ed., Belfast, 1977, pp. 91-110, and ATTILIO MELLONE, *Luce*, in *ED*, pp. 706-713.

[20] «Ma però che qui è fatta menzione di luce e di splendore, a perfetto intendimento mostrerò [la] differenza di questi vocabuli, secondo che Avicenna sente. Dico che l'usanza de' filosofi è di chiamare "luce" lo lume in quanto esso è nel suo fontale principio; di chiamare "raggio" in quanto esso è per lo mezzo, dal principio al primo corpo dove si termina; di chiamare "splendore" in quanto esso è in altra parte alluminata ripercusso». [«But since light and reflected light have been mentioned here, I will, in order to be perfectly clear, clarify the difference between these terms according to the opinion of Avicenna. I say that it is customary for philosophers to call luminosity light as it exists in its original source, to call it radiance as it exists in the medium between its source and the first body which it strikes, and to call it reflected light as it is reflected into another place that becomes illuminated»]. *Convivio* III, xiv, 5. According to Vasoli and other scholars, in this passage Dante quotes Avicenna through other sources. Many agree that Bartholomew of Bologna's *De luce* was one of the most likely Dante's direct sources. Among these scholars are LEONARDO OLSCHKI, *Sacra dottrina e Teologia mystica. Il canto XXX del Paradiso*, «Giornale dantesco», XXXVI, n.s., IV, 1933, pp. 3-25, who shows the parallelism between the *De luce* and the third treatise of *Convivio*; CELESTINO PIANA, *Le questioni inedite «De glorificatione Beatae Mariae Virginis» di Bartolomeo da Bologna O. F. M. e le concezioni del Paradiso Dantesco*, «L'Archiginnasio», XXXIII, 1938, pp. 247-262. Celestino Piana noticed that Dante's description of the Empyrean corresponds to that Bartholomew of Bologna exposed in two *quaestiones* on the glorification of Mary. See also EGIDIO GUIDUBALDI, *Bartolomeo da Bologna*, in *ED*, pp. 526-527, and MARIA CORTI, *Metafisica della luce come poesia*.

In general, medieval Christian thinkers were reluctant to accept the identification of God with light, for the implicit risk of emanatism involved by assuming that light propagated through a chain of participative hierarchical illuminations. This assumption contradicted the notion of creation ex nihilo and prevented a logical separation between creator and creatures. For these reasons, authors such as Bonaventure adapted Neo-Platonic ideas on light to the theological framework of revelation, by distinguishing divine and material lights, and admitting only an analogical relation between them.[21] Light could be used solely in metaphorical sense to talk about creation and its ontological and gnosiological orders.

The vast use of images and metaphors of light in the *Commedia* testifies to the flourishing rhetoric of brightness that characterized medieval writers, even of different orientations.[22] Light metaphors and images recurred also in the writings of Thomas Aquinas, who strongly opposed the tendency of Franciscan Scholasticism to attribute formal substantiality to light. The Dominican theologian, following Aristotle, defended the accidental nature of light by arguing that substantial forms were not sensitively perceivable. Divine light is purely intellectual and, he argues:

it should be noticed that intellectual light is nothing else than a manifestation of truth, according to Ephesians 5:13: "All that is made manifest is light." Hence to enlighten means nothing else but to communicate to others the manifestation of the known truth (*S. Th.* I^a, q. 103, a. 1).

considerandum est quod lumen, secundum quod ad intellectum pertinet, nihil est aliud quam quaedam manifestatio veritatis; secundum illud ad *Ephes.* V, "omne quod manifestatur, lumen est." Unde illuminare nihil aliud est quam manifestationem cognitae veritatis alteri tradere

Dante's use of images and themes related to light certainly has a structural relevance, but it does not imply a Neo-Platonic vision of illumination. Rather, his conception of light seems closer to Aquinas's conception of light as man-

---

[21] See BONAVENTURE, *In II sent.* D. 13, a. 12, q. 11, ad 4. Quaracchi edition, p. 323 and ff. The distinction between divine light (*lux*) and material light (*lumen*) became quite common among Christian theologians, and it is particularly relevant in the *Commedia*. Such distinction already appeared in the PSEUDO-GROSSATESTA's *Summa philosophiae*. In the tradition of Neo-Platonic mysticism, as in Dionysius, divine light is not accessible, but defined as obscure luminosity. The idea of the 'point' which constitutes the center of Dante's speculation on light in the *Commedia*, finds an antecedent in Jewish mysticism, as illustrated in the *Book of Splendor* (*Zepher-a-Zohar* XIIIth c.). The 'point' is the origin of luminous energy beyond which resides the unknowable infinite.

[22] On Dante's use of metaphors of light in the *Commedia* see, SILVIA FINAZZI, *La metafora scientifica e la rappresentazione della corporeitas luminosa*, in *La metafora in Dante*, pp. 167-192.

ifestation of truth. Moreover, in the *Commedia* light is essentially instrumental. It is a means to actualize potentiality in matter, and a cognitive medium (both these functions are connected to angelic operations). It is not, as for the metaphysicians of light, a substantial form.

As in other areas, Scholastic authors tried to make the science of *perspettiva* compatible with Christian doctrine. While the elaboration of a scientific theory of vision based on physical laws provided ground for a theory of contemplative and beatific visions, the rhetoric of light developed in the womb of Augustinian Neo-Platonism provided a language able to represent ineffable divine reality as vision. Ideas and imagery of light had become common knowledge by Dante's times, and it is not by chance that they play an essential role in his poetic universe. As with other aspects of contemporary theological and scientific culture, Dante adapted them to his needs, rendering it difficult to classify his original thought within any specific doctrinal field.

The use of technical language and concepts such those Dante employed in the *Convivio* and even in the *Rime*, where a distinction between eternal immaterial light and corporeal luminosity already exists, is evidence that even before the *Commedia* Dante had familiarity with the basic notions of *perspettiva*.[23] In *Paradiso*, his acquaintance with the geometrical laws of optics is evident as we follow in the pilgrim's footsteps through the pattern of visions that leads to the extreme experience of the Trinitarian image in canto XXXIII, 54. Before this climax, the unbearable brightness of the 'point' at the center of the Empyrean hides the vision of the three circles. At his highest intellectual capacity, Dante's sight pierces through the intense light to achieve the vision of the sacred image («ché la mia vista, venendo sincera, / e più e più intrava per lo raggio /de l'alta luce che da sé è vera» [«because my sight, becoming pure, was able / to penetrate the ray of Light more deeply- / that Light, sublime, which in Itself is true»], *Par.* XXXIII, 52-54). Dante's eyes 'enter' the ray, becoming one with it until his sight is «consumed» («tanto che la veduta vi consunsi!» [«so long that I spent all my sight on it!»], *Par.* XXXIII, 84) until his power of vision entirely surpasses its physical boundaries and becomes absorbed in the purely intellectual light of the 'point'. Once his sight dissolves in the brightness of the ray, Dante can see the universal conflation of all forms into unity. Only then can he observe the moment that is opposite and complementary to that of creation, the moment in which he saw: «sustanze e accidenti e lor costume / quasi

---

[23] For the existence of a distinction between «lux» and «lumen» in the *Rime*, see GIAMPIERO W. DOEBLER, *Non mi può far ombra: Le distinzioni fra luce e lume nelle Rime di Dante*, «Tenzone», VII, 2006, pp. 29-50.

conflati insieme, per tal modo / che ciò ch'i' dico è un semplice lume» («substances, accidents, and dispositions / as if conjoined-in such a way that what / I tell is only rudimentary», *Par.* XXXIII, 88-90).

In the lower heavens, Dante had been able to observe how, from the absolute unity and perfection of God's mind, the action of hierarchies produced the infinite dispersion of reality. Now he can see the corresponding opposite moment in which love bounds together the scattered forms of reality («ciò che per l'universo si squaderna» [«what, in the universe, seems separate, scattered»], *Par.* XXXIII, 87) as words arranged in ordered strings of meaning to form a 'volume'. The «quasi» of line 89 signals, however, that Dante does not see the final return to unity of forms and accidents – which are still distinguishable to him. Rather, he discerns the «modo», namely the manner in which divine love holds them all together.[24]

The superior capacity of vision and understanding Mary's intercession had granted him, he can only dimly reproduce in words. Once resolved in language, the highest light becomes just a simple reflection «un semplice lume». And yet this dim light is enough to enlighten the readers and allow them to proceed with Dante to see the circles of Trinity shining in the depths of brightness:[25]

> Ne la profonda e chiara sussistenza
> de l'alto lume parvermi tre giri
> di tre colori e d'una contenenza; (*Par.* XXXIII, 115-117)

> In the deep and bright
> essence of that exalted Light, three circles
> appeared to me; they had three different colors,
>   but all of them were of the same dimension.

The circles mutually reflect light in a play of mirrors that at last reveals to Dante the seal of God in human effigy:

---

[24] On the 'modo' or order of creation, see chapter III, I c and f.

[25] The use of «lume» in line 116 is not congruent with the distinction Dante maintains between divine and material light, unless we interpret the line in terms of Dante's desire to stress the consistency of light when it becomes perceivable by the human eye, rendering an 'appearance', i.e., an image of the divine transcribed in decipherable visual language. This interpretation seems supported by the repeated idea of the circles as 'appearances' in *Paradiso* XXXIII, 127-129, where the Trinitarian image is again described as «reflected light»: «Quella circulazion che sì concetta / pareva in te come lume reflesso, / da li occhi miei alquanto circunspetta.» [«That circle-which, begotten so, appeared / in You as light reflected-when my eyes / had watched it with attention for some time»]. For a theological and interpretation of light in canto XXXIII see DIEGO FASOLINI, *'Illuminating' and 'Illuminated' Light: a Biblical-Theological Interpretation of God-as-Light in Canto XXXIII of Dante's Paradiso*, «Literature and Theology», XIX, 4, 2005, pp. 297-310.

> dentro da sé, del suo colore stesso,
> mi parve pinta de la nostra effige:
> per che 'l mio viso in lei tutto era messo (*Par.* XXXIII, 130-132).

> within itself and colored like itself,
> to me seemed painted with our effigy,
> so that my sight was set on it completely.

Beginning and end, center and periphery, coincide with the revelation of 'our' likeness to God in Incarnation. This final image perhaps hints at the limits of our understanding that, even when transhumanized in angelic intellectual perception, remains within the boundaries of the Word.

In the theological cosmology of the *Commedia*, the Empyrean 'point' is the unique source of pure immaterial light. No other body, whether corporeal or not, emanates light directly. The angels reflect it on the celestial bodies and from them it passes on to all creatures.[26] All reflected light, as seen in the above quoted passages from canto XXXIII, is properly 'illumination' «lume». In the setting of the poem, the angels propagate light by reflecting it. They are mirrors. Once their glowing rays penetrate the material spheres, the optical phenomena of reflection follow the geometrical laws of perspective in channeling the virtues and forms that flow down from the primary luminous source to activate the potencies of the material world.

The intellectual and physical mechanics of light reflection required an instrumental rather than an ontological function of light, in strict relation with angelic operations. In this respect, Dante seems closer to Aquinas than to the philosophers of the 'metaphysics of light'. From these philosophers, however, Dante derived other important elements besides the above-mentioned rhetoric of light, and particularly the cognitive properties of illumination. The relation between light and knowledge was a main concern in most Christian treatises on optics. Crucial to express it was the interpretation of Biblical passages concerning the creation of light, and particularly *Genesis* 1:3, upon which the exegetes based their interpretation of the Biblical creation of the angels: «Dixitque Deus: "Fiat lux". Et facta est lux». («And God said: Be light made. And light was made»).

A fundamental text in any medieval debate on light and creation, besides Basil of Cesarea's *Hexaëmeron*, was Augustine's *De Genesi ad litteram*.[27] Au-

---

[26] Dante had already illustrated this process in *Convivio*, III, xiv, 4. See discussion at p. 65.

[27] Basil of Cesarea describes light as a created substance that makes things visible. Light and night are opposite but belonging to the same act of creation, before which pure darkness embraced all elements. See BASIL OF CESAREA, *Sulla Genesi (Omelie sull'esamerone)*, M. Naldini ed., Milano,

gustine's interpretation of Genesis 1:3 seems particularly significant to retrace a possible ideological tradition behind Dante's final vision of Incarnation. In commenting on this Biblical verset, Augustine observed that if «be light made» referred to the creation of spiritual light, then this was to be understood not as God's co-eternal light, but as the generated Wisdom that illuminates spiritual and rational beings:

Si autem spiritalis lux facta est, cum dixit Deus: *Fiat lux*; non illa vera Patri coaeterna intellegenda est, per quam facta sunt omnia, et quae illuminat omnem hominem; sed illa de qua dici potuit: *Prior omnium creata est sapientia*. Cum enim aeterna illa et incommutabilis, quae non est facta, sed genita Sapientia, in spiritales atque rationales creaturas, sicut in animas sanctas se transfert, ut illuminatae lucere possint; fit in eis quaedam luculentae rationis affectio, quae potest accipi facta lux, cum diceret Deus: *Fiat lux*: si iam erat creatura spiritalis, quae nomine coeli significata est, in eo quod scriptum est: *In principio fecit Deus coelum et terram*; non corporeum coelum, sed coelum incorporeum coeli corporei, hoc est, super omne corpus, non locorum gradibus, sed naturae sublimitate praepositum (*Genesis ad litteram*, I, 17.32).

But if it was a spiritual light that was made when God said *Let light be made* (*Gen.* 1:3), it is not that true light, co-eternal with the Father, that is to be understood, through which all things were made, and which enlightens every human being, but that about which it could be said: *Before all things there was created wisdom* (*Sir* 1:4). When that eternal and unchangeable Wisdom, you see, which was begotten, not made, transfers itself into spiritual and rational creatures, as it does into holy souls, so that being thus enlightened they can themselves become sources of light, there is produced in them a kind of infection of shining, glowing intelligence; and this can be taken as made light, made when God said *Let light be made*, provided there was already a spiritual creation, which was signified by the word "heaven", where it is written, *In the beginning God made heaven and earth*. This was not a corporeal heaven but the incorporeal heaven of the corporeal heaven, set that is above every kind of body, not by degrees of space, but by the sublimity of its nature.

The *Fiat lux* of *Genesis* 1:3 is not the coeternal and absolute light of God Creator, but the light of the generated Word that instills a «luculentae rationis affectio»: a state of illuminated reason that guides spiritual and human creatures in their desire to know God. Augustine interlaced this created light of *Genesis* with the one mentioned in the *Ecclesiasticus*, thus providing Old-Tes-

---

Mondadori (Fondazione Lorenzo Valla), 1990, II, 7-8, pp. 59-69 and AUGUSTINE, *La genesi alla lettera*, Roma, Città Nuova (*Opera omnia di Sant'Agostino*), vol. IX/2, 1989; English translation from AUGUSTINE, *On Genesis*, Introduction, transl., and notes by E. Hill, O.P. and J.E. Rotelle, O.S.A., New York, New City Press («The Works of Saint Augustine. A Translation for the 21st Century», I/13).

tamentary ground for interpreting Christ as the pure light shining from the center of creation (an idea that Bonaventure developed in his *Conferences on Hexaëmeron*). In this way, he implicitly linked the «*fiat lux*» to the «*ego sum lux mundi*» of John 8:12 through the created Wisdom of *Ecclesiasticus*.[28] This connection between light and knowledge, and its Christological implications, emerge in Dante's sequence of images preceding the final vision of light in canto XXXIII.

The transposition of eternal light into enlightening wisdom characterized other authors engaged in speculation on optics, such as the already mentioned Bartholomew of Bologna, who diffusely commented on the Gospel passages in which the metaphor of light is developed in Christological perspective. Bartholomew interpreted the Redeemer as the primary source of pure light, that of divine cognition. Placed at the center of the universe, Christ is like a sun emanating rays that first touch upon those closest to him (the apostles), and then reach to the extreme limits of universe.[29] As in Augustine, this illumination penetrates both spiritual and rational beings. In Bartholomew's expanded similitude, the rays simply convey divine illumination, and the splendor is the luminosity the enlightened bodies emanate by reflection. Whether or not Bartholomew was Dante's direct source, his Christocentric vision of the 'sun' radiating wisdom seems analogous to what we find in the *Commedia*. The essential feature of light in Dante is indeed its connection with wisdom, as developed within an attested tradition, and its instrumental function with respect to angelic operations.

III. THE ANGELIC HIERARCHY

The concept of hierarchy was a cornerstone of Dionysian angelology and remained a key notion in Dante, who adapted it to a different context where angelic orders and heavenly spheres were connected. Dante maintained the Dionysian definition of hierarchy as an ordering of the angelic substances based on their 'nominal' characteristics, but he revised his mechanism of hier-

---

[28] The *Ecclesiasticus* (*Siracides* or *Ben Sira*) is one of the Deuterocanonical books of the Christian Bible. It is considered the last canonical representation of Jewish wisdom even if the book is not part of the Hebrew Canon. It contains a history of Salvation and identifies wisdom with Mosaic Law. For the relevance of this Book in Dante's *Commedia*, see ANTONIO ROSSINI, *Il Dante sapienziale*, pp. 171-176.

[29] BARTHOLOMEW OF BOLOGNA, *Tractatus de luce Fr. Bartholomaei di Bononia inquisitions et textus*, I. Squadrani ed., «Antonianum», VII, 1932, pp. 201-238, 337-376, 465-494, and in particular *Caput* III, p. 235.

archical illumination to make angels immaterial mirrors of divine enlightenment that connected material and spiritual worlds.

In the Dionysian system, the concept of hierarchy provided the structure whereby angels and human souls completed their return to unity with God through an anagogical process. Dionysius thus defined the celestial hierarchy:

> Est quidem hierarchia ordo divinus et scientia et actio, deiforme, quantum possible, similans et ad inditas ei divinitus illuminationes proportionaliter in dei similitudinem ascendens (*CH* III 1. 164 C).[30]
>
> In my opinion a hierarchy is a sacred order (*taxis*), a state of understanding (*episteme*) and an activity (*energheia*) approximating as closely as possible to the divine. And it is uplifted to the imitation of God in proportion to the enlightenment divinely given to it.

As Louis Bouyer notes, the term 'hierarchy' was Dionysius's invention. It defined the celestial ordering of immaterial substances, but it did not indicate a closed order.[31] Rather, this order was open towards inferior beings, for each angelic substance was subject to change in the action of transmitting and assimilating the illumination it received from the superior orders, and ultimately from God. Enlightenment and imitation connoted the process through which each hierarchy fulfilled its end in achieving participation and perfect union with God. Order (*taxis*), understanding (*episteme*), and activity (*energheia*) were the three constitutive elements of the notion of hierarchy. The first indicates the ranking of angelic substances in three hierarchies, each containing three orders; the second indicates the kind of science transmitted to angels; the third indicates the imitative activity through which they participate in the divine. These characteristics belonged to each order and corresponded to the three moments of purification, illumination, and union (or perfection) in which the angels' anagogical process occurs.[32]

---

[30] The Latin translation is from Johannes Scotus Eriugena (ca. 810-ca. 877), *Iohannis Scoti Eriugenae Expositiones in ierarchiam coelestem*, J. Barbet ed. Turnholti, Brepols, 1975. Greek quotations are based on PSEUDO-DIONYSIUS, THE AREOPAGITE, *La hiérarchie celeste*, Introd. by R. Roques; critical text edited by G. Heil; translation and notes by M. de Gandillac, Paris, Les Éditions du Cerf, 1958. Unless otherwise indicated, English translations are from *Pseudo-Dionysius. The Complete Works*, C. Luibheid, P. Rorem, R. Roques, with introductions by J. Pelikan, J. Leclercq, K. Froehlich, New York, Paulist Press, 1987.

[31] LOUIS BOUYER, *La spiritualità dei Padri* (III-IV secolo), in *Storia della spiritualità*, vol. 3/B. New edition expanded and updated by L. Dattrino and P. Tamburrino, Bologna, Edizioni Dehoniane, 1986, p. 136.

[32] Most interpreters and commentators used to assign each moment to the three orders of each hierarchy. The lowest would be associated with purification, the median with illumination, and the highest with purification. Albert the Great, however, demonstrated that this classification did not

The Areopagite was more interested in defining a structure that exemplarily indicated the process through which creatures participated and reunited with their creator than in angelic ministries. As seen in section I of this chapter, the thirteenth-century commentators of the *De coelesti hierarchia*, however, had prevalently interpreted angelic hierarchies in a soteriological perspective, and had grafted the exegetical tradition that focused on the moral aspects of angelic activity onto Dionysian mysticism. In attributing a role in the government of the material world to angels, Dante took a step forward in this direction that required a partial reformulation of the constitutive elements of Dionysius's definition of hierarchy.

a. *The Divine* Taxis

In defining hierarchical ordering, Dionysius applied to angelology his theology of divine names, whereby angelic arrangement found its justification in the Biblical – and therefore sacred – origin of angelic names.[33] These names designated the nature of angels and their specific operations, which in turn determined the position of angels in the nine orders grouped in three hierarchies.[34] By providing Biblical foundation to angelic functions, Dionysius established the ordering as «sacred» and therefore immutable. In Dante, however, the orders needed to be functional to the correspondence between angelic and heavenly spheres he had established in the *Commedia*. This conscious adaptation of the Dionysian ordering to his needs emerges in *Paradiso* XXVIII, 130-139, where Beatrice rejects Gregory the Great's classification (which Dante had previously adopted in *Convivio* II, v, 5-11), and indicates Dionysius's as the one who saw the truth because he arranged the angels 'like she does' («com'io»). Dante presents here a reversed perspective where theological doctrine is in agreement with the *Commedia* rather than the opposite,

---

belong to Dionysius's thought, for he attributed the three elements of purification, illumination, and unification to all hierarchies. See DIEGO SBACCHI, *La presenza di Dionigi Areopagita* (pp. 59-64).

[33] In the *De coelesti hierarchia* Dionysius specifies that Seraphim means ardor, Cherubim power of knowledge, Thrones detachment from anything base, Domination lordship as freedom from servility, Virtues virility in pursuing and transmitting virtue, Powers order in receiving and administering divine dispositions, Principalities Godlike princeliness in divine order. Dionysius does not specify the meaning of Archangels and Angels. He associates the former with both the functions of Principalities and Angels in administering and ordering things that pertain to the mundane sphere. As the lowest order, and the one closest to human beings, Angels directly intervene and enlighten those below them to direct them toward salvation. *CH*, *passim*.

[34] Starting from the highest, the first hierarchy includes Seraphim, Cherubim, and Thrones. The second includes Dominations, Virtues, and Powers; the third Principalities, Archangels, and Angels.

thus reaffirming the preeminence of his poetic truth over theology and philosophy.[35] Beatrice's indirect reference to Saint Paul's vision as the source of Dionysius's authority strengthens this perspective, and implicitly reinforces Dante's own authority as both seer and scribe.

In the systematic Dionysian *taxis*, the arrangement of angels corresponded to a hierarchy in contemplation, which determined the distribution of angelic functions according to a criterion synthesized by Aquinas in his *Summa Theologiae* I^a, q. 108, a. 6: «the highest hierarchy contemplates the ideas of things in God Himself; the second in the universal causes; and third in their application to particular effects».[36] In *Convivio*, probably through the mediation of Bonaventure, Dante associated the order of contemplation of the divine to the three persons of the Holy Trinity.[37] The first hierarchy from above contemplated God, the second Christ, and the third the Holy Spirit.[38] According

---

[35] Peter Hawkins observes the same pattern in *Purgatory* XXIX, where Dante makes John's Apocalypse agree with the *Commedia* rather than the other way around. PETER S. HAWKINS, *All Smiles: Poetry and Theology in Dante*, «PMLA», CXXI, n. 2, 2006, pp. 371-387: 372-373.

[36] «prima hierarchia accipit rationes rerum in ipso Deo; secunda vero in causis universalibus; tertia vero secundum determinationem ad speciales effectus». *S. Th.* I^a, q. 108, a. 6r. The correspondence between hierarchical arrangement and governing functions is evident in Aquinas' commentary on Dionysius's ordering: «Et quia Deus est finis non solum angelicorum ministeriorum, sed etiam totius creaturae, ad primam hierarchiam pertinet consideratio finis; ad mediam vero dispositio universalis de agendis; ad ultimam autem applicatio dispositionis ad effectum, quae est operis executio; haec enim tria manifestum est in qualibet operatione inveniri. Et ideo Dionysius, ex nominibus ordinum proprietates illorum considerans, illos ordines in prima hierarchia posuit, quorum nomina imponuntur per respectum ad Deum, scilicet Seraphim et Cherubim et Thronos. Illos vero ordines posuit in media hierarchia, quorum nomina designant communem quandam gubernationem sive dispositionem, scilicet Dominationes, Virtutes et Potestates. Illos vero ordines posuit in tertia hierarchia, quorum nomina designant operis executionem, scilicet Principatus, Angelos et Archangelos». [«And because God is the end not only of the angelic ministrations, but also of the whole creation, it belongs to the first hierarchy to consider the end; to the middle one belongs the universal disposition of what is to be done; and to the last belongs the application of this disposition to the effect, which is the carrying out of the work; for it is clear that these three things exist in every kind of operation. So Dionysius, considering the properties of the orders as derived from their names, places in the first hierarchy those orders the names of which are taken from their relation to God, the "Seraphim," "Cherubim," and "Thrones"; and he places in the middle hierarchy those orders whose names denote a certain kind of common government or disposition – the "Dominations," "Virtues," and "Powers"; and he places in the third hierarchy the orders whose names denote the execution of the work, the "Principalities," "Angels," and "Archangels"»]. *S. Th.* I^a, q. 108, a. 6.

[37] Bonaventure expounds in his *Coll. in Hex.* XXI, 1-3 the order of contemplation according to Father, Son, and Holy Spirit that Dante follows in *Convivio* II, v, 7-18. See also, as Busnelli and Vandelli suggested, Bonaventure's *Coll. in Hex.* XXI, 17-20. Vasoli indicates also Vincent of Beauvais, *Speculum Historiale*, I, 12. See Vasoli commentary on *Convivio* II, v, 8.

[38] The same order of contemplation appears in the passages where Dante describes the hierarchies and their functions in *Convivio* II, v, 5-18. In this order, there is an echo of Dionysius's attribution to all divine minds of a threefold distinction: «in tria dividuntur secundum se super mundane ratione omnes divini intellectus, in essentiam et virtutem et operationem» [«there is within all divine

to Bonaventure's interpretation of Dionysian hierarchies, in turn influenced by Hugh of Saint Victor's commentary on the *De coelesti hierarchia*, hierarchies receive a triple order of illumination, connected to the divine science of the Trinity (directed to God, Nature, Human Beings), and which induced in the angels the desire to participate in the design of salvation in accord with their different functions.[39] The Franciscan theologian posited three principles whereby the hierarchies were in relation to the Trinity. The first principle is conformity to their exemplar, expressed by their nine-tiers ordering, which is the same as Dionysius. The second principle is integrity, which corresponds to three types of science, mystic, speculative, and moral respectively corresponding to the three types of life, contemplative, mixed, and active. The third principle is multiformity of the heavenly monarchy, which corresponds to three aspects of angelic operations as related to superior things, to themselves, and to human beings. According to the last principle, the first hierarchy, composed of Seraphim, Cherubim and Thrones is ordered in terms of the process of conversion toward God in the three steps of purification, illumination, and union (*ordinata per conversionem*). Dominations, Virtues, and Powers compose the second hierarchy, which relates to Wisdom as principle of order and governance (*ordinata per potestatem*). The third hierarchy, composed by Principalities, Archangels, and Angels, communicates with the human world and receives divine science with respect to the effects of the Holy Spirit (*tria beneficia*). Its task is to guide and assist humankind, as Principalities govern over princedoms, archangels reveal events that affect all humankind, and angels assist human beings individually.[40]

---

minds the threefold distinction between being, power, and activity»] (*CH.* XI, 2, 284 D). As Mellone observes, while Bonaventure (and Vincent of Beauvais in his *Speculum historiale* I.12) linked the hierarchies to the persons of the Trinity in terms of 'analogical relation' (*In Hexaëmeron, Coll.* XXI, 17-20), Dante developed it in terms of object of angelic beatific vision. Another difference with Bonaventure concerns the order of contemplation within each hierarchy. Although Dante does not specify it for all hierarchies, it appears different from the one present in Bonaventure, as in *Convivio* the superior orders of each hierarchy seem to contemplate respectively the Father, the Son, and the Holy Spirit with respect to themselves. ATTILIO MELLONE, *Angelo*, p. 269. See also EDMUND G. GARDNER, *Dante and the Mystics*, p. 203. The hypothesis that Bonaventure was the main source of Dante's Trinitarian order of contemplation originated with LUIGI M. CAPELLI, *Le gerarchie angeliche e la struttura del Paradiso dantesco*, «Giornale dantesco», VI, 1898, pp. 241-259.

[39] As Luscombe notes, Hugh of Saint Victor «advanced the notion that the divine Trinity itself constituted a hierarchy, not in the sense that inequalities are found in the divine Persons but in the sense that their inter-communication impressed a divine likeness on created beings which caused them to be formed into hierarchies». DAVID E. LUSCOMBE, *Hierarchies in the Writing of Alain de Lille, William of Auvergen, and St. Bonaventure*, p. 16.

[40] BONAVENTURE, *Coll. In Hex.* XX-XXI. For the Bonaventurian scheme of illumination of the hierarchies see D. LUSCOMBE, *Hierarchies in the Writing of Alain de Lille, William of Auvergen, and St. Bonaventure*, p. 24.

The influence of Bonaventure's interpretation of the hierarchies emerges in *Convivio* II, V, 7-18), where Dante follows the same hierarchic order of contemplation of the three Persons of the Trinity, although adopting the Gregorian arrangement. Dante stresses the triadic essence of angelic contemplation as a three-times-three process, perhaps hinting at the miraculous meaning of number nine he had celebrated in the *Vita Nuova* and associated to angels and to Beatrice.[41] In the above mentioned passage from *Convivio*, the orders speculate according to what proceeds from, separates from, and reunites with the Trinitarian person their hierarchy contemplates («procede», [...] «si parte», [...] sé unisce».[42] Such ordering reflects the mystic pattern of *exitus* from and *reditus* to God, which indirectly associates the threefold arrangement of orders in each hierarchy to the different moments of the anagogical process of reunification with God.

In *Paradiso*, however, the articulated partition of the contemplative order remains implicit. The different power of vision of angels remains the only factor that determines the object of their contemplation:

> e dei saper che tutti hanno diletto
> quanto la sua veduta si profonda
> nel vero in che si queta ogne intelletto (*Par.* XXVIII, 106-108).

> and know that all delight to the degree
> to which their vision sees – more or less deeply –
> that truth in which all intellects find rest.

---

[41] See Chapter IV, III.

[42] «E con ciò sia cosa che ciascuna persona nella divina Trinitade triplicemente si possa considerare, sono in ciascuna gerarchia tre ordini che diversamente contemplano. Puotesi considerare lo Padre non avendo rispetto se non ad esso: e questa contemplazione fanno li Serafini, che veggiono più della Prima Cagione che nulla angelica natura. Puotesi considerare lo Padre secondo che ha relazione al Figlio, cioè come da lui si parte e come con lui sé unisce: e questo contemplano li Cherubini. Puotesi ancora considerare lo Padre secondo che da lui procede lo Spirito Santo, e come da lui si parte e come con lui sé unisce: e questa contemplazione fanno le Potestati. E per questo modo si puote speculare del Figlio e dello Spirito Santo: per che convengono essere nove maniere di spiriti contemplativi a mirare nella luce che sola se medesima vede compiutamente». [«Since each person of the divine Trinity can be considered in a threefold manner, there are in each hierarchy three orders that contemplate in different ways. It is possible to consider the Father with regard to but him alone, and this contemplation the Seraphim perform, who perceive more of the First Cause than any other angelic nature. It is possible to consider the Father with respect to the relation he has to the Son, that is, how he is separated from him and how united with him; and this the Cherubim contemplate. It is further possible to consider the Father with respect to how the Holy Spirit proceeds from him, and how it is separated from him and how united with him; and this contemplation the Powers perform. In this same way it is possible to contemplate the Son and the Holy Spirit: consequently it is appropriate that there should be nine classes of contemplative spirits, to gaze upon the light which can only be completely beheld by itself»]. *Convivio* II, V, 9-11.

The angelic vision is deeper according to the nature and merit of the angels, and the relation to the persons of the Trinity is transformed in terms of the different kinds of science they can receive through divine illumination.

The hierarchical order of contemplation and the corresponding order of illumination of angelic substances are essential to understand how Dante structured the relation between unity and multiplicity. In *Paradiso* XXVIII, 64-78, Beatrice explains to Dante the relation between the One – the bright point «acuto si' che 'l viso ch'elli affoca / chiuder conviensi per lo forte lume» («so acute / a light, that anyone who faced the force / with which it blazed would have to shut his eyes», *Par.* XXVIII, 17-18) at the center of all spheres- and the multiplicity of the angelic choirs.[43] Although angels do not possess creative power, as instruments of divine illumination they are crucial presences in the creative passage from unity to multiplicity, and in the order with which multiplicity conflates back to unity. As created substances, and not emanations from the One, through the hierarchical *taxis* angels participate in the divine, thus making the co-existence of unity and multiplicity possible.

As seen in previous section II, at the climax of his journey Dante uses the harvesting metaphor of the «volume» to describe the discernible bonds tying the dispersed multiplicity of creation to the luminous point that represents its origin and end:

> Nel suo profondo vidi che s'interna,
> legato con amore in un volume,
> ciò che per l'universo si squaderna:
> sustanze e accidenti e loro costume
> quasi conflati insieme, per tal modo
> che ciò ch'i' dico è un semplice lume (*Par.* XXXIII, 85-90).

> In its profundity I saw-ingathered
> and bound by love into one single volume-
> what, in the universe, seems separate, scattered:
> substances, accidents, and dispositions
> as if conjoined-in such a way that what
> I tell is only rudimentary.

Line 90 introduces a fascinating zeugma, typical of Dante's rhetorical art. Three different elements are called into question, writing, language, and light, and their properties overlap. «What I tell» refers to Dante's art of writing, and here has the solemnity and authority of *dictum*, which is however presented

---

[43] The relation between unity and multiplicity constitutes an important theological and philosophical theme in the poetics of the *Commedia*, to which Chapter III will return.

neither as vision (of letter signs) nor as sound (of spoken language), but as «simple reflection». The metaphorical light refers to the poem, which although not divine, by analogy bears the illuminating power of divine messages. Dante's poem enlightens his readers with the simple illumination of human art, which is however an illumination inspired by the vision of truth. At the end of his journey, angelic mediation is no longer necessary. Thanks to Mary's intercession, vision becomes 'face to face', conveying a surplus of grace that enables Dante's sight to jump into the highest dimension of the visible divine. What he sees is an image of order, reflected in the arrangement of the hierarchies.

b. *The Angelic Episteme: Order and Science*

As seen in the previous section, in Dionysian angelology order, divine science, and operations were constitutive elements of the definition of hierarchy, and were correlated. Angelic ordering reflected the different degrees of vicinity to God of angels, determining in turn their different capacities of vision and therefore the specific science each order received. Angels acted based on this science, and their arrangement corresponded to the different types of wisdom they possessed and transmitted in accord with the harmony of creation based on the *analogia entis*. Angelic order reflected the order of cosmos. In the *Commedia* Dante developed this aspect of the Dionysian doctrine by establishing a concrete correspondence between angelic choirs and celestial spheres; their ordering became an aspect of the general design of cosmic order.[44] The criterion of hierarchical arrangement, therefore, could not be the arbitrary choice of exegetes, but must be of divine inspiration, for it determined the modes of divine intervention in the universe, and had ontological and cognitive implications.

Dionysius's ordering of the nine angelic choirs prevailed on all others in authority. The widespread belief that Saint Paul had instructed the Areopagite on the 'true' hierarchical «*taxis*» contributed to its sacrality. Previous orderings were therefore considered «errors», and thus Dante rejected the Gregorian hierarchical ordering in favor of the Dionysian one in *Paradiso* XXVIII:

> E Dïonisio con tanto disio
> a contemplar questi ordini si mise,
> che li nomò e distinse com'io.

---

[44] On the importance and nature of order in the *Divina Commedia*, see GIUSEPPE MAZZOTTA, *Dante's Vision and the Circle of Knowledge*, Princeton, Princeton University Press, 1993 (pp. 197-218). See also MARC COGAN, *The Design in the Wax. The structure of the Divine Commedia and its Meaning*, Notre Dame, University of Notre Dame Press, 1999, pp. 149-247.

> Ma Gregorio da lui poi si divise;
> onde, sì tosto come li occhi aperse
> in questo ciel, di sé medesmo rise (*Par.* XXVIII, 130-135).

> And Dionysius, with much longing, set
> himself to contemplate these orders: he
> named and distinguished them just as I do.
> Though, later, Gregory disputed him,
> when Gregory came here-when he could see
> with opened eyes-he smiled at his mistake.

Gregory the Great's epiphanic smile is the sign of a truth he finally attains in *Paradiso*; the same truth Dante now contemplates on the names and the order of the angels. This truth does not simply concern the structure of the hierarchies, but also the now visible purpose of the ordering itself, which in Dante is no longer simply related to the angels' anagogical process of participation and reunification with God, but consists in the simultaneous act of being attracted by and attracting toward the bright center around which the angels circle:

> Questi ordini di sù tutti s'ammirano,
> e di giù vincon sì, che verso Dio
> tutti tirati sono e tutti tirano (*Par.* XXVIII, 127-129).

> These orders all direct-ecstatically-
> their eyes on high; and downward, they exert
> such force that all are drawn and draw to God.

The specular position of the active and passive forms of the verb «tirare» (to draw) in line 129, captures angelic activity in its intimately related speculative and active moments («tutti tirati sono e tutti tirano»). The insistence on the action of drawing in the reiterate use of «tirare», stresses the angels' effort (*energheia*) in being all uplifted and in uplifting all inferior substances toward their common end of participative reunification with God. The «tutti [...] tutti» emphasizes the universality of the movement toward God that angels guide. Dante's combination of the two moments, to draw and to be drawn, is connected to the two different angelic operations, the contemplative, which correspond to their beatific vision, and the active reflection of divine illumination, which represents their intervention in and on the material spheres. Both actions in Dante presuppose direct access to divine enlightenment, and a hierarchical order that depends on the different capacities of vision of the angels.

The action of 'drawing' occurs through illumination and it brings attention to the knowledge it conveys. As we have seen, Dionysius defined *episteme* («a state of understanding») as one of the constitutive elements of the definition of hierarchy, but did not specify its content, thus giving origin to diverging interpretations. In commenting on the *De coelesti hierarchia*, Albert the Great interpreted the meaning of the science conveyed by divine enlightenment by distinguishing between two types of illumination, beatific and hierarchical. While beatific vision was direct, hierarchical illumination arrived to angels in gradations, reflected by the higher orders onto the inferior ones. According to the Dominican teacher, this illumination conveyed a transmissible science of God's nature, the order of the universe, and the mysteries of the Church.[45] Angels, and through angels humankind, acquired this knowledge according to their capacity to receive illumination, which guided them in their operations. Although Albert the Great's distinction between beatific and hierarchical visions, and his accent on the practical purpose of hierarchical science, remained characteristic traits of the *Commedia*, important differences emerge with respect to Dante's interpretations of the Dionysian *episteme* and its function in defining the role of hierarchies. While admitting direct beatific vision, Albert the Great maintained mediated hierarchical illumination, and did not attribute to the angels any cosmological function. Dante instead assumed direct illumination and interpreted Dionysius's «state of understanding» in terms of operative science that included the providential action of angels as celestial movers. In the *Commedia* all angels and Blessed enjoy beatific vision, but only angels receive via direct illumination the science of their ministries, and reflect divine light directly on the sphere they move.

The revision of the hierarchical order of angelic illumination was essential to connect heavens and angels. Dante exposed the association of the dynamics of light with the mirroring function of the angels in *Convivio* III, XIV, 4, suggesting a possible early approach to the doctrine of the Areopagite:[46]

---

[45] According to Albert the Great, only the second type of illumination is hierarchical: «deus offert se beatis omnibus immediate, quantum pertinet ad obiectum beatitudinis, quia nihil potest beatificare nisi ipse, in quo est perfectio bonitatis. [...] Quantum vero pertinet ad revelationem occultorum divinorum, quae sunt in natura eius et quae pertinent ad dispositionem universi et mysteria ecclesiae, non vident eum omnes immediate». [«God presents Himself in an unmediated way to all blessed as object of beatitude, for nothing can beatify but God Himself, who is perfection of Goodness. [...] As for the revelation of the divine mysteries concerning His nature, the order of universe, and the secrets of the Church, they do not see him in an unmediated way»]. ALBERT THE GREAT, *Super Dionysium De caelesti hierarchia*, P. Simon and W. Kübel eds., Aschendorff, Monasterii Westafalorum, 1993, p. 53.

[46] In this case, Dante's later adoption of Dionysius's hierarchical ordering, and the explicit reference to his authority in *Paradiso* XXVIII would not be motivated by a lately discovered of the *De*

lo primo agente, cioè Dio, pinge la sua vertù in cose per modo di diritto raggio, e in cose per modo di splendore riverberato; onde nelle intelligenze raggia la divina luce sanza mezzo, ne l'altre si ripercuote da queste intelligenze prima illuminate (*Convivio* III, xiv, 4).

Here we must further know that the first agent, namely God, instills his power into things by means of direct radiance or by means of reflected light. Thus the divine light rays forth into the Intelligences without mediation, and is reflected into the other things by these Intelligences which are first illuminated.

This passage reveals that Dante already in *Convivio* assumed that angelic essences receive direct illumination from God («sanza mezzo»), while on «other things» illumination arrived through angelic reflection.[47] The same assumption is maintained in the *Commedia* where the mechanics of light transmission structures the way in which the angelic orders move the celestial spheres.

The assumption of direct enlightenment constituted a crucial difference between Dante and Dionysius. The strict hierarchical scheme of the latter presented problematic aspects, and was hardly compatible with the structure of the *Commedia*, where each order directly illuminates and moves one of the spheres. How could the hierarchical transmission of God's light be compatible with the fact that, according to Biblical authority, all blessed souls and angels enjoyed direct vision of God? This question was one of Aquinas's concerns in expounding the Dionysian structure of hierarchical illumination in *quaestio* 103 of his *Summa theologia* I. In replaying to the objection that all angels have beatific vision, Aquinas distinguishes between direct illumination from God, and indirect through the hierarchies:

Ad primum ergo dicendum quod omnes Angeli, tam superiores quam inferiores, immediate vident Dei essentiam; et quantum ad hoc, unus non docet alium. De hac

---

coelesti hierarchia, but by a change in Dante's opinion on the hierarchical system that best fitted his vision of cosmo-theological order.

[47] The two modalities of illuminations, direct and indirect, reflect a similar distinction between the principle of government and its execution. According to the first, God governs directly, according to the second, God governs through His instruments. For an exposition of this principle, see Thomas Aquinas, *S. Th.* I^a, q. 103, a. 6: «gubernatione duo sunt consideranda, scilicet ratio gubernationis, quae est ipsa providentia; et executio. Quantum igitur ad rationem gubernationis pertinet, Deus immediate omnia gubernat, quantum autem pertinet ad executionem gubernationis, Deus gubernat quaedam mediantibus aliis» [«In government there are two things to be considered; the design of government, which is providence itself; and the execution of the design. As to the design of government, God governs all things immediately; whereas in its execution, He governs some things by means of others»]. On angelic enlightenment as intellectual operation, and its hierarchical transmission see also *S. Th.* I^a, q. 106, a. 1. Departing from the Areopagite, Dante assumed that the Blessed, similarly to the angels, receive direct enlightening. This theme will be further discussed in chapter IV, section I.

enim doctrina propheta loquitur, unde dicit non docebit vir fratrem suum, dicens, cognosce dominum. Omnes enim cognoscent me, a minimo eorum usque ad maximum. Sed rationes divinorum operum, quae in Deo cognoscuntur sicut in causa, omnes quidem Deus in seipso cognoscit, quia seipsum comprehendit, aliorum vero Deum videntium tanto unusquisque in Deo plures rationes cognoscit, quanto eum perfectius videt. Unde superior Angelus plura in Deo de rationibus divinorum operum cognoscit quam inferior; et de his eum illuminat. Et hoc est quod dicit Dionysius, IV cap. de *Div. Nom.*, quod Angeli existentium illuminantur rationibus. (*S. Th.*, q. 103, a. 1).

All the angels, both inferior and superior, see the Essence of God immediately, and in this respect one does not teach another. It is of this truth that the prophet speaks; wherefore he adds: "They shall teach no more every man his brother, saying: 'Know the Lord': for all shall know Me, from the least of them even to the greatest." But all the types of the Divine works, which are known in God as in their cause, God knows in Himself, because He comprehends Himself; but of others who see God, each one knows the more types, the more perfectly he sees God. Hence a superior angel knows more about the types of the Divine works than an inferior angel, and concerning these the former enlightens the latter; and as to this Dionysius says (*Div. Nom.* IV) that the angels "are enlightened by the types of existing things".

As Albert the Great before him, Aquinas interpreted Dionysian hierarchical illumination in a restrictive sense. All angels receive direct illumination, and this pertains to the science of divine truths each order can receive according to its capacity of vision. At the same time, angels also hierarchically receive and transmit illumination. Higher orders reflect onto the lower orders the knowledge of sacred truths that their superior capacity of vision allows them to achieve. Their different visions determine their distinct tasks in the government of the world. In this interpretation, both direct and indirect enlightenment become compatible. Dante maintained hierarchical communication of light, for this was necessary to assure the supervision of higher onto inferior angelic orders in the organizational design of their governing functions, so as to assure the anagogical process of angels and souls.

The combination of both indirect and direct illuminations not only rendered Dionysius compatible with the Scriptures, but also, and more importantly, it limited the excessive importance that a strict hierarchical system placed on the role of intermediation of angels, which indirectly reduced the centrality of Incarnation. Possibly related to this issue, is Bonaventure's breach in the rigid Dionysian hierarchy of light transmission. In discussing the classification of the hierarchies according to the Trinitarian principle of government in his *Conference* XXI, 21, he provided Dante with an authoritative antecedent by assuming direct illumination of angelic orders.[48] The previously quoted passages from *Convivio* III, XIV, 4, and the *Commedia* suggest that Dante elaborated the enlightening structure of the poem in line with an established hermeneutic tradition that found in Albert the Great, Aquinas, and Bonaventure its major representatives.

Dante's departure from a criterion of purely hierarchical transmission of divine light had important implications in the *Commedia*, where direct illumination assures the acquisition of specific ministerial science to hierarchies. While in the Dionysian system hierarchical illumination elicited imitation in human beings so that ecclesiastical and secular orders mirrored the celestial one, the same does not happen in Dante's poem. The consequence of this change is visible in the ecclesiological sphere. In admitting direct divine illumination to all separate substances – although limited to particular forms of divine science –, Dante implicitly lessened the authority of the Church's mediation. Unlike the Areopagite's, his system rendered the continuity between angelic and ecclesiastic hierarchies problematic, and critically weakened the 'sacred' authority of the organization of the Church and its ministries.

c. *Angelic* Energheia *and Imitative Theology*

A fundamental aspect of the participative theology of the Areopagite consisted in the leading role of imitation in the anagogical process that through purification, enlightenment, and participation drove the hierarchies toward unification with God.[49] The desire of the angels to become as similar as pos-

---

[48] In his reformulation of Dionysius's strict hierarchical order of enlightenment, Bonaventure thought of direct illumination of all orders, including the Blessed (absent in the above quoted passage from *Convivio* but present in the *Commedia*), while human beings would receive it through angelic mediation. «Nota autem, quod prima hierarchia non originatur nec illuminatur nisi a solo Deo; media autem illuminatur a Deo et a suprema; infima autem illuminatur a Deo et a suprema et media; ecclesiastica autem ab omnibus», [«It should be noted that the first hierarchy originates and is illuminated by God only. The intermediate hierarchy receives illumination from God and the superior hierarchy, the inferior hierarchy from God, the superior, and the intermediate hierarchies. The ecclesiastical hierarchy receives illumination from all»]. BONAVENTURE, *Coll. In Hex.* XXI, 21. This passage shows that Bonaventure assumed a double order of contemplation, direct and indirect. Bonaventure, however, seems to assume a distinction between «hierarchical spirits» and angels, which is not apparent in Dante. See BONAVENTURE, *Coll. in Hex.* XXI, 16.

[49] «Quia et secundum seipsum unisquisque et caeleris et humanus animus speciales habet et primas et medias et ultimas ordinationes et virtutes ad dictas per unumquemque hierarchicarum illuminationum proprias anagogas proportionaliter manifestatas, per quas unumquodque in participatione fit, sicut idipsum et fas est et possible, supercognitissimae purgationis plenissimi luminis, anteperfectae perfectionis». [«Each intelligent being, heavenly or human, has his own set of primary, middle, and lower orders and powers, and in accordance with his capacities these indicate the aforementioned upliftings, directly relative to the hierarchic enlightenment available to every being.

sible to their creator was the effect of the illumination they received and that they, luminous transparencies, communicated to creation, thus connecting sensible and divine realities.

The Dionysian tripartite system offered an interpretation of reality as «collection of images», or symbols, which not only demonstrated the existence of God but also allowed the process of ascent to and descent from God. Through angelic orders, God's science descended in the form of intellectual light from the highest to the lowest spiritual creatures, thus stimulating in them the desire to return and reunite with Him.[50] The *Mystica theologia*, the shortest but not least influential of Dionysius's treatises, expounded the hermeneutics of this circular process. The descending movement of divine light, originating like creation from a gratuitous act of divine love, was intrinsic to the act of creation and corresponded to positive (*cataphatic*) theology. The ascending movement corresponded instead to negative (*apophatic*) theology. Marta Cristiani argues that while it is possible to find both the ascending and descending moments in the *Commedia*, apophatic theology, as well as the idea of God as perfect obscurity beyond any luminous divine manifestation, do not however characterize the poem.[51] Rather, it is a hymn to light showing that poetry supersedes the ineffability to which any theological and rational discourse is confined. The idea of divine obscurity implicit in apophatic theology is not however incompatible with Dante's poetry of light if we consider the cosmological map of the *Commedia* as depicting the universe of revelation, not encompassing all of the possible infinite dimensions of the divine. Dante's vision represents what human intellect at its highest – almost angelic – perfection can see and know. In Dante's confessed incapacity of expression surfaces a reality that remain hidden beyond the luminous point. What Dante cannot say, lies beyond what he sees in his final apotheosis; beyond the revealed image at the center of light.[52]

It is in accordance with this arrangement which each intelligent entity – as far as he properly can and to the extent he may – participates in that perfection preceding all perfection»]. *CH* X, 3, 273C.

[50] Nygren notices that the Dionysius developed in Christian terms the ancient idea of love as *Eros*; the desire to recompose a multifaceted reality in a unity. See ANDERS NYGREN, *Eros e Agape. La nozione cristiana dell'amore e le sue trasformazioni*, Bologna, Edizioni Dehoniane, 1990, pp. 597-602.

[51] Marta Cristiani excludes a substatial influence of the Pseudo-Dionsysius on Dante for two reasons: the role Dante assigns to angels as movers of the heavens; and the absence of Dionysius's negative theology in the structure of the poem. «[...] è ben evidente, ci sembra, che gli splendori del cielo di Dante – simbolismo "positivo" e "simile" al suo oggetto in un ricorso costante e sistematico a immagini luminose – escludono la dialettica interna affermazione-negazione, centro del sistema dionisiano». [«It seems evident that the splendors of Dante's heaven – "positive" symbolism, and "similar" to its object through a constant and systematic use of luminous images, excludes the internal positive-negative dialectics that is at the core of the Dionysian system»]. MARTA CRISTIANI, *Dionigi l'Aeropagita (Pseudo)*, p. 461.

[52] A possible example of how Dante alludes to and brings into the poem "what cannot be said" is his use of geographical names. William Franke has shown that the names designing the topogra-

In Dionysius, unlike in Neo-Platonism, the circular process of *exitus* and *reditus* is not a process of expansion and contraction of the same ontological reality. The universe is not strictly emanation from the One, but is the result of creation of distinct beings that reunite (*énosis pròs tòn theón*) with their creator by exceeding their ontological boundaries. Through the hierarchies, all spiritual creatures – including the human soul – undergo a process of participation in the divine that culminates in their 'divinization', which consists in acquiring likeness to God in proportion to their capacities. At the core of this process of deification (*théopoiesis*) is the imitative mechanism of angels (*imitatio dei*).[53]

This idea also surfaces in the *Commedia*, and encompasses Dante's entire journey through the three otherworldly realms. By moving from the farthest recesses of Hell to the ineffable – but poetically uttered – vision of Trinity through the circles of heaven, Dante endures his personal and exemplary experience of 'divinization' (transhumanization). His journey, however, is not linear, for it ends with Dante's return to a spiritually and morally renovated life, ready to fulfill the assigned task of reporting what he saw. His 'true' vision, consigned to the language of poetry, capable of reproducing what could not otherwise be told, pursues the same end as that of superior creatures. The circularity of motion presented in the poem, however, is not similar to the one found in Dionysius, where it remains circumscribed to the realm of intellectual creatures. In the *Commedia*, the desire that vision elicits in the angels immediately becomes an act of imitation whose effects pass on to human beings through the angelic influence on celestial bodies. Participation is the climax of the angelic anagogical process, which relies on and aims at similitude; it is the consequence and the cause of illumination, involving intellectual and moral spheres as it extends its imitative effects to the human beings. Ultimately, even Dante's poem is enlightened participation and imitation. By showing to his readers what to pursue and what to avoid in view of eternal life, it becomes, analogically, an instrument of grace, similar to an angelic operation in providing humankind with moral illumination. The idea of imitation, central to Dionysius's contemplative angelology, thus extends in Dante to the human and poetic spheres to become the moral fulcrum of the *Commedia*.

---

phical map of *Paradise* represent the spatial 'otherness' beyond the imaginative figuration of the poem: «Within the topographical descriptions of Italy in the *Paradiso*, it is especially the proper names for geographical places that suggest something real that remains stubbornly external to the poem and to the possibility of representation altogether». WILLIAM FRANKE, *The place of the proper name in the topographies of the Paradiso*, forthcoming in his book *The Veil of Eternity: Language and Transcendence in Dante's Paradiso*.

[53] The idea of assimilative knowledge belonged to the Biblical and Patristic traditions, particularly to the tradition of the Greek Fathers, which applied this notion to divine revelation. For the influence of this tradition in Dionysius see LOUIS BOUYER, *La spiritualità dei Padri*, pp. 151-155.

Paradiso XXIX illustrates how the idea of imitation encompasses the corporeal world and the moral sphere. The vision of light is knowledge of God, more or less profound according to the different intellectual natures of the angels. The profundity of vision is spatially represented by the angels' different distances from God, which correspond to their varying degrees of perfection achieved through their desire of imitation. The different degrees of intensity of vision elicit in the angels an incommensurable fervor:

> Onde, però che a l'atto che concepe
> segue l'affetto, d'amar la dolcezza
> diversamente in essa ferve e tepe (*Par.* XXIX, 139-141).
>
> and this is why (because affection follows
> the act of knowledge) the intensity
> of love's sweetness appears unequally.

The opening sequence of causal clauses («onde», «perciò») renders with poetic complexity the logical and theological succession of understanding («atto che concepe») and love («affetto»). The tercet recalls the Dionysian definition of hierarchy while at the same time placing contemplative action at the beginning of cosmic movement. If love is the universal engine that secures the regular movement of creation, understanding is what sets the cosmic engine in motion. In seeking similitude with God, the hierarchies stimulate a similar desire of likeness in others. Angels perfect their natures in this imitative process, which makes them «fellow workers of God».[54]

Est enim uncuiquique hierarchiam sortientum perfectio, hoc est secundum propriam analogiam in dei imitationem ascendere et omnium divinius, ut eloquia aiunt, dei cooperatorem fieri et ostendere divinam in seipso actionem, secundum quod possible est, relucentem (*CH* III, 2, 165B).[55]

---

[54] «Interpretatio igitur hierarchiae est ad deum, quantum possible, similitudo et unitas, ispum habens omnis sanctae et scientiae et actionis ducem et ad suum divinissimum decorum immutabiliter quidem diffiniens» [«The goal of the hierarchy, then, is to enable beings to be as like as possible to God and to be at one with Him. A hierarchy has God as its leader of all understanding and action. It is forever looking directly at the comeliness of God. A hierarchy bears in itself the mark of God»]. (*CH* III 2, 165A). Dionysius refers to the angels' capacity to perfection their nature as their *dynamis*, or potency. He also calls *Dynamis*, the second order of the second hierarchy, the virtues. See *CH* VIII, 1. The concept of potency in Dionysius finds three distinct applications. It refers to God, to the angels, and it designates the human intellective capacity. See SALVATORE LILLA, *Dionigi l'Aeropagita e il platonismo cristiano*, Brescia, Morcelliana, 2005, pp. 199-226.

[55] In the effort of enhancing a similitude with God, the angels participate in the divine attributes of beauty, goodness, and virtue: «Non ergo unusquisque hierarchiae dispositionis ordo secundum propriam analogiam reducitur ad divinam cooperationem, illa perficiens gratia et deodata vir-

Perfection [of a hierarchy] consists in uplifting to the imitation of God, each order according to its own capacity, and – and this is the most divine thing – in becoming what the Scripture calls "fellow workman of God", thus revealing how divine activity manifests itself in each order within the limits of the capacities attributed to that order.

As noted above, it is possible to see a reflection of this process of angelic «deification» in the pilgrim's «transhumanization» occurring in the *Commedia*.[56] Transposed in Christian terms, human efforts to become Christ's collaborators correspond to the angelic effort to become «collaborators of God», in coherence with a theology of Pauline inspiration.[57] It is possible to suggest that Dante analogically transposed the Dionysian doctrine of angelic imitation of God in human terms, as «*imitatio Christi*».[58] As the angels perfect their nature by becoming collaborators of God through participation and divine enlightenment, so human beings perfect their nature by receiving angelic illumination, which elicits in them the desire to imitate Christ, and in so doing they exemplarily illuminate others. If angels are the closest creatures to God, human beings are the closest

---

tute, quae divinitatis naturaliter et supernaturaliter insunt et ab ea superessentialiter acta et ad possibilem deum diligentium hierarchice imitationem hierarchice manifestata?» [«And so it comes about that every order in the hierarchical rank is uplifted as best it can toward cooperation with God. By grace and a God-given power, it does things which belong naturally and supernaturally to God, things performed by him transcendentally and revealed in the hierarchy for the permitted imitation of God-loving minds»]. I have slightly modified the English translation (*CH* III 3, 165D).

[56] The idea of human divinization is present in the Areopagite: «et nullum ab eorum quae sunt, simile nominari proprie et omnino valet. Verumtamen quaequmque et intellectualium et rationalium ad unitatem eius et qualiscumque virtus universaliter et ad divinas ipsius illuminationes, quantum possible, incessabiliter extenditur secundum virtutem, si iustum dicere, divina imitatione et divina univocatione facta est». [«No being can in any way or as a matter of right be named like to it [God]. Yet every being endowed with intelligence and reason, which, totally and as far as it can, is returned to be united with him, which is forever being raised up toward his divine enlightenments, which if one may say so, tries as hard as possible to imitate God – such a one surely deserves to be called divine»]. *CH* XII 3, 293B.

[57] In this parallel between angels and humans the presence of Saint Paul, who called Christians «co-workers» of Christ in 1 *Cor.* 3:9, surfaces again. The Greek text has «synergoi», literally «co-workers». The *Vulgata* translates the word as «adiutores».

[58] The analogy between angelic imitation of God and human imitation of Christ is coherent with the importance Dante attributed to the role of good will in the divine order of things. The idea that the pursuit of virtue is a necessary step in the path to beatitude finds roots in the philosophical tradition derived from Aristotelian ethics, and was a tenet of monastic spirituality. The same idea, connecting virtue and imitation, is present in Severinus Boethius. For Boethius, virtuous life was a necessary premise to happiness, which consisted in achieving as much as possible closeness to God: «Securo igitur concludere licet dei quoque in ipso bono nec usquam alio sitam esse substantiam» [«But we have determined that true happiness is the perfect good; therefore true happiness must dwell in the supreme Deity»]. SEVERINUS BOETHIUS, *The Consolation of Philosophy*, Oxford, Oxford University Press, 2000 (*De Philosophiae Consolatio*), III, IX, 10. On this theme in Dante's spiritual journey as transhumanization in order to 'deificate' see STEPHEN BOTTERILL, *Dante and the Mystical Tradition: Bernard of Clairvaux in the Commedia*, Cambridge, Cambridge University Press, 1994, p. 229.

creatures to Christ. In their imitative process, humans find the way to become His collaborators on earth. This collaboration has ethical and political consequences in Dante, for to be Christ's collaborator means to pursue virtue, ethical and intellectual, as well as individual and civic.

d. *The Idea of Merit*

An important element of Dionysius's definition of hierarchy is the concept of merit, defined as the angels' disposition to receive divine illumination as determined by nature and grace. In the *Commedia*, the vision of light is the first and fundamental moment that gives origin to the process of acquisition of the divine science of angelic operations. Angels penetrate this vision more or less profoundly according to their different intellectual natures, and accordingly divine light permeates all universe but glows «in one part more and in another less» [«in parte più e meno altrove», *Par.* I, 3]. The angels' capacity to receive illumination depends however not only on nature but also on grace.

The dialectics of merit and illumination assumes particular relevance in Dante's angelology, for it connects the cognitive properties of light to the ranking of angelic functions. Dante illustrates this Dionysian point in two passages, in *Paradiso* XIX, 136-138, and *Paradiso* XXVIII, 100-102.

In *Paradiso* XXIX, 134-136 Dante speaks of God as the «prima luce» (first light) received in as many ways as is the number of angels:[59]

> La prima luce, che tutta la raia,
> per tanti modi in essa si recepe,
> quanti son li splendori a chi s'appaia (*Par.* XXIX, 136-138).

> The First Light reaches them in ways as many
> as are the angels to which It conjoins
> Itself, as It illumines all of them;

An important aspect of the theological and philosophical debate on angelic nature surfaces in the quantitative suggestion of line 138: «tanti [...] quanti». Aquinas had observed that since matter is the principle of differentiation in a species, there must be as many species of angels as their number.[60] The kaleidoscopic multiplication of angelic essences does not contradict the fact that an indeterminate number of angels can belong to each given order, for they are assigned to their positions based on their merit. The different modes in which

---

[59] See also *Par.* XXIX, 142-145.
[60] AQUINAS, *S. Th.* I\ua, q. 50, a. 4.

light propagates from the point correspond to the modalities of divine intervention in the cosmos and implicitly relates to the different operations angels perform within each order, in accord with the specific functions assigned to the hierarchy to which they belong. Merit, illumination, and operations are thus intimately and individually connected. In this way, Dante linked the cognitive power of illumination with merit and ranking positions in the hierarchies, and opened to a plurality of different, specific, angelic interventions within each order, corresponding to the illuminations the «splendori» receive.

The second mentioned tercet shows the relation Dante established between merit and the incessant movement of angelic essences in their effort of acquiring likeness to their creator:

> Così veloci seguono i suoi vimi,
> per somigliarsi al punto quanto ponno;
> e posson quanto a veder son soblimi (*Par.* XXVIII, 100-102).

> [...] They follow
> the ties of love with such rapidity
> because they are as like the Point as creatures
> can be, a power dependent on their vision.

This tercet describes the motion of Seraphim and Cherubim. The angels' desire of likeness expressed in their mystical participative movement toward God becomes in Dante physical motion around the «point». By insisting on the idea of capacity («ponno»; «posson»), Dante stresses the connection between the angels's different speed of motion and their different capacities of vision. The term «soblimi», of mystical resonance, significantly rhymes with «vimi» (ties, bonds) and their respective positions at the end of lines 100 and 102 emphasize the relation between angelic vision and the «ties of Love», which bound together the 'volume' of Creation. The word 'sublime' hints at something that transcends angelic natural capacity: the increased power of vision God granted to loyal angels after Lucifer's fall by an additional gift of grace. Nature and grace thus determine their different powers of vision. In Dante's notion of merit, vision elicits motion and both determine the relative positions of angels in the hierarchies, and therefore the various functions each order performs.

At the climax of his journey, Dante insists on the concept that vision elicits love, and that love drives the angels to reverberate divine splendor onto the lower hierarchies, including the human one. In *Paradiso* XXVIII, 109-114 Dante illustrates again this fundamental Dionysian notion and makes evident the centrality of the concept of merit.

> Quinci si può veder come si fonda
> l'esser beato ne l'atto che vede,
> non in quel ch'ama, che poscia seconda;
> e del vedere è misura mercede,
> che grazia partorisce e buona voglia:
> così di grado in grado si procede (*Par.* XXVIII, 109-114).
>
> From this you see that blessedness depends
> upon the act of vision, not upon
> the act of love – which is a consequence;
> the measure of their vision lies in merit,
> produced by grace and then by will to goodness:
> and this is the progression, step by step.

Beatitude, of angels as well as of souls, primarily derives from the act of seeing, to which love follows, thus eliciting in spiritual creatures the desire to imitate God. It is worth recalling that vision involves acquisition of divine science, and it is accompanied by, but distinct from, grace. Merit («mercede») measures the angels' natural capacity of receiving divine light, and it «partorisce» («begets») grace and good will, thus steering the imitative activity of hierarchies through which «di grado in grado» («from step to step») the ascending process takes place.

In Dante's exemplary ultramundane adventure in the *Commedia*, it is possible to see a transposition of the Dionysian idea of angelic merit in terms of human will. Similarly to angels, humankind receives enlightenment from Intelligences, which kindles in them the desire to cooperate with grace. The more intense the ardor consequent to vision, the more their merit increases.[61] The interaction of vision and cooperation with grace results in a further augmentation of merit, and grants what Dante names 'illuminating grace'.[62] The

---

[61] This relation is illustrated also in *Paradise* XXIX, 64-66: «e non voglio che dubbi, ma sia certo, / che ricever la grazia è meritorio / secondo che l'affetto l'è aperto». [«I would not have you doubt, but have you know / surely that there is merit in receiving / grace, measured by the longing to receive it»].

[62] Following the exposition of Thomas Aquinas, there are three distinctions of Grace: 1. *gratia gratum faciens* and *gratia gratis data*; 2. *operans* and *cooperans*; 3. *preveniens* and *subveniens*. The first one, properly sanctifying Grace, is a *habitus* Christians receive at baptism. The *gratia gratis data* is a gift from God that makes those who receive it cooperate in bringing others to salvation (this type of Grace seems to be the one bestowed on Beatrice). The distinction between *gratia preveniens* and *subveniens* is similar to the difference between the desire of a given good and the capacity to act in order to realize it. Illuminating grace, as a grace of the mind, is part of the actual grace (*operans* and *cooperans*), which God bestows on us so that we may act effectively on our will in the effort of uplifting ourselves toward Him. «The illuminating grace of the intellect (*gratia illuminationis, illustrationis*) first presents itself for consideration. It is that grace which in the work of salvation suggests good thoughts to the intellect.

pilgrim's moral progression as he moves across the heavens exemplarily illustrates this process, for the light mirrored in Beatrice's eyes increasingly enhances his capacity to receive the vision of attainable beatitude.[63]

Paradiso XXIX illustrates the gradual development of the angels' capacity to receive illumination and their consequent desire to cooperate with grace. In this canto, Beatrice illustrates the doctrine of the creation of angels, and that of the origin and nature of the increased beatitude they gained after Lucifer's rebellion. Loyalty, humility, and obedience made the angels who did not rebel worthy of grace beyond their nature[64]

> per che le viste lor furo essaltate
> con grazia illuminante e con lor merto,
> sì c'hanno ferma e piena volontate (*Par.* XXIX, 61-63);
>
> through this, their vision was exalted with
> illuminating grace and with their merit,
> so that their will is constant and intact.

Illuminating grace, as God's gratuitous gift, exalted the angels' power of sight by enhancing their natural merit so that they had «ferma e piena volontate» («full and steadfast will»). It is possible to see a connection between the tercet quoted above and the idea of sublimity expressed in the previously quoted tercet from *Paradiso* XXVIII, 102, for there is no sublime vision without merit and grace. Thus, the 'sublime' could be interpreted as the capacity

---

This may happen in a twofold manner, either mediately or immediately». J. POHLE, *Actual Grace*, in *Catholic Encyclopedia*, vol. VI, Online edition, 2003. See THOMAS AQUINAS, *Summa Theologiae* I, Iª, q. III, and J. VAN DER MEERSCH, *Grace*, in *Dictionnaire de Théologie Catholique*, A. Vacant, E. Mangenot and E. Amann eds., Paris, Letouzey et Ané, 1903-1976, VI, pp. 1553-1687.

[63] The same idea of merit as applied to human beings returns in Dante's explanation of the additional grace humankind will enjoy after the Final Judgment, when the souls reunite with their bodies at resurrection: «La sua chiarezza séguita l'ardore; / l'ardor la visïone, e quella è tanta, / quant' ha di grazia sovra suo valore. / Come la carne glorïosa e santa / fia rivestita, la nostra persona / più grata fia per esser tutta quanta; / per che s'accrescerà ciò che ne dona / di gratüito lume il sommo bene, / lume ch'a lui veder ne condiziona; / onde la visïon crescer convene, / crescer l'ardor che di quella s'accende, / crescer lo raggio che da esso vene». [«Its brightness takes its measure from our ardor, / our ardor from our vision, which is measured / by what grace each receives beyond his merit. / When, glorified and sanctified, the flesh / is once again our dress, our persons shall, / in being all complete, please all the more; / therefore, whatever light gratuitous / the Highest Good gives us will be enhanced- / the light that will allow us to see Him; / that light will cause our vision to increase, / the ardor vision kindles to increase, / the brightness born of ardor to increase»]. *Par.* XIV, 40-51.

[64] «Quelli che vedi qui furon modesti / a riconoscer sé da la bontate / che li avea fatti a tanto intender presti», [«Those whom you see / in Heaven here were modestly aware / that they were ready for intelligence / so vast, because of that Good which had made them»]. (*Par.* XXIX, 58-60). Dante seems to follow here the traditional doctrine on the fall of the rebel angels. For the additional grace angels received after Lucifer's fall see AQUINAS, *S. Th.* I, q. 62, a. 9.

of the mind to go beyond its natural power of understanding and soar into a superior dimension. The reference to good will seems instead connected to the «buona voglia» of *Paradiso* XXVIII, 113, and indicates the strong relation between grace and good will in Dante's notion of merit and the connection between cognitive and ethical moments in the angelic act of seeing.

e. *The Divine Mirrors*

The relation between acquisition of science and ministerial function reflects the philosophical continuity between the contemplative and practical spheres Dante postulated. The hierarchical order and the mechanism of light transmission, as Dante reinterpreted them, are structural elements that allowed for transforming cognitive into operative functions. By acting as *specula* (mirrors), not simply metaphorically but also literally, angels assure the communication of the science they receive.[65]

Dionysius presented the literal and metaphorical function of «mirrors» as the consequence of the angels' desire of imitation, through which they become 'images' of God:[66]

Interpretatio igitur hierarchiae est ad deum, quantum possibile, similituddo et unitas, ipsum habens omnis sanctae et scientiae et actionis ducem et ad suum divinissimum decorum immutabiliter quidem diffiniens: quantum vero possible, reformat, et suos laudatores agalmata divina perficit, specula clarissima et munda receptive principalis luminis et divini radii et inditae quidem claritas sacre repletas eamque iterum copiose in ea quae sequuntur, declarantia, secundum divinas leges (*CH*, III, 2, 165A).

The goal of a hierarchy, then, is to enable beings to be as like as possible to God and to be at one with him. A hierarchy has God as its leader of all understanding and action. It is forever looking directly at the comeliness of God. Hierarchy causes its members to be *images* of God in all respects, to be clear and spotless *mirrors*, reflecting the glow of primordial light and indeed of God himself. It ensures that when its members have received this full and divine splendor they can then pass on this light generously and in accordance with God's will to beings further down the scale.

---

[65] For an analysis of the main philosophical and literary sources of images of mirrors in Dante DIEGO SBACCHI, *La presenza di Dionigi Areopagita*, pp. 93-113.

[66] In a similar way, but with greater stress on the image than on the imitation of God, Dionysius defines the angel in his *The Divine Names* IV, 22 as follows: «The angel is image of God. He is a manifestation of the hidden light. He is a mirror, pure, bright, untarnished, unspotted, receiving, if one may say so, the full loveliness of the divine goodness and purely enlightening the within itself as far as possible the goodness of the silence in their inner sanctuaries». For the possible influence of images of light derived from Dionysius's *Divine Names* and its commentary by Aquinas, see MARCO ARIANI, *"e sì come di lei bevve la gronda / de le palpebre mie"*, pp. 131-152: 144-149.

In this passage, «image» and «likeness» are connected and yet distinct concepts. As 'images', angels become mirrors reflecting onto others the glow of the divine; in pursuing likeness, they aim at becoming similar to God. Being mirrors pertains to their nature, but likeness is the result of their choice of action and requires the aid of grace. Like human beings, angels were created in the image and likeness of God, and like human beings, they conquer similitude through their actions.

The notion of «speculum», conveying the idea of an indirect knowledge of the divine truths as reflected in a mirror, derived from Saint Paul's 1 *Corinthians*: «Videmus nunc per speculum in aenigmate tunc autem facie ad faciem» («Now we see through mirrors, then we shall see face to face»).[67] The image of the mirror, on which Dionysius insists also in his *Divine Names*, and which was common among Neo-Platonic authors, captures both the continuity between Dante and Dionysius's angelologies, and their common roots in the Pauline theology.[68] The contact with Saint Paul is particularly relevant for the stress Pauline theology placed on pursuing virtue, and its impact on the ethical economy of the *Commedia*.

An important corollary of Dante's theory of angelic 'mirrors' concerns its metaphysical implications. As seen in chapter I, Dante insists on the pure immateriality of angels. How then could spiritual essences be transparent 'surfaces' with the power to absorb and reflect a luminous ray? The answer could be found in the fact that angels possess a fully actualized potentiality in receiv-

---

[67] SAINT PAUL, 1 *Corinthians*, 13:12. The image is present also in the Old Testament, *Book of Wisdom* 7:26, and ORIGEN, *In Iohannis Evangelium*, XII, 153 (249, 32). The image of the mirrors connected to the issue of the transmission of divine knowledge recurs in other Dante's works: *Epistle* V, 23; *Monarchia* II, ii, 8; *Questio* 61. For an analysis of the relation between the image of mirror and *analogia entis* in Dante, see FRANCESCO MAZZONI, *Il trascendentale dimenticato*, in *Omaggio a Beatrice*, R. Abardo ed., Firenze, Le Lettere, 1997, pp. 120-123.

[68] The notion of light reflection and mirror was common among Neo-Platonic authors, and images of light reflection in mirrors appear in various authors, from Macrobius's commentary on *Somnium Scipionis* (I, xiv, 15) to Alain de Lille' *Anticlaudianus*: «Presentat Fronesi speculum, quo cuncta resultant / Que locus empireus in se capit, omnia lucent, / Que mundus celestis habet, sed dissona rerum / Paret in hiis facies. Hic res, hic umbra uidetur, / Hic ens, hic species, hic lux, ibi lucis imago». [«In this mirror is reflected everything which the fiery region encompasses: in it shines clear everything which the heavenly universe holds, but the appearance of these things differs from the real objects. Here one sees reality, here a shadow; here being, here appearance; here light, there an image of light»]. The difference between Alain and Dante is evident in the passage. While Alain's physical mirror reflects only shadows of true reality, in Dante angels are mirrors that reflect divine image and convey forms. ALAIN OF LILLE (Alanus ab Insulis, XIIth c.), *Anticlaudianus*, R. Bossuat ed., Paris, Vrin, 1955, VI, 119-130. English translation from Alan of Lille, *Anticlaudianus*, J.J. Sheridan ed., Toronto, Pontifical Institute of Medieval Studies, 1973, p. 160. Sbacchi notices that the word «speculo» and its variations such as «speglio», are not present in the poetic tradition of Dante's times. Dante derived this terminology from the lexicon of the mystics, which in turn derived from Dionysius. See DIEGO SBACCHI, *La presenza di Dionigi Areopagita*, pp. 93-113.

ing divine light, which is the principle of their operations. In postulating the transmission of purely intellectual light to material spheres, Dante implicitly introduced a certain degree of 'opaqueness' in the angels, so that the physical laws of reflection could apply materially to the celestial phenomena, and analogically to the intellectual illumination of the Empyrean.

The mechanics of light transmission is vital to a definition of the cosmological function of Dante's Intelligences, for angelic reflection triggers the transmission of virtues to the celestial bodies, thus actualizing their capacity to influence the sublunar world. In *Purgatorio*, where the operations of angels in assisting the atoning souls and the two wayfarers play a visible role, Dante presents a technical explanation of the mechanics of light diffusion, thus implicitly assimilating the effects of angelic enlightenment on the material spheres to those on the human souls. In the inter-realm where the poet illustrates the contact between the divine and the human, angels communicate by singing, talking, and reciting the beatitudes that correspond to the virtues necessary to defeat the vice punished in each terrace. An unbearable splendor, which intensifies as Dante progresses toward the Eden, accompanies these communicative acts. In *Purgatorio* XV, 16-24, Dante describes the intense and unbearable luminosity emanating from the angel of the terrace of the envious by using a reference to a scientific simile of the mechanics of light refraction:

>   Come quando da l'acqua o da lo specchio
> salta lo raggio a l'opposita parte,
> salendo sù per lo modo parecchio
>   a quel che scende, e tanto si diparte
> dal cader de la pietra in igual tratta,
> sì come mostra esperïenza e arte;
>   così mi parve da luce rifratta
> quivi dinanzi a me esser percosso;
> per che a fuggir la mia vista fu ratta (*Purg.* XV, 16-24).

>   As when a ray of light, from water or
> a mirror, leaps in the opposed direction
> and rises at an angle equal to
>   its angle of descent, and to each side
> the distance from the vertical is equal,
> as science and experiment have shown;
>   so did it seem to me that I had been
> struck there by light reflected, facing me,
> at which my eyes turned elsewhere rapidly.

The «art» in line 21 refers to the *perspettiva*, whose laws Dante applies here to explain his reaction to the angel's luminosity.[69] Light arrives from the celestial being as refracted («rifratta») in a mirror, and the poet describes it from the point of view of the physical impression of the receiver. It is an imprecise perception («mi parve») of brightness that his eyes are not yet able to tolerate. At this stage of his journey, Dante's physical and intellectual capacities are not sufficient to make him able to bear the intense radiance of the angel. These capacities, however, increase as the pilgrim progresses toward the summit of the holy mountain where he assists to Beatrice's descent.

The metaphorical expansion of the Pauline notion of mirror achieves its apex in the third canticle, and testifies to the influence of Dionysian mystical language. While in *Purgatorio* language – accompanied by exemplary visions – is still a viable means of communication, in *Paradiso* the expressive strength of the rational word progressively fades away to yield to the mystical, analogical language of the «divine names» and to the imagery of light. In *Paradiso* XXIX, 142-145 the theme of the mirror appears linked to that of creation from unity to multiplicity, recalling the powerful images of universal order that introduced the figure of the Areopagite in *Paradiso* X:

> Vedi l'eccelso omai e la larghezza
> de l'etterno valor, poscia che tanti
> *speculi* fatti s'ha in che si spezza,
> uno manendo in sé come davanti (*Par.* XXIX, 142-145).[70]

> By now you see the height, you see the breadth,
> of the Eternal Goodness: It has made
> so many *mirrors*, which divide Its light,
> but, as before, Its own Self still is One.

This first light propagates in rays that the angelic mirrors refract, while maintaining its essential unity. The insistence on light, vision, splendor, and the spectrum of terms referring to the same concepts, are characteristic of Dante's description of angels and intensifies their image as instrumental vehicles of light. This intellectual light is «valor», virtue or goodness that expresses itself in the creation of the angelic *specula*. In Dante thus the angelic function

---

[69] Gilson suggests the influence of Albert the Great's mediation on the description of the optical phenomenon in the above quoted tercets, particularly of his *De causis proprietatum elementorum*, *De meteoris*, and *De natura locorum*. See SIMON GILSON, *Medieval Optics*, pp. 112-116.

[70] See for the same concept the quoted passage from *Convivio* III, xiv, 4-5. The image of angelic mirrors returns also in *Par.* XIII, 55-60. My Italies.

as mirrors is no longer limited to the anagogical participative process as in Dionysius, but it becomes necessary also to the governing function of the material world that angels perform in the *Commedia*.

Conclusions

This chapter has shown that Dionysian angelology, although providing a viable model of cosmic harmonization based on the contemplative and illuminative faculties of angels, did not, however, offered a complete theory of angelic operations able to explain the ways in which angels effectively influenced and guided the actions and the choices of human beings. We have seen that another tradition, parallel to the Dionysian, chiefly represented by Gregory the Great, provided Dante with an approach to angelology that focused on the practical importance of angelic operations. This tradition, of Biblical and Pauline foundations, merged with the Dionysian, and provided ground for Dante's design of a 'cosmological' angelology in which Christian theology and Aristotelian metaphysics found an equilibrium that served the tropological purpose of the *Commedia*.

The analysis of Dionysian angelology has shown its problematic aspects when adapted to a system in which angels played a cosmological role by connecting the motion of celestial spheres to the immaterial regions of the Empyrean through illumination. While keeping Dionysius's hierarchical structure, Dante needed to adapt fundamental Dionysian concepts related to the very notion of hierarchy in order to set his essentially providential angelology at the core of the cosmology of the *Commedia*. Particularly problematic in Dionysius's strict hierarchical system was the stress it placed on angelic intermediation, in a way that diminished the central role of Incarnation. A possible way of reducing this risk was to develop and insist on the material offices operated by the angels and on their instrumentality within a soteriological perspective. In order to take this direction, in the *Commedia* Dante reinterpreted key notions in the Dionysian system such as light, hierarchy, merit, and angelic mirroring action to adjust them to the function of movers of the celestial spheres within the eschatological perspective of the poem. The following chapter examines this function.

CHAPTER THREE

MOVING THE STARS:
ANGELS AND THE ORDER OF THE COSMOS

This chapter explores the cosmological and ontological implications of the most important of angelic operations: movers of the heavenly spheres. Chapter II has illustrated how Dante reinterpreted the Dionysian hierarchical arrangement of the angelic orders to structure the anagogical scheme of the *Commedia*. As seen in Chapter I, a crucial aspect of his reinterpretation consisted in connecting angelic contemplative and operative functions. The analysis of the cosmological cantos of the third canticle reveals the structural and poetic necessity of angelic hierarchies in reconciling providential destiny and free will by connecting material and spiritual spheres.

The first section of this chapter explores the relation between angelology and cosmology. It first sets the cosmological questions presented in the *Commedia* within the broader context of the medieval debate on the harmonization of ancient science and Christian doctrine that developed during the twelfth and the thirteenth centuries. It then analyses the metaphysical and theological implications of attributing the role of movers to the Dionysian hierarchies, and places Dante's solution within the context of the medieval debate on movers. After discussing the eschatological character of Dante's cosmology and showing that the moving operations of the angels are structurally necessary to his design of a cosmology that is both 'scientific' and providential, it discusses the role of angels in governing the cosmos, and analyses Dante's distinction between two ordering criteria: the Holy Spirit, as cause and origin of order, and Providence, as its governing instrument. In addition, this section discusses love and knowledge as two dialectically related moments on which the dynamics of Dantean cosmos relies. The section then proceeds by investigating how these elements vitally interact in the production of the material reality that the hierarchies supervise, and highlights the importance of angelic operations in the metaphysics of the *Commedia* by exploring the order-ontology relation in the proemial cantos I and II of *Paradiso*. The second section of the chapter examines the ethical consequences of the doctrine of astrological influence caused by angelic orders in regulating the motion of the celestial spheres. It discusses the relation between angels and astrology in the *Commedia* within the context of the theological debate on astrological determinism, and illustrates the complementarity of inclinations and free will in the moral cosmogony of the *Commedia*.

I. COSMOLOGY AND ANGELOLOGY

In attributing the role of movers to the angels, Dante reconciled the Aristotelian physical explanation of celestial motion with the Neo-platonic theological justification of its causes. In so doing, he established a philosophical and physical structural relation between angelology and cosmology. This section analyzes the philosophical and scientific aspects of the cosmological functions of angels.

a. *The 'Book' of Stars*

Starting in the eleventh century, a wave of new translations from Arab and Greek texts reintroduced in the West a long-lost scientific culture that the Arab philosophers had saved, developed, and transmitted.[1] These texts greatly influenced the formation of a 'modern' scientific cosmology, and included Aristotle's scientific works such as *Physics*, *Metaphysics*, *On Heavens* (*De coelo*), the vastly influential *Liber de causis* (initially attributed to Aristo-

---

[1] The literature on the translations and transmission of scientific texts that greatly influenced the development of medieval science between the eleventh and the fourteenth century is vast. Among the texts relevant to place Dante in context, see EDWARD CRANZ, *The Reorientation of Western Thought c. 1100 A.D.*; TULLIO GREGORY, *Riscoperta della natura e nuove scienze nel secolo XII*, in *Speculum naturale. Percorsi del pensiero medievale*, Roma, Storia e Letteratura, 2007, pp. 15-33. See also EDWARD GRANT, *The Foundations of Modern Science in the Middle Age*, Cambridge, Cambridge University Press, 1996. For the influence of Aristotelian physics and cosmology on the theory of elements in Dante, see PATRICK BOYDE, *Dante Philomythes and Philosopher: Man in the Cosmos*, Cambridge, Cambridge University Press, 1981, particularly pp. 57-131; CESARE VASOLI, *Dante, Alberto Magno e la scienza dei peripatetici*, in *Dante e la scienza*, P. Boyde and V. Russo eds., Ravenna, Longo, 1995, pp. 55-70. Among the numerous works Bruno Nardi dedicated to the influence of Aristotelian and Arab philosophy on Dante, and that are relevant for the angelic question see at least *Dal Convivio alla Commedia*, in BRUNO NARDI, *Dal Convivio alla Commedia*, pp. 37-150, and *Dante e la cultura medievale*, Bari, Laterza, 1942, where in the essay on *Tutto il frutto ricolto del girar di queste spere*, pp. 245-264, Nardi illustrates important changes in Dante's opinions on movers, and their relation to the question of the origin of creation from the *Convivio* to the *Commedia*. On the impact of the translations of astronomical texts on medieval thought, see also GRAZIELLA FEDERICI VESCOVINI, *"Arti" e filosofia nel secolo XIV. Studi sulla tradizione aristotelica e i moderni*, Firenze, Nuovedizioni Enrico Vallecchi, 1983.

tle), and a consistent corpus of astrological treatises such as Claudio Tolomeus' *Almagest*, Alfraganus (Ahmad ibn Muhammad ibn al Farghānî)' *Liber de aggregatione scientiae stellarum et principiis coelestium motuum*, and Alpetragius (Nur el-Din al-Betrugi)' *Liber de motibus coelorum*. Together with texts on medicine and optics (*perspettiva*), these works became the foundation of a new idea of nature and humanity, and contributed to form the intellectual backdrop against which Dante wrote the *Commedia*.[2]

Before the principles of Aristotelian physics begun to reshape philosophical and theological thought, the medieval vision of the cosmos was rooted on the scriptural exegesis of creation. This vision relied on a handful of texts inherited from antiquity and taught in the schools in Dante's time, such as Macrobius' *Somnium scipionis* (*The Dream of Scipio*), Martianus Capella's *De nuptiis mercurii et philologia* (*The marriage of Mercury and Philology*), and Plato's *Timaeus* in Calcidius' version and commentary. Moreover, Platonic and Neo-Platonic influences, mediated through the Greek Fathers and Byzantine theology, shaped cosmological representations before the twelfth century. Their mystical imprint never completely disappeared, even in authors such as Aquinas and Bonaventure. Aristotelian thought, however, provided the ground for developing the great 'scientific' theology of the thirteenth century.

Before the eleventh and the twelfth centuries, nature and sacred Scriptures were conceived of as two divine 'books' of parallel authority. Nature was the book God had written with his finger before the Fall. After Adam and Eve's expulsion from Eden, the Scriptures became necessary, and remained the enlightening medium to the divine even after the Incarnation of the Word.[3] Analogical hermeneutical principles applied to the reading of both Nature and Scriptures; both were collections of *signa*, symbols signifying the meanings God had encrypted on the visible side of creation that grace enabled to read.[4] Dante expressed this idea in the *Convivio*:

---

[2] For Dante's knowledge of these texts see MASSIMO MIGLIO, *Alfragano*, in *ED* I, pp. 122-123; ID., *Alpetragio*, in *ED* I, pp. 180-181, and BRUNO NARDI's seminal articles on *Dante e Pietro d'Abano* and *Dante e Alpetragio*, in *Saggi di filosofia Dantesca*, Firenze, La Nuova Italia, 1967, pp. 40-62 and pp. 139-166. In the second article, Nardi suggests Dante's possible knowledge of the Arab astronomer through Albert the Great's commentary on Aristotle's *De coelo*. See also in the same volume, Nardi's study on the influence of the *Liber de causis*, in *Le citazione dantesche del "Liber de causis"*, pp. 81-109.

[3] On the impact of Aristotelianism on the medieval idea of nature, see TULLIO GREGORY, *Nature au Moyen Âge* and, with particular attention to physics and astrology, *Riscoperta della natura e nuove scienze nel secolo XII*, in *Speculum naturale. Percorsi del pensiero medievale*, Roma, Edizioni di Storia e Letteratura, 2007, pp. 1-14 and pp. 15-33.

[4] An authoritative and widely taught text that illustrated the analogical symbolic interpretation of nature and Scripture, was HRABANUS MAURUS' *De universo*. Dante placed Maurus among the wise souls in the heaven of the Sun in *Par*. XII, 139.

Li numeri, li ordini, le gerarchie narrano li cieli mobili, che sono nove, e lo decimo annunzia essa unitade e stabilitade di Dio. E però dice lo Salmista: "Li cieli narrano la gloria di Dio, e l'opere delle sue mani annunzia lo firmamento" (*Convivio* II, V, 13).[5]

The moving heavens, which are nine, declare the numbers, the orders, and the hierarchies, and the tenth proclaims the very unity and stability of God. Therefore the Psalmist says: "The heavens declare the glory of God, and the firmament proclaims his handiwork".

Following the Psalmist, Dante presents the heavens as a 'narrative' of the works of creation, a 'book of stars' in which it is possible to read, as in a book, the structure («li numeri») and the organization («li ordini, le gerarchie») of God's «handiwork».

For the medieval mind, many texts inherited from antiquity such as Pliny's *Natural History*, or the anonymous *Physiologus*, codified by Isidore of Seville in his *Etymologie* and *De natura rerum* contained the hermeneutical keys to the interpretation of the forms and symbols composing the universe.[6] Other texts presented a cosmological vision modeled on the idea that the universe was a mirror image of the divine mind (*imago dei*) and that ontological continuity existed between the celestial macrocosm and the human microcosm. Possible sources of inspiration for Dante, were Bernardus Silvestris' *De mundi universitate*, Alain de Lille's *Anticlaudianus* (1182-84), and Restoro d'Arezzo's *La composizione del mondo* (1282); the latter based on the *Liber aggregationibus* by Alfraganus.[7] By Dante's time, many ideas illustrated in these treatises had become common knowledge, and had been absorbed by the current theological and philosophical views on universe.

The introduction of Aristotelian physics stimulated the study of nature on its own terms, as a reality rooted in the spiritual ground but governed by its own principles, and which could be the object of independent investigation. The theologians of the twelfth and the thiteenth centuries gradually integrated

---

[5] The reference in the quotation is to *Psalm* 18, 1: «Caeli enarrant gloriam Dei, et opera manum eius annuntiat firmamentum». The *Psalm* places the accent on the capacity of the heavens to 'narrate' the story of creation.

[6] For a discussion on the importance of these texts in medieval thought, see TULLIO GREGORY, *Nature au Moyen Âge*, pp. 2-3.

[7] The critical literature about the possible sources of Dante's cosmology is too vast to summarize. A brief but exhaustive discussion on Dante's direct and indirect knowledge of some of the major authors and his approach to their texts is in CESARE VASOLI, *Dante e l'immagine del mondo nel Convivio*, in *L'idea e l'immagine dell'universo nell'opera di Dante*, Centro Dantesco dei Frati Minori Conventuali Ravenna, Ravenna, Longo, 2008, pp. 83-102. See also, on Dante and Ristoro d'Arezzo, ADRIANA OLIVIERO, *La composizione dei cieli in Restoro d'Arezzo e Dante*, in *Dante e la scienza*, Longo, Ravenna, 1995, pp. 351-362.

the well-ordered system of the laws of Aristotelian causality into the non-systematic biblical cosmology that the Fathers had developed in their hexaemeral literature. The book *Lambda* (XII) of *Metaphysics*, the *Physics*, the treatise on *De coelo et mundo*, and the *Liber de causis* together with their Arab and Latin commentaries, formed the nucleus of the new natural science and became the palette that philosophers, theologians, and artists used to illustrate their new views on cosmology.[8]

The cornerstone of this new approach to natural science was the firmly held belief that celestial bodies influenced the processes of generation and corruption that distinguished finite from infinite reality.[9] Based on the principle of astral causality, medieval thinkers, and Dante among them, believed that stars and planets influenced the infinite combinations of the four elements (earth, water, air, and fire) determining individual and universal events, conditioning individual characters, and designing the path of universal his-

---

[8] The anonymous *Liber de causis* (*Book of Causes*), for a long time attributed to Aristotle, had a deep influence on medieval cosmological speculation. Dante quoted it in several passages of the *Convivio* and also in the *Epistle* XIII to Cangrande. See BRUNO NARDI, *Le citazioni*. The *Liber de causis* is in reality an excerpt from Proclus's *Elements of Theology*. Thomas Aquinas discovered its true origin and dedicated a vast commentary to it. On Aquinas' interpretation on the *Liber de causis* and its influence on his philosophy, see CRISTINA D'ANCONA COSTA, *Recherches sur le Liber de causis*, Paris, Vrin, 1995.

[9] «In dubium nulli sapienti vertitur quin omne motus naturalis inferiorum corporum ex motu celestis corporis causentur», [«no erudite scholar doubts that the motion of celestial bodies causes every natural motion in the bodies below»]. AQUINAS, *Responsio ad magistrum Joannem de Vercellis de 43 articulis*, in *Opera omnia*, Studium Fratrum Predicatorum, t. XLII, Rome, 1979, pp. 327-335. Quoted from TULLIO GREGORY, *Natura e Qualitas planetarum*, in *Speculum naturale. Percorsi del pensiero medievale*, Roma, Edizioni di Storia e Letteratura, 2007, p. 72. Dante too attributes this widely shared belief to the philosophers. In speaking about the similarity of heaven and sciences, he affirms: «E la terza similitudine si è lo inducere perfezione nelle disposte cose. Della quale induzione, quanto alla prima perfezione, cioè della generazione sustanziale, tutti li filosofi concordano che li cieli siano cagione, avegna che diversamente questo pongano», [«The third similarity consists of bringing about perfection in those things disposed thereto. Concerning the bringing about of perfection, insofar as the first perfection is concerned, namely substantial generation, all philosophers agree that the heavens are the cause, although they explain it differently»]. *Conv.* II, xiii, 5. For the importance of the doctrine of celestial causality in thirteenth – century thought and in Dante, see the studies dedicated to this theme by BRUNO NARDI, in *Saggi di filosofia Dantesca*, and *Dante e la cultura medievale*. Two other texts are particularly relevant for the reception of the principle of astral causality in the Middle Ages: the *Speculum astronomiae*, which Lynn Thorndike attributed to Albert the Great (but the debate on this attribution is still open), and Albumasar's (Ja'far ibn Muhammad Abū Ma'shar al-Balkhī), *Introductorium in Astronomiam*. See LYNN THORNDIKE, *The History of magic and Experimental Science*, v. 4, New York, Columbia University Press, 1958, pp. 692-719, and PAOLA ZAMBELLI, *The "Speculum astronomiae" and its Enigma*, Dordrecht-Boston-London, 1992. On the impact of the translation of the ninth-century Persian astronomer Albumasar, see the seminal work by RICHARD LEMAY, *Abu Ma'shar and Latin Aristotelianism in the Twelfth Century. The Recovery of Aristotle's Natural Philosophy through Arabic Astrology*, Beirut, 1962. For a discussion of the reception of celestial causality in the theology of the twelfth and the thirteenth centuries, see TULLIO GREGORY, *Natura e Qualitas planetarum*, in *Speculum naturale* (47-68). See also Cesare Vasoli's commentary on *Convivio* II, xii.

tory. The study of the stars – their motions and their nature – became therefore of vital importance for theologians, and for the harmony between superior and inferior worlds that Dante sought in the *Commedia*.[10]

The assimilation of Aristotelian scientific thought into Christian Biblical cosmology caused a vast debate among theologians and astronomers, who aimed at coping with the non-secondary issues posed in reconciling Aristotelianism with the doctrine of creation and its eschatological character.[11] The biblical exegesis of creation needed to integrate explanations of sidereal motion and celestial causality within the context of Aristotelian-Arab physics and metaphysics. In this perspective, God identified with the first cause, as the origin and principle of universal motion, while angels – generally identified with the Aristotelian Intelligences – performed a function of intermediation between intelligible and sensible worlds.

If the new science provided a sound basis for a physics of creation, it did not provide, however, an ontological explanation of creation that was compatible with the Scriptures. While this new science provided a rational justification of sidereal motion, it did not solve the problem of the creation of multiplicity from the One. Moreover, it did not provide scientific ground for the mystical cycle of *exitus* from and *reditus* to God, meaning the anagogical process of ascent and descent that was at the core of the Christian – and Dante's – vision of the cosmos.

These issues, however, were not foreign to the Platonic and Neo-Platonic traditions, which provided a viable bridge across Aristotelian and Biblical cosmologies. As seen in the preceding chapters, the Dionysian celestial hierarchy and the doctrines on light elaborated by Franciscan Scholastics within the Augustinian line were two major elements of Neo-Platonic influence that became integrated into the cosmogony of the Scriptures. Both these elements, as shown in chapter II, constituted an integral part of the structure of the *Commedia*. In the spirit of his times, Dante used a syncretistic approach, which combined different elements from diverse traditions. He re-

---

[10] «Necesse est theologum scire scientia» [«Theologians must know science»], wrote Roger Bacon. Like him, many others repeted the same concept in their writings. Tullio Gregory provides and comment on numerous passages from the twelfth-century texts testifying to the theologians' new interest in astrology. See TULLIO GREGORY, *Astrologia e teologia nella cultura medievale*, in *Mundana Sapientia. Forme di conoscenza nella cultura medievale*, Roma, Edizioni di Storia e Letteratura, 1992, pp. 291-328.

[11] On the early debates among the theologians and their defense of astrology as compatible with Christian doctrine, see TULLIO GREGORY, *Astrologia e teologia*, MARIE-THÉRÈSE D'ALVERNY, *Astrologues et théologiens au XIIe siecle*, in *Mélanges offerts a M-D Chenu*, Paris, Vrin, 1967, pp. 31-50; JOHN TOLAN, *Reading God's Will in the Stars. Petrus Alfonsi and Raymond de Marseille defend the New Arabic Astrology*, «Revista Espanola de Filosofia Medieval», 7, 2000, pp. 13-30.

conciled these elements with his grand cosmological design, and transformed angelic hierarchies in the necessary interface between the spiritual and material dimensions of the cosmos. Dante's assumption that angels could perform simultaneous cognitive and ministerial functions distanced him from the mainstream angelology of the twelfth century, and allowed him to make angels the unifying factor of spiritual and sensible worlds. In presiding over cosmic motion and anagogical deification, angels ensured that the 'great sea of being' would flow back toward its natural port, thus fulfilling providential design.

b. *The Blessed Movers: Angels and Intelligences*

Dionysian angelology provided a theological foundation for Dante's design of cosmic harmonization in the *Commedia*. Although Dionysius did not present a philosophically refined doctrine of angelic operations that could satisfy Dante's needs, his system remained nonetheless essential for the tropological purpose of the *Commedia*. To meet his structural goal, in the *Commedia* Dante combined the Dionysian hierarchical structure (with the important departures seen in the preceding chapters) with the Aristotelian doctrine of Intelligences. He therefore established a correspondence between the nine angelic hierarchies and the nine spheres in the cosmos, by connecting enlightening and moving functions. In attributing to the angels the function of divine movers, Dante introduced an astrological element in his angelological-cosmological system that enhanced the role of angels as providential instruments of the order of creation. The assumption that angels were the movers («beati motori») of celestial bodies was not, however, straightforward. Integrating Aristotelian Intelligences and angelic figures of Biblical tradition was a step not all theologians were ready to take.

The identification of angels and Intelligences came as a natural result of the general Christianization of ancient philosophy that began in the early centuries of the Christian era. Although this identification was common, theologians disagreed on attributing the role of celestial movers to separate substances. To better understand the terms of one of the liveliest theological debates of the thirteenth century and its influence on Dante, it is necessary to take a step back and consider the philosophical implications of the Aristotelian doctrine of the Intelligences.[12]

---

[12] For an analysis of the syncretistic approach to angels in late antiquity and the Middle Ages, and the influence of Aristoteliansim on Dante's angelology, see STEPHEN BEMROSE, *Dante's angelic Intelligences*, Roma, Edizioni di Storia e Letteratura, 1983.

Plato's doctrine of the ideas had produced a fracture between the universe of «true» exemplars (ideas or forms) and sensible reality, which was modeled on them. His literary invention of the demiurge in his *Timaeus*, the skilled architect who occupied the intermediate space between the superior world of ideas and the inferior world of things, had not healed this fracture. Although Aristotle's metaphysics defined finite substances as a «synolon» (unity) of matter and form, the issue of how ideas became embedded in matter remained a central issue. From the point of view of the principles (*archai*) and causes (*aitiai*) of things Aristotle distinguished three principles: form (*logos* or *eidos*), its opposite or 'privation' (*steresis*), and matter (*hule*).[13] As for cognitive act, substances are a synolon of matter and form. Form, however, is also the actualizing principle of matter and foundation of being, and in this sense, it can be thought of in its own terms (a controversial point indicating a possible persistence of Platonic elements in Aristotle's metaphysics). Matter always needs form to exist.[14] Privation is the principle of change that triggers the passage of form into matter (for example, when dying a cloth white, a transformation occurs by eliminating the opposite 'not-white'), and characterizes the corporeal world where substance only exists as synolon of matter and form. Change allows for the actualization of matter, so that privation becomes in general terms a principle of motion upon which contingent reality relies.[15] For this reason, movement and the science that studies movement (physics), are of central importance in Aristotle's philosophical system: «Nature is a principle of motion and change, and it is subject of our inquiry. We must therefore see that we understand what motion is; for if it were unknown, nature too would be unknown».[16] The explanation of motion presupposed a necessary hierarchical order of causes, which in turn led to the assumption that the motion of the celestial bodies – composed of ether, a superior incorruptible fifth element – caused the movement of the inferior world, thus influencing all events involving generation and corruption of the four elements.[17]

---

[13] ARISTOTLE, *Metaphysics*, Λ, XII, 1069b, 30. English translation by W.D. ROSS, *The Complete Works of Aristotle*, J. Barnes ed., 2 vols., Princeton, Princeton University Press, 1984, pp. 1552-1728.

[14] Matter is indeed pure potency. It exists only in God's mind. It does not seem to exist contradiction with the Aristotelian principle and what Dante says in *Paradiso* XXIX, 22-24.

[15] In his *Physics*, Aristotle defines motion as any type of change, or more precisely, as the actualization of a potentiality. *Physics* 201a, 10-11, 27-29, b 4-5. In *The Complete Works of Aristotle*, pp. 315-446, translated by R.P. Hardie and R.K. Gaye.

[16] ARISTOTLE, *Physics*, III, 200b-10.

[17] For the Aristotelian principles of motion Dante applied to angels, see BRUNO NARDI, *"Sì come rota c'igualmente è mossa"*, in *Nel mondo di Dante*, Roma, Edizioni di Storia e Letteratura, 1944, pp. 337-350.

Aristotle introduced the immaterial movers of the celestial bodies as necessary links in the physical and ontological chain of beings to explain the celestial causality that conditioned this process of generation and corruption.[18] Following the geocentric cosmological structure of eight heavens that he derived from Eudoxus and Callippus, Aristotle postulated the existence of a number of movers equal to the number of the movements of each sphere.[19] The circular (perfect) motion of stars and planets assured the universal harmony of the cosmos in a well-ordered organism governed by the physical laws of necessity and causality.

The movers were external, non-formal, causes of the celestial bodies. Although composed of a finer matter than the corruptible bodies, planets and stars had less nobility than purely spiritual creatures. For this reason, superior and nobler entities must be the external causes of their motion. Aristotle likened the Intelligences to divinities that, by means of the movements they impressed and regulated, directed the cosmos toward the common end that the first cause (the immovable mover) had established through uniform universal movement.

In the second century AD, Claudius Tolomeus (the discovery was attributed to him much later by John Philoponus) extended the spheres to nine.[20]

---

[18] Aristotle assumed a finite number of Intelligences, equal to the number of sidereal movements (55 or 47), based on Eudoxus and Callippus' calculations. Dante hypothesized an infinite number of angels but a limited one of angelic orders corresponding to the number rather than to the movements of the spheres. He assumed, however, that the movement of heavens was the 'perfection' of the movers, following in this opinion Aristotle, Averroes, and with alterning views, Aquinas. The role of movers, however, was the main but not the unique operation Dante attributed to angels, who served on different tasks, related to but distinct from their moving function. For these other operations, see Chapter IV of this work. For the debate on the number of movers in Dante's times and the relation between Plato's secondary divinities – who produced corruptible things – and Aristotle's Intelligences, see BRUNO NARDI, *Dal Convivio alla Commedia*, pp. 37-42. In the *Convivio* Dante seemed inclined to accept the doctrine of the 'secondary causes' in Creation. In the *Commedia*, however, angelic orders only induce but do not 'fabricate' contingent things, for they are efficient and not formal causes. The Parisian bishop Etienne Tempier condemned this doctrine in 1277 («Quod cum intelligentia sit plena formis, imprimit illas formas in materiam per corpora coelestia tanquam per instrumenta», [«that being an intelligence full of forms, it impresses these forms on matter through celestial bodies as its instruments»]), but, as Cesare Vasoli notices, this did not prevent theologians from widely accepting it. See CESARE VASOLI, *Il canto II del Paradiso*, in *Lectura Dantis Metelliana*, Roma, Bulzoni, 1992, pp. 27-51; 48.

[19] ARISTOTLE, *Meth*. XII, 1074a, 5-10. Aristotle estimated the total number of movements in 55, including those of the moon and the sun. See BRUNO NARDI, *Dal Convivio alla Commedia*, pp. 37-42. See discussion on the number of movers and its philosophical implications in Dante in Chapter I.

[20] John Philoponus, a sixth-century astronomer from Alexandria, based his attribution on Tolomeus' *Planetary Hypotheses*, a work that circulated mainly among Islamic astronomers, and less known to the West. Tolomeus' most influential works in the Middle Ages were the *Almagest*, first translated into Latin from the original Greek by Gerard of Cremona in 1175, and his treatise on electional astrology, *Tetrabiblos* paraphrased by Proclus in the fifth century. The introduction of the

The observation of two different movements in the celestial bodies – the diurnal one from East to West, and one of a degree every one hundred years from west to East in the zodiac – had led to suppose the existence of a ninth sphere to justify the observed, second movement of the fixed stars. Although initially introduced for this reason, the ninth sphere soon assumed theological character, as theologians identified it with the see of the divine mind, the Empyrean. The Aristotelian-Ptolemaic cosmology, enriched by the contributions of Arab and Latin commentators to Aristotle's scientific works such as Albert the Great and Thomas Aquinas' commentaries on *De coelo*, remained substantially unaltered throughout the Middle Ages. In the attempt to reconcile Aristotelian and scriptural cosmological structures some theologians, such as Aquinas, maintained Neo-Platonic elements that had been absorbed into Christian thought and that persisted, with the implied issues, also in the *Commedia*. This reconciliatory approach presented two opposite threats to its sustainers: the risk of emanatism implicit in Neo-Platonism, and the risk of astral determinism that could derive from assuming causality *strictu sensu*. The *Commedia* represents an attempt to overcome both threats by linking a transcendental principle of creation and order to a cosmos structured on Aristotelian foundations. The key factor in making Dante's poetical attempt successful was the attribution of operative function to angelic hierarchies.

The question of the movers acquired such importance that in 1271 the Dominican general, John of Vercelli, sent a questionnaire on 43 themes to three major theologians of his order: Albert the Great, Thomas Aquinas, and Robert Kilwardby.[21] More than half of the questions (23) concerned the celestial

---

ninth sphere in medieval astronomy initially explained the movement of the sphere of the fixed stars with no theological implications. For an account on the history of the ninth sphere in medieval thought see EDWARD GRANT, *Planets, Stars, and Orbs: The Medieval Cosmos, 1200-1687*, Cambridge, Cambridge University Press, 1996, pp. 315-323. For the theological doctrine of the Empyrean, or ninth orb, and its characteristic in the *Divine Commedia*, see BRUNO NARDI, *La dottrina dell'Empireo e la sua genesi storica e nel pensiero dantesco*, in *Saggi di filosofia Dantesca*, pp. 167-215, and CHRISTIAN MOEVS, *The Metaphysics of Dante's Commedia*, pp. 15-35.

[21] For a detailed account of the different answers to five crucial questions included in John of Vercelli's questionnaire, see TIZIANA SUAREZ-NANI, *Les anges et la philosophie*, pp. 120-142. The questions enlighten the different positions of the three theologians on themes that are particularly relevant for the *Commedia*: «An Deus moveat aliquod corpus immediate» [«Whether God moves any body without mediation»]; «An omnia que moventur naturaliter, moveantur ministerio angelorum movente corpora celestia» [«Whether all things that are moved in nature, are moved angels' ministry as movers of the celestial bodies»); «An angeli sint motores corporum caelestium» [«Whether angels are movers of celestial bodies»]; «An infallibiter sit probatum angelos esse motores corporum caelestium apud aliquos» [«Whether anyone has infallibly demonstrated that angels are the movers of celestial bodies»]; «An infallibiter sit probatum angelos esse motores corporum caelestium, supposito Deum non esse immediatum motorem illorum corporum» [«Whether it can be infallibly demonstrated that angels are the movers of the celestial bodies, assuming that God is not the immediate

bodies and their motions. The episode testifies to the importance of the issue on celestial movers, and to the fact that a well established doctrine did not yet exist in this field. The answers of the three theologians show an array of different positions among the masters of the new Scholastic theology.[22] While Albert the Great rejected the identification of angels and Intelligences, both Aquinas and Kilwardby accepted it, but for different reasons. For Albert the angels did not have a role in the natural order of things; they retained their original biblical function of divine messengers and officers and could not be identified with the Intelligences, for Intelligences were pure motionless forms eternally in act and connected to the bodies they moved. The angels instead, being equally incorporeal and immaterial, could move and respond to God's commands, and therefore maintained a residual potency actualized when called to perform divine missions. As Tiziana Suarez-Nani observes, Albert's apodictic refusal to attribute to the angels the role of movers aimed at delimiting the domain of philosophical speculation, which could not extend to an investigation of divine reality.[23] The issue about celestial motion should therefore remain within the field of natural philosophy. In maintaining a distinction between philosophical inquiry and theological speculation, the master of Cologne relegated the angels to outside the domain of physical order. Similarly, Robert Kilwardby tried to provide a natural explanation of motion, but unlike Albert the Great and Aquinas, explained universal dynamism by attributing to the celestial bodies a natural inclination toward movement of divine origin.

Aquinas's position differed from those of the other two theologians. In particular, he did not share Albert's opinion that the extrinsic movers must be 'philosophical' Intelligences of a different nature and function from angels.[24] He did not attribute the function of movers to all angels, however,

---

mover of these bodies»]. Robert Kilwardby studied in Paris. He was Regent at Oxford University and provincial prior Dominican Order in England prior his appointment as archbishop of Canterbury in 1272. He died in Viterbo in 1279.

[22] The texts containing the answers of the three theologians are the following: AQUINAS, *Responsio ad magistrum Joannem de Vercellis de 43 articulis*, in *Opera omnia*, Studium Fratrum Predicatorum, t. XLII, Rome, 1979, pp. 327-335; ROBERT KILWARDBY, in MARIE-DOMINIQUE CHENU, *Les réponses de Saint Thomas et de Kilwardby à la consultation de Jean de Verceil*, in *Mélanges Mandonnet*, vol. I, Paris, 1930, pp. 191-222; ALBERT THE GREAT, in *Alberti Magni Opera Omnia*, t. XVIII, 1, Monasterii Westfalorum, 1975, pp. 45-64. See also TIZIANA SUAREZ-NANI, *Les anges et la philosophie*, pp. 108-142, and for a comparison of medieval theological cosmologies see PIERRE DUHEIM, *Le Système du Monde. Histoire des Doctrines cosmologiques de Platon à Copernic*, 10 vols., Paris, Hermann, 1913-1959, in particular Tome III for Franciscan and Dominican cosmologies, and Tome VI for the 1277 condemnation of the propositions about the movers.

[23] TIZIANA SUAREZ-NANI, *Les Anges et la philosophie*, pp. 125-127.

[24] Aquinas dealt with questions related to cosmology and the movers in many of his works,

but only to the Virtues, which superseded to the orders predisposed to the government of human affairs.

Albert the Great and Kilwardby's different objections to the identification of angels and Intelligences reveal their intention to keep philosophy and theology separate, each governed according to its own principles. Aquinas' position reveals instead an inclination to reconcile philosophical and theological truths, which must find reciprocal support. Following Aristotle, Aquinas argued that to a hierarchy of beings corresponded a hierarchy of causes, and that an intellectual principle superior in nature to the material spheres, must cause sidereal motion. By identifying this principle with the angels, Aquinas created that bond between creation and cosmology, philosophy and theology that also constitutes an essential trait of Dante's *Paradiso*.

A critical issue in the debate on the movers concerned the mechanics of transmission of the movement from the transcendent immovable being in the Empyrean to the nine material spheres. To assume that the heavens would move by their own will, each desiring to imitate the sphere placed above them, implied the assumption that the heavens were animated, a doctrine at odds with their supposed instrumentality in the providential architecture of celestial influence. The doctrine of animation had very ancient origin, and Arab philosophers had further developed and adapted it to a monotheistic context. Avicenna, the great Arab physician and philosopher of the tenth century, had introduced two categories of angels, corresponding to inferior and superior Intelligences, in order to explain the movement of the spheres. Superior angels were immobile causes of movement; the formal principles of the orbs were instead inferior angels, who activated motion by eliciting the desire to imitate the spheres above them in the celestial bodies. Both Averroes and Aquinas criticized Avicenna's theory, although Aquinas never firmly rejected the hypothesis of animated heavens. Several passages of *Paradiso* present, as discussed later, ambiguity on this issue.[25] The concep-

---

showing an evolution in his positions. For an analysis of the various texts and their comparison see THOMAS LITT, *Les corps célestes dans l'univers de Thomas d'Aquin*, Louvain, Publications Universitaires, 1963, pp. 99-109. See also BRUNO NARDI, *Tutto il frutto ricolto*, and ETIENNE GILSON, *The Christian Philosophy of Thomas Aquinas*, Notre Dame, University of Notre Dame Press, 1956, pp. 160-173.

[25] On this issue, critics have different opinions. According to Tiziana Suarez-Nani, Aquinas retained Avicenna's distinction in combining the idea of angelic hierarchies as superior movers but attributed to the Virtues the material function of activating cosmic dynamics. Others, such as Litt, affirm that Aquinas remained in doubt about the existence of the heavenly souls. See THOMAS LITT, *Les corps célestes*, pp. 108-109. For a discussion of a possible Avicennian influence on Aquinas on this issue, see TIZIANA SUAREZ-NANI, *Les anges*, pp. 116-120.

tual framework of *Paradiso*, however, leaves little doubt that for Dante angelic hierarchies initiated the cosmic dynamics and that the spheres were merely their instruments.

Although both Aquinas and Dante identified the angels with the celestial movers, there is a main difference between their angelologies. While Aquinas attributed the function of movers to the order of Virtues, Dante followed the Dionysian taxonomy of angelic functions, and made all orders move the heavens according to their correspondent degrees of nobility. Aquinas attributed to the highest hierarchy (Seraphim, Cherubim, and Thrones) an essentially contemplative function, whereas the lowest hierarchy (Principalities, Archangels, and Angels) had ministries relative to the government of the sublunar world. The ministry of the intermediate hierarchy (Dominations, Virtues, and Powers) was to actuate the dispositions of divine providence. The task of movers thus properly pertained to this hierarchy, and specifically to the Virtues. Aquinas based his choice on Dionysius' explanation of the name «virtue» as meaning universal dynamic force, based on the authority of Luke's Gospel.[26] By distancing himself from Aquinas, and giving to all angelic orders the role of chronocrators, Dante connected in a specular way angelic and physic orders, thus making the hierarchies the link of the spiritual and the material sides of the cosmos. In so doing, angels are the hinge Dante used to establish the cor-

---

[26] «Secundo autem, ab operante et exequente distribuitur et multiplicatur ad varios effectus. Quod quidem fit per ordinem virtutum, quarum nomen, ut Dionysius ibidem dicit, significat quandam fortem virilitatem in omnes deiformes operationes, non relinquentem suimet imbecillitate aliquem deiformem motum. In quo patet quod principium universalis operationis ad hunc ordinem pertinet. Unde videtur quod ad hunc ordinem pertineat motus caelestium corporum, ex quibus, sicut ex quibusdam universalibus causis, consequuntur particulares effectus in natura: et ideo virtutes caelorum nominantur Luc. 21-26, ubi dicitur: virtutes caelorum movebuntur. Ad eos etiam spiritus pertinere videtur executio divinorum operum quae praeter naturae ordinem fiunt, nam ista sunt sublimissima in divinis ministeriis: propter quod Gregorius dicit quod virtutes dicuntur illi spiritus per quos signa frequentius fiunt. Et si quid aliud universale et primum est in ministeriis divinis exequendis, conveniens est ad hunc ordinem pertinere». [«Then, secondly, there is a distribution and multiplication in the form of diverse effects on the part of the agent and executor. In fact, this is done by the order of Virtues, whose name, as Dionysius says in the same place, suggests "a strong forcefulness in regard to all Godlike operations, one which does not abandon its Godlike movement because of any weakening in itself." It is evident from this that the source of universal operation belongs to this order. Hence it appears that pertinent to this order is the motion of the celestial bodies, from which bodies as universal causes, the particular effects in nature follow. So, they are called "the powers of the heavens" where it is said: "the powers of the heavens shall be moved" (Luke 21:26). Also pertinent to these spirits is the execution of divine works which are done outside the order of nature, for these are most sublime among the divine ministrations. For which reason, Gregory says, "those spirits are called Virtues through which miracles are frequently wrought" [In Evangelium, homil. 34]. And if there be anything else that is universal and primary in the carrying out of divine ministrations, it is proper to assign it to this order»]. AQUINAS, *Contra Gentiles* III, 80, 11. English translation by Vernon J. Bourke available online at the site of the Thomas Institut (http://www.thomasinstituut.org).

respondence between physical and metaphysical, eternal and temporal, upon which he based the cosmological and poetic structure of the *Commedia*.

c. *Dante's Eschatological Cosmology: the Order of the Holy Spirit*

Rethinking the cosmos based on philosophical principles had crucial theological implications, particularly with regard to the relation between cosmology and creation. Often magmatic for lack of established doctrine, the opinions of thirteenth – century theologians reflected two different approaches. Some, such as Albert the Great and his Arab predecessors like Avicenna and Averroes, maintained a separation between physical and spiritual realms. In Albert the Great's view, recourse to divine intervention was necessary only when a natural explanation of phenomena was impossible. Others, such as Aquinas, tried to reconcile natural and supernatural dimensions, and considered philosophy as auxiliary to theology («ancilla theologiae»): the former researched the causes and the origin of motion in nature; the latter situated them in the supernatural sphere.[27] The origin of motion was a metaphysical question, however, and when the mechanics of celestial movement was set in a Christian context, it became impossible to separate it from the issue of the relation between the One and the multiplicity. Transposed in Christian terms, speculation on the causes of heavenly motion directly concerned the origin of creation.

Neo-Platonism, with the works of Plotinus and his disciple Porphyry, laid the ground upon which, through the filter of the Arab philosophers, theologians of the thirteenth century attempted a reconciliation of Plato and Aristotle's cosmological views. A fundamental element in this attempt was the finding of the nexus that linked contingent beings to their eternal point of origin without incurring in emanatism. The Platonic doctrine of the *anima mundi* (soul of the world) played a major role in searching for what «binds all things in one volume», and it is particularly relevant to understanding the delicate equilibrium of Aristotelianism and Neo-Platonism at work in the *Commedia*.[28]

---

[27] «Non enim [theologia] accipit sua principia ab aliis scientiis, sed immediate a Deo per revelationem. Et ideo non accipit ab aliis scientiis tanquam a superioribus, sed utitur eis tanquam inferioribus et ancillis», [«For it [theology] accepts its principles not from other sciences, but immediately from God, by revelation. Therefore it does not depend upon other sciences as upon the higher, but makes use of them as of the lesser, and as handmaidens»]. AQUINAS, *S. Th.* I, q. 1, a. 5.

[28] An important account of the historical development of the idea of the *anima mundi* from antiquity to Middle Ages, with particular reference to William of Conches and the School of Chartres, is in TULLIO GREGORY, *"Anima mundi." La filosofia di Guglielmo di Conches e la scuola di Chartres*, Firenze, Sansoni, 1955, pp. 123-174. On the importance and limits of the Neo-Platonic tradition of *anima mundi* on Dante, see also GIULIANA CARUGATI, *"Pulcherrima mundi figura"*, in *Il ragionare della carne. Dall'anima mundi a Beatrice*, Lecce, Manni, 2004, pp. 27-46.

The hypothesis of an inherent vital principle of cosmological motion was an attempt to explain the origin of the world with an immanent cause. Developed by Stoicism on the basis of an ancient, pre-Socratic, animistic tradition, the idea of a 'mind' spreading its *logoi spermaticoi* (animated principles) to originate life and movement, passed to the Middle Ages through classical sources, mainly Virgil's *Aeneid*, Calcidius' commentary to Plato's *Timaeus*, Macrobius' commentary to the *Somnium Scipionis*, and Boethius' *De consolatione philosophiae*.[29] Since animism conflicted with the Christian principle of a transcendent God, Neo-Platonism situated the *anima mundi* above sensible reality, as a superior entity able to access ideas and imprint them in all beings. This unifying and unified principle controlled the dispersion of multiplicity by embracing all forms of life and moving them toward a finalistic return to the One. With the development of Trinitarian theology, the Greek Fathers tended to identify the *anima mundi* with the Holy Spirit.[30] Many theologians, such as Theodoric of Strasburg and William of Conches (although with alternating opinions), followed in their steps. The Latin fathers however, starting with Augustine, had reservations about this identification.[31] In the twelfth century, with the reaffirmation of the religious legitimacy of astrology, Raymond of Marseille -the author of the *Tables of Marseille* and one of the first defenders of

---

[29] In particular, PUBLIO VIRGILIO MARONE, *Aeneid* VI, 724-33; MACROBIUS, *In Somium Scipionis*, I, XIV, 15; BOETHIUS, *De consolatione philosophiae*, III, m. IX.

[30] As the Western Christian civilization appropriated the symbolic representations of ancient and oriental cosmographies, a production of archetypal images of the cosmos common to the religious and scientific milieux flourished. In the iconography of the twelfth and the thirteenth centuries, the zodiac became associated with the representation of divine cosmology, where the heavens figure as steps to climb toward God. The association of stars and planets to the angels and to the gifts of the Holy Spirit are present in the iconography of the cosmos already in the twelfth century. See JURGIS BALTRUSAITIS, *L'image du monde celeste du IXe au XIIe siècle*, «Gazette des Beaux-Arts», XX, 1938, pp. 138-148; ID., *Cercles atrologiques et cosmographiques à la fin du Moyen Âge*, «Gazette des Beaux-Arts», XXI, 1939, pp. 65-84. On the presence of angels in the medieval cosmological iconography, see also BARBARA BRUDERER EICHBERG, *Les neuf chœurs angéliques*, pp. 39-42.

[31] As Tullio Gregory shows through the analysis of William of Conches' *Philosophia*, there were at least three different and prevalent opinions about the nature of the *anima mundi*. The first identified *anima mundi* and Holy Spirit (e.g. Theodoric of Chartres in his *De sex dierum operibus* and Raymond of Marseilles, the author of the astronomical Marseille tables). Some of the supporters of this position, such as Abelard, regarded the *anima mundi* as Holy Spirit only in a metaphorical sense, as a dispenser of charismatic gifts. The second opinion considered the *animal mundi* as an immanent power God had placed in all creatures. Finally, a third opinion saw the vivifying principle of the soul of the world as an incorporeal substance that operated in different ways in all bodies, according to their different qualities. Although inclined to accept the identification of *animal mundi* and Holy Spirit, William of Conches changed his mind consequently to the fierce opposition of William of Saint Thierry, and Innocent II's condemnation of this doctrine. William of Conches did not take any position in his *Philosophia*, but embraced the identification of *animal mundi* with the Holy Spirit in his commentary to Plato's *Timaeus*, where he defined *animal mundi* as a natural fecundatory force (vigor) that regulated heavenly motion. See TULLIO GREGORY, *Anima mundi*, p. 141.

astrology – attributed to the Holy Spirit a cosmological function, thus connecting the motion of the celestial spheres to the providential order of creation.[32]

A major issue with the identification of the vital principle with the Holy Spirit was that the soul of universe strictly depended on creation, while the Holy Spirit as the third person of the Trinity was eternal. Abelard's reply to this objection shows the amplitude and subtlety of the debate on this subject. Moreover, it serves to understand the conceptual contexts within which Dante weaved Neo-Platonic and Aristotelian elements in his work. Peter Abelard introduced an important distinction between the Holy Spirit as informing divine Love and as principle of order in creation. The Holy Spirit was the «soul of the world» – «spiritus quippe nomen est naturae, anima vero officii» [«for indeed the term spirit refers to its nature, while anima refers to its office»] – only as governing criterion of universal order.[33] This distinction did not overcome the suspicion of heresy that led to its condemnation by William of Saint Thierry, Bernard of Clairvaux, and ultimately by Innocent II. Abelard's opinion, although rejected by the Church, appealed however to those concerned with the cosmological aspect of divine order, for it resurfaced in the works of theologians who tried to defend the religious legitimacy of astrology. Dante's idea of Providence as the governing instrument of the Holy Spirit operating through the angelic Intelligences can be situated in this line of thought.

In several occasions throughout the circles of *Paradiso*, Dante seems to distinguish between the ordering function of the Holy Spirit, and the unifying power of Love-caritas. In the heaven of the Moon, after Beatrice turns her response to Dante's question about the moon spots into a doctoral dissertation on the ontological principles of reality, Piccarda Donati explains to Dante that the disposition of the blessed in *Paradiso* follows the order dictated by the Holy Spirit:

> Li nostri affetti, che solo infiammati
> son nel piacer de lo Spirito Santo,
> letizian del suo ordine formati (*Par.* III, 52-54).

> Our sentiments, which only serve the flame
> that is the pleasure of the Holy Ghost,
> delight in their conforming to His order.

---

[32] For Raymond of Marseille's arguments in defense of the astrologer as prophet and exegete, see TULLIO GREGORY, *Astrologia e teologia*, p. 295, n. For Raymond of Marseille and his *Liber cursuum planetarum*, see RICHARD LEMAY, *Abu Ma'shar and Latin Aristotelianism in the Twelfth Century*, pp. 141-157, and CHARLES H. HASKINS, *Studies in the History of Medieval Science*, Frederick Ungar, New York, 1967, pp. 96-98.

[33] PETER ABELARD, *Introductio ad theologiam* II, 17; PL 178, 1082, as quoted in TULLIO GREGORY, *Anima mundi*, p. 147.

All souls are inflamed (a verb that evidences the effects and the essence of divine love) with the «pleasure» of the Holy Spirit, where the choice of the word «pleasure» hints at the self-complacency of the Trinitarian person in the perfection of the well-ordered world of creation. The blessed share this pleasure («letizian»), for order is the formal principle of paradisiacal souls («del suo ordine formati»). In another passage (*Par.* XIX, 101) Dante returns to this concept by calling the blessed the «fires» of the divine spirit, and stresses the strict connection existing between spiritual substances and love as a formal element of their natures.[34]

While love is the origin and cause of creation, Providence is the formal ordering principle ruling the universe, and distributes offices and ministries. So we read in *Paradiso* XXVII, when the blessed souls remain silent to let Saint Peter talk to Dante:

> La provedenza, che quivi comparte
> vice e officio, nel beato coro
> silenzio posto avea da ogne parte, (*Par.* XXVII, 16-18)

> After the Providence that there assigns
> to every office its appointed time
> had, to those holy choirs, on every side,
> commanded silence,

The «here» («quivi») refers not only to the heaven of the fixed stars where Dante hears Peter's invective against the disorder of the Church, but also to all heavens, where Providence distributes («comparte») their different offices among the blessed and the angels. As Peter Damian had already revealed to Dante in canto XXI, the blessed, like the angels, are servants of *caritas* and obey the «counsel» that rules the universe:

> Ma l'alta carità, che ci fa serve
> pronte al consiglio che 'l mondo governa,
> sorteggia qui sì come tu osserve (*Par.* XXI, 70-72).

> But the deep charity, which makes us keen
> to serve the Providence that rules the world,
> allots our actions here, as you perceive.

«Consiglio» (counsel) may have here the general meaning of wisdom, but it is also one of the gifts of the Holy Spirit, which metonymically indicates the governing role associated with its seven charismatic gifts. If Providence is the

---

[34] «Poi si quetaro quei lucenti incendi / de lo Spirito Santo» [«After the Holy Ghost's bright flames fell silent»], *Par.* XIX, 100-101.

instrument of the Holy Spirit in governing the order of *Paradiso*, then the angels are instruments of Providence and as such, they are related to the gifts of the Holy Spirit, as the analysis of *Purgatorial* angels in chapter IV will show. They fulfill their task as chronocrators by actualizing the virtues present in the celestial bodies. Through them, Providence orders the dispersed multiplicity and binds it together in the unitary design of salvation made possible by Christ's redemptive sacrifice.

The role assigned to Providence and its instruments reflects the essentially eschatological character of Dante's cosmology, and the importance of the angelic functions in the spiritual structure of the *Commedia*.[35] The eschatological character of Dante's cosmology emerges in *Paradiso* XXIII, where the poet assists to the triumphs of Christ and of Mary, thus anticipating the vision of the mystic rose in *Paradiso* XXX. In this canto, as in *Paradiso* II, the poet experiences a holistic view of universal harmony. It is a view from below, while the poet is still immersed in the ethereal matter of the heaven of the fixed stars and suspended in the constellation of Gemini. As Dante admires the triumph of Christ, Beatrice directs his attention to the host accompanying Christ:

> e Beatrice disse: «Ecco le schiere
> del triunfo di Cristo e tutto 'l frutto
> ricolto del girar di queste spere!» (*Par.* XXIII, 19-21).
>
> And Beatrice said: «There you see the troops
> of the triumphant Christ-and all the fruits
> ingathered from the turning of these spheres!».

The «schiere» are composed of blessed and angels, and the saved souls testify to the triumph of Christ's redemptive work. Lines 20-21 place the revolutions of the spheres in a perspective of salvation, for their ultimate end is the holy fruit of beatitude.[36]

---

[35] The eschatological character of cosmology, centered on the salvific role of Incarnation, that characterizes Dante's poem, was already present in John Scotus Eriugena's writings. As Tullio Gregory notes: «La *recapitulatio* o l'*instaurare omnia in Christo* viene cosí a costituire la prospettiva fondamentale dell'escatologia eriugeniana, lungo una linea di pensiero che da Paolo a Ireneo si era venuta svolgendo attraverso tutta la patristica greca». [«The *recapitulatio* or the *instaurare omnia in Christo* constitutes the fundamental perspective of Eriugena's eschatology, in a line of thought that from Saint Paul to Ireneus had developed in all Greek Patristic»]. TULLIO GREGORY, *L'escatologia di Giovanni Scoto*, in *Mundana sapientia*, p. 234.

[36] In the same perspective speaks also Folchetto of Marseille in *Paradiso* IX, 106-108: «Qui si rimira ne l'arte ch'addorna /cotanto affetto, e discernesi 'l bene / per che 'l mondo di sù quel di giù torna» [«For here we contemplate the art adorned / by such great love, and we discern the good / through which the world above forms that below»].

ANGELS AND THE ORDER OF THE COSMOS

Immediately before passing from the ninth heaven of the Primum Mobile to the immaterial and a-temporal space of the Empyrean, Beatrice expounds the general principles governing the motion of the Primum Mobile to the wayfarer. We are at the confine that separates the material and immaterial worlds, ready to cross the invisible membrane that separates matter from eternal and pure spirituality. This is Dante's final holistic vision of the universe in which cosmology and metaphysics come into unity, drawn in a superior theological synthesis:

> [...] Le cose tutte quante
> hanno ordine tra loro, e questo è forma
> che l'universo a Dio fa simigliante (*Par.* I, 103-105).
>
> [...] All things, among themselves,
> possess an order; and this order is
> the form that makes the universe like God.

The order is the «form» of the universe, the specific arrangement of «all things». It is possible to think of this disposition linking all substances so that they have order «among themselves» in analogy with the concept of form that applies to language. In book Z of *Metaphysics* Aristotle explained that in taking two letters, a and b, and their combination into a syllable (ab, or ba), their form consists in the order in which the letters are arranged so that 'ab' is a different 'substance' from 'ba'.[37] Order is therefore an informing criterion, the formula of *Paradiso* as well as of corporeal reality that makes things similar to God. In their act of seeing, the angels acquire the knowledge of order and its purpose:

> Qui veggion l'alte creature l'orma
> de l'etterno valore, il qual è fine
> al quale è fatta la toccata norma (*Par.* I, 106-108).
>
> Here do the higher beings see the imprint
> of the Eternal Worth, which is the end
> to which the pattern I have mentioned tends.

This tercet expresses a teleological vision («fine») to which order gives meaning and operation («norma»). At the core of the cosmological construction of the *Commedia*, we find that the providential structure dictated by

---

[37] «Since that which is compounded out of something so that the whole is one, not like a heap but like a syllable-now the syllable is not its elements, ba is not the same as b and a». ARISTOTLE, *Meth.* Z (VII), 1041b.

God's love relies on the operative function of the angelic hierarchies, for they are the instruments that transmit and preserve order in the universe. In this perspective, the correspondence between choirs and spheres – assured by the combination of Dionysian and Aristotelian ministries – was necessary to structure the all-encompassing, eschatological construction of *Paradiso*. Creation and cosmology, unity and multiplicity, thus find a theological synthesis rooted in metaphysics.

Trinitarian doctrine and Incarnation were central to Dante's eschatological cosmology.[38] In presenting the ordering principles of his cosmology in the two proemial cantos of *Paradiso*, Dante anticipates the harvesting episode of canto XXXIII, where he sees the unity of the three circles embodying the three moments of creation (power), incarnation (wisdom), and love (the providential order presided by the Holy Spirit). Before introducing the philosophical and theological arguments that structure his path through the heavens in *Paradiso* II, Dante places at the centre of his cosmological discourse the crucial event in providential history: the mystery of Incarnation. Inflamed by the desire to see the splendor of the «eternal daisy», Dante marvels at his own perception, uncertain as to how he can be at the same time corporeal and suspended in the ether:

> S'io era corpo, e qui non si concepe
> com'una dimensione altra patio,
> ch'esser convien se corpo in corpo repe,
> accender ne dovria più il disio
> di veder quella essenza in che si vede
> come nostra natura e Dio s'unio (*Par.* II, 37-42).

> If I was body (and on earth we can
> not see how things material can share
> one space-the case, when body enters body),
> then should our longing be still more inflamed
> to see that Essence in which we discern
> how God and human nature were made one.

---

[38] Dante's eschatological vision centered on Incarnation seems to follow a line of thought that starting from Saint Paul, through Ireneus and the Greek Fathers continued in the Middle Ages, particularly in the works of John Scotus Eriugena (also the most authoritative translator of Dionysius). As Tullio Gregory observes, in Eriugena «È dunque la mediazione di Cristo – non la mediazione intelligibile operata dall'uomo – a dare al *reditus* il suo fondamento, realizzando nell'incarnazione quell'unione che è garanzia del finale ritorno», [«it is therefore Christ's mediation – not the intelligible mediation man operates – to give foundation to the *reditus*, thus realizing in Incarnation that union that guarantees the final return»]. TULLIO GREGORY, *Mundana Sapientia*, p. 234.

The realization of the coexistence of material and spiritual dimensions, impossible to conceive rationally, should elicit in all human beings the desire to ascend spiritually and contemplate the mystery of Christ's human and divine nature. In the poet's progression toward the Empyrean, the individual experience of the pilgrim connects with the universal destiny of human beings, exemplarily embedded in the implicit double subjectivity of the «I» and «we». Incarnation appears strictly related to the foundational motive of the angelic anagogical voyage toward divine similitude, paralleled by and connected to the human imitative journey toward Christ. While the first is a spiritual theophanic process, the second type of imitation is a *conversio mori*, a necessary step toward fulfilling the human destiny of salvation. If this imitation finds its constituent ground in the superior spiritual sphere, then it occurs within the ethical sphere of human operations. The cosmological structure of the *Commedia* thus connects the ethical and spiritual aspects of theological imitation.

d. *Love, Knowledge, and Dante's Epic Journey.* Paradiso I

The opening canto of *Paradiso* takes the reader into the rarified atmosphere of the outer space where the luminous bodies not only testify to a reality beyond the visible, but also are the visible instruments of the divine love that, from the infinite depths of Empyrean, administers the manifolds aspects of sensible reality. While love is the governing principle of the universe, we also know (see chapter II) that Dante places love second to knowledge in the physical and spiritual economy of the universe – in the same way that the grammar of the Word comes before the synthax of creation. Angels become agents of motion only after they contemplate the divine mind (although to speak in temporal terms is incorrect; for in the Empyrean all occurs outside place and time). Dante elaborated the relation between contemplation and motion by combining the Dionysian anagogical process (consisting of the three moments of enlightment, participation, and unification) with the metaphysics of causality. Enlightenment is thus not only «revelation» of the divine mind, but also impregnates Intelligences with divine forms – as Dante read in proposition 10 of the *Liber de causis* – which they transfer to the spheres they move. The angels neither create, nor «possess» forms. They convey the images apprehended through speculation by reflecting them on the inferior spheres as mirrors of the divine intellect. Intellectual light, although dissimilar from a material entity of any sort, has nonetheless a real transformative power since it conveys the actualizing force of the divine.[39] As

---

[39] See chapter III for a brief discussion of the influence of the metaphysics of light in Dante. See also CHRISTIAN MOEVS, *The metaphysics*, pp. 19-21.

discussed in chapter II, the influence on Dante's conception of light and of the laws of reflection and refraction of medieval optical theory is evident. Dante transforms the angelic function of mirrors from a symbolic and metaphorical function into a concrete material one.[40] The highest orders perfectly mirror the sacred images of the divine mind on the uniform and transparent matter of the Primum Mobile, and imperfectly on the lower heavens, whose thicker ethereal matter causes light to refract and diffuse. The optical effects correspond to the different stages in which the passage of forms from the Empyrean to the heavens occurs. Their mirroring function illustrates the role of angels as efficient causes in the creation. In this action, knowledge (of forms derived from contemplation) and love (the angelic desire of likeness to God that initiates the cosmological motion) indissolubly combine in initiating the harmonious dynamic of the universe.[41] This dynamics is the expression of the order that finds its origin in love. If love is the ordering principle of cosmos, knowledge is at the very origin of its existence.

The two related principles of love and knowledge thus constitute the nucleus of universal harmony. They are the two forces that drive universe toward its ultimate destiny, symbolized in Christ and in the mystery of Incarnation. Knowledge and knowledge-driven love determine the spiritual and the physical motions in which the order of cosmos articulates, revealing the strict connection between creation and cosmology.[42] Creation and its resolution from

---

[40] Already in *Convivio* Dante assumed that the movers were the origin of the virtues of the stars, and that they transmitted them through light: «li raggi di ciascuno cielo sono la via per la quale discende la loro vertude in queste cose di qua giù. E però che li raggi non sono altro che uno lume che viene dal principio della luce per l'aere infino alla cosa illuminata, e luce non sia se non nella parte della stella, però che l'altro cielo è diafano, cioè trasparente, non dico che vegna questo spirito, cioè questo pensiero, dal loro cielo in tutto, ma dalla loro stella. La quale per la nobilità delli suoi movitori è di tanta vertute, che nelle nostre anime e nell'altre nostre cose ha grandissima podestade, non ostante che essa ci sia lontana [...]» [«it is necessary to know that the rays of each heaven are the paths along which their virtue descends upon these things here below. Since rays are nothing other than the shining which comes from the source of the light through the air to the thing illuminated, and there is no light except from the body of the star, because the rest of the heaven is diaphanous (that is, transparent), I do not say that this spirit (that is, this thought) comes from their heaven as a whole but from their star. This star, by reason of the nobility of those who move it, is of such great virtue that it has immense influence upon our souls and upon all things belonging to us, notwithstanding that its distance from us [...]»]. *Convivio* II, vi, 9-10.

[41] Order and harmony, intended in musical terms, are correlated. Correspondence also exists between angelic choirs and music, showing another aspect of the relation Dante established between angels and order. For an analysis of music and angelic orders, see FRANCESCO CIABATTONI, *Dante's Journey to Polyphony*, Toronto, Toronto University Press, 2010.

[42] For a discussion on the dialectic love-knowledge in *Paradiso*, particularly with respect to Dante's theory of creation in *Par.* XXVII, see GIUSEPPE MAZZOTTA, *Cosmology and the Kiss of Creation (Paradiso I, 27-29)*, «Dante Studies», CXXIII, 2005, pp. 1-21, and *La metafisica della creazione*, in *L'idea e l'immagine dell'universo nell'opera di Dante*, Ravenna, Longo, 2008, pp. 61-81.

time into eternity are the two poles around which the eschatological cosmology of the *Commedia* revolves.

Dante experiences his voyage to deification in analogy with the angels and the souls. The reason he – «not Aeneas, not Paul» (*Inf.* I, 32) – benefits of this special grace, resides in his poetic voice. Dante is the one for whom (and because of whom) Beatrice stood above all other poets of his time; the one – as Rachel says to Beatrice – «that-for your sake-he's left the vulgar crowd» («uscì per te de la volgare schiera», *Inf.* II, 105). This poetic superiority finds its apex in the *Commedia*, where the distinctions in genres proposed in the *De vulgari eloquentia* among love, virtue, and arms are annihilated. The poem is epic, lyric, and educative at the same time; it mirrors the image of the cosmos and is part of its ethical order. It is the reason Dante is granted special grace to enjoy the «true» vision of the otherworld.[43] His spiritual sailing of the metaphysical waters of Paradiso is the result of the intimate interactions of love and knowledge, faith and art.

While Dante opened the *Convivio* with the famous Aristotelian quote about humans' natural desire for knowledge, in the *Commedia* the poet inscribed this universal desire within a sacred dimension, making knowledge the root of the natural desire to reunite with God.[44] The themes of love and knowledge permeate the entire *Commedia*, and are at the very core of Dante's final voyage through the heavens in *Paradiso*. The long nautical metaphor that opens canto II addresses and warns the readers about the difficulty of the last leg of the journey they are about to undertake following Dan-

---

[43] «The *Divina Commedia*, which is a cosmological epic, can legitimately be called the poetic model of the cosmos as much as it gives the specular image of the cosmos. It contains and mirrors the design of Creation while being part of Creation. In this way, it leads the soul of the reader to the discovery of divine wisdom». GIUSEPPE MAZZOTTA, *Cosmology and the Kiss of Creation*, p. 12.

[44] «naturalissimo è in Dio volere essere – però che, sí come ne lo allegato libro si legge, "prima cosa è l'essere, e anzi a quello nulla è" – l'anima umana essere vuole naturalmente con tutto desiderio; e però che il suo essere dipende da Dio e per quello si conserva, naturalmente disia e vuole essere a Dio unita per lo suo essere fortificare». [«The human soul by nature desires with all its will to exist; and since its being depends on God and is preserved by him, it naturally longs and desires to be united with God in order to strengthen its being»], *Convivio* III, ii, 7. The book Dante quotes in this passage is the *Liber de causis*. There is an innate desire of knowledge in man, determined by his desire to reunite with God: «Sì come dice lo Filosofo nel principio della Prima Filosofia, tutti li uomini naturalmente desiderano di sapere. La ragione che puote essere ed è, che ciascuna cosa, da providenza di propria natura impinta, è inclinabile alla sua propria perfezione; onde, acciò che la scienza è ultima perfezione della nostra anima, ne la quale risiede la nostra ultima felicitade, tutti naturalmente al suo desiderio siamo subietti», [«As the Philosopher says at the beginning of the First Philosophy, all men by nature desire to know. The reason for this can be and is that each thing, impelled by a force provided by its own nature, inclines towards its own perfection. Since knowledge is the ultimate perfection of our soul, in which resides our ultimate happiness, we are all therefore by nature subject to a desire for it»]. *Convivio* I, I, 1.

te's *navicella*.⁴⁵ The image of «perigliose acque» that opened the first canto of *Inferno* returns here to signal another kind of danger: the danger of an intellectual challenge that can destroy those not equipped with the scientific and theological knowledge necessary to deal with the complexity of the subject. The sea is no longer behind the pilgrim, but extends before him as an ocean he must sail to conquer the vision of beatitude. The novelty of Dante's voyage is not only in the astonishing sequence of visions, but also in the gallery of characters and in the use of poetic language to voice what had been so far unspeakable. It consists in the challenging intellectual voyage through the partially new waters of his original cosmological structure, where knowledge surpasses the limits of specific scientific fields to harmonize in superior poetic wisdom.

The opening lines of canto II hint at the narration of another perilous expedition beyond the limits of human knowledge: Ulysses' short journey on the unexplored waters beyond Hercules' pillars.⁴⁶ Unlike Ulysses's, however, Dante's navigation is not doomed to cognitive wreckage. The parallel between the two characters surfaces through textual allusions in *Paradiso* II, 1-3, where we find echoes of the tercets of *Inferno* XXVI.⁴⁷ Here Dante's company, meaning his readers, are in a «piccioletta barca» a small boat. They follow at a distance the poet's solitary navigation on his «legno», a metonymy indicating Dante's vessel, similar to Ulysses's denotation of his ship:

> ma misi me per l'alto mare aperto
> sol con un legno e con quella compagna
> picciola da la qual non fui diserto (*Inf.* XXVI, 100-102).

> Therefore, I set out on the open sea
> with but one ship and that small company
> of those who never had deserted me.

---

⁴⁵ For an analysis of the nautical metaphor of canto II and its mythological subtexts, see MICHELANGELO PICONE, *Canto II*, in *Lectura Dantis Turicensis. Paradiso*, G. Güntert and M. Picone eds., Firenze, Cesati, 2002, pp. 35-52.

⁴⁶ Maria Corti notices the parallel exploration of the two semantic fields of 'flight' and 'sailing' in *Inferno* XXVI and *Paradiso* II, and explores the textual connection between the two cantos in *Metafisica della luce come poesia*, in ID., *Percorsi*, pp. 151-153. See also MARIA CORTI, *Le metafore della navigazione, del volo e della lingua di fuoco nell'episodio di Ulisse (Inferno, XXVI)*, in ID., *Scritti su Cavalcanti e Dante*, Torino, Einaudi, 2003, pp. 348-364.

⁴⁷ «O voi che siete in piccioletta barca, / desiderosi d'ascoltar, seguiti / dietro al mio legno che cantando varca, / tornate a riveder li vostri liti: / non vi mettete in pelago, ché forse, /perdendo me, rimarreste smarriti», [«O you who are within your little bark, / eager to listen, following behind / my ship that, singing, crosses to deep seas»], *Par.* II, 1-3.

Similarly to Dante's audience, only a few (the «compagna picciola» that reminds of the readers in their «piccioletta barca») followed Ulysses in his «wild flight» («folle volo», *Inf.* XXVI, 125). As Dante, Ulysses sails on his «legno»; as Ulysses' crew, Dante describes those willing to follow him in his enterprise in a world beyond human boundaries (to the «mondo sanza gente», *Inf.* XXVI, 117) as being full of desire to know («desiderosi d'ascoltar» *Par.* II, 2).[48] In *Paradiso* II, Dante warns those who have not nurtured their intellects with wisdom, and are unprepared to sail the uncharted depths of the heavens, to return to their «liti», the familiar shores of their known world. These shores recall the ones Ulysses contemplated before reaching the point of no-return («L'un lito e l'altro vidi infin la Spagna», *Inf.* XXVI, 103). The connection between Ulysses as anti-type of Dante, and his daring voyage as pre-figuration of Dante's, emerges even more forcefully in canto XXVII of *Paradiso*, where the poet sets his cosmology within the grand design of creation.

As Giuseppe Mazzotta argues, Dante evokes in this canto both Ulysses and Francesca as two figures characterizing the dialectic between knowledge and desire that Dante places at the heart of his vision of the cosmos.[49] In the heaven of the moon, the shadow of Francesca surfaces behind the figures of Piccarda Donati and Costanza of Altavilla, who breached their spiritual vows by yielding to violence, just as Francesca yielded to the violence of passion in breaching her wedlock. Francesca also represents the weakness of reason when it remains bound to the senses. In this meaning, her allegorical figure re-emerges in Beatrice's warning to Dante:

> certo non ti dovrien punger li strali
> d'ammirazione omai, poi dietro ai sensi
> vedi che la ragione ha corte l'ali (*Par.* II, 55-57).
>
> struck by the arrows of amazement once
> you recognize that reason, even when
> supported by the senses, has short wings.

---

[48] Ulysses' oration instills in his companions a frantic desire to experience the unknown waters: «Li miei compagni fec'io sì aguti, / con questa orazion picciola, al cammino, /che a pena poscia li avrei ritenuti», [«I spurred my comrades with this brief address / to meet the journey with such eagerness / that I could hardly, then, have held them back», *Inf.*, XXVI, 121-123.

[49] «The two figures are joined by a self-evident complementariness: one's desire to know counters the other's knowledge of desire [...] In effect, Ulysses and Francesca come through as two failed metaphysicians, who are caught by misunderstanding and deceptions constitutive of their thoughts and actions». GIUSEPPE MAZZOTTA, *Cosmology and the Kiss of Creation*, pp. 6-7. See also on Francesca and her failure in self-knowledge, CHRISTIAN MOEVS, *The Metaphysics*, pp. 157-158.

Dante should not feel surprise at the novelty of the truths Beatrice unveils here. She reminds him of the limits of reason in understanding what only noetic intuition can grasp. The wings of reason, rooted in the senses, are too short to fly to *Paradiso*. The metaphor of line 57 takes the reader back to other wings that cannot fly: the wings of Paolo and Francesca in *Inferno* V: helplessly still and upright in an indelible image of impotency. There Virgil had compared the two lovers to two doves whose wings had become incapable of voluntary flight against the fury of the infernal wind. The metaphor hints at the sin of the lustful, punished for having submitted reason to passion, and thus turned by *contrappasso* in passive fliers. They are forced to submit to the *turbo* as they had submitted to devious inclinations.⁵⁰ The same fault in the sense-reason relation reappears in the lowest heaven of *Paradiso*, with a different nuance but similar meaning. While the lustful had not used reason to follow the senses, the blessed of the heaven of the moon enjoy a diminished degree of beatitude because they let violence (and fear) prevail over reason. Both Piccarda and Costanza discerned the good, but their will failed them. Human reason is doomed by its intimate relation with the senses. As the problematic status of sense perception in *Paradiso* shows, sight and hearing are often deceived, blurred, and unable to lead to certainty. Their qualities are poetically misplaced by the recurrent use of zeugma and synesthesia. As Dante experiences in canto XVIII, in the presence of the singing lights forming the words of the eagle, sight and hearing exchange objects, and are both insufficient for the intellectual challenges of *Paradiso*. Similarly, in the heaven of the Moon, Dante is deceived about the moon spots because his reason still follows his visual senses.⁵¹

---

⁵⁰ *Inf.* V, 39.

⁵¹ The motif of the wings is recurrent in *Paradiso* and marks canto II. Dante often mentions his own wings as a metaphor of his intellectual capacity to fly through the heavens. The metaphor of wings and flight had a philosophical meaning in the poetic debate that developed among the poets of the 'Tuscan school', like Guittone d'Arezzo and Bonagiunta Orbicciani, and the poets of the Dolce Stil Nuovo, the New Sweet Style. When Dante was still a young poet in Florence, Bonagiunta and Guido Guinizzelli debated about whether poetry should have subtle philosophical content or not. Bonagiunta attacked Guinizzelli for his excessive subtlety: «Così passate voi di sottigliansa / e non si può trovar chi ben ispogna, / cotant'è iscura vostra parlatura», [«So you exceed so much in subtleties that it is impossible to find anyone able to explain what you say»] Bonagiunta Orbicciani son. 19A, 9-11. In his response, Guinizzelli used the image of the flying birds to indicate the «altezza d'ingegno», the intellectual superiority of the 'new' poetry, whose lofty philosophical contents Bonagiunta had criticized: «Volan ausel' per air di strane guise /ed han diversi loro operamenti, / né tutti d'un volar né d'un ardire», [«Birds fly in the air in different ways, and differently operates. Not all equally flying and equally daring»]. Guido Guinizzelli, son. 19b, 9-11. The question of the vows illustrates the inconstancy and the weakness in will of the souls appearing in the heaven of the Moon. It is related to the major theme of free will, already touched in *Purgatorio* XVI and will assume particular relevance in the cosmological context of *Paradiso*, as we will see in the next sections. The son-

Unlike Ulysses, Dante is not alone in his journey. His guides are Minerva, the goddess of wisdom, and Apollo, the god of poetic grace and of the divine enigmatic word, who possesses the mind of those called to voice the spirit of the god. While Dante and Ulysses share a similar desire to explore the literary unknown, the poet's enterprise is not doomed. Enlightened by grace, his adventure will be as successful as that of Jason, the mythical sailor who left Argo to conquer the Golden Fleece with the help of Medea's magic. Jason, the successful Argonaut, appears as metamorphosis of Ulysses, his complement in defining Dante's own *quête*. In combining the two myths of Ulysses and Jason, Dante rewrites in Christian terms the myth of the conquest of knowledge, where the hero returns holding his Golden Fleece, his symbol of victory: the poem itself.

e. *The Desire of Motion*. Paradiso I

As Dante's '*navicella*' sets sail, the readers face an intellectual challenge for the poet introduces the first fundamental principle of the cosmological structure of the *Commedia*: the principle of causality of motion. In transposing the Aristotelian vertical hierarchy of causes within the Christian universe, Dante identified God as the ultimate source of universal motion. This was the prevalent position among theologians of his time. Nevertheless, the issue was to specify in what sense God caused the movement of celestial bodies. According to the principles of Aristotelian physics, a mover must have some connection with the object it moves. It was not conceivable, however, that God – the perfect quietness whose being is thoroughly actualized – would directly initiate and maintain sidereal motion.

Avicenna had postulated that the desire of the spheres to become similar to God, namely to acquire a condition of perfect uniformity, initiated the circular movements of the stars. This hypothesis, however, implied the animation of the celestial bodies, and both Averroes and Aquinas rejected it. Dante transposed Avicennas' idea of motion-causing desire from the spheres to the angels, and made desire depend upon the degree of knowledge angels acquired in contemplation. The closer the angelic orders were to God, the clearer was their vision (and their reflecting power), from which desire derived. In applying the dialectic of desire to angels and enlightment, Dante dis-

---

nets are quoted from GIANFRANCO CONTINI's edition of *Poeti del Duecento*, Milano-Napoli, Ricciardi, 1960. On the flight of the mind in Neo-Platonic and Islamic sources, see DANIELA BOCCASSINI, *Il volo della mente. Falconeria e sofia nel mondo mediterraneo: Islam, Federico II, Dante*, Ravenna, Longo, 2003, pp. 281-334.

placed the principle of movement from the material heavens to the space-less and time-less Empyrean, ensuring its perfect transcendentality.[52]

Explaining the process that originates the heavenly movements was such a crucial issue that Dante placed it at the very beginning of *Paradiso*, as a key to understanding his cosmological design. As the poet reflects upon the process of «trasumanar» he is about to undergo, he addresses the love that rules over the heavens («amor che 'l ciel governi», *Par.* I, 74), presenting it as a sort of 'passive' principle in being «desiderato» (desired):

> Quando la rota che tu sempiterni
> desiderato, a sé mi fece atteso
> con l'armonia che temperi e discerni, (*Par.* I, 76-78)

> When that wheel which You make eternal through
> the heavens' longing for You drew me with
> the harmony You temper and distinguish,

With these lines Dante introduces a distinction between love – as the divine source of the regulated harmony of the universe – and desire, which moves all creatures to reunite with their origin in the Empyrean. Love is the opening of God from the infinite to the finite, reaching all creation («s'aperse in nuovi amor l'etterno amore», *Par.* XXIX, 18). Desire unifies; it draws universe toward its point of origin like a spiritual gravitational force. While love characterizes the moment of expansion of the divine (creation and the angelic transmission of light are the two major instances of this

---

[52] Christian Moevs, argues that Dante's conception of the Empyrean marks another of Dante's departures from mainstream theology: «*In radical distinction from the Empyrean of the Scholastics, the Empyrean of the* Commedia *is absolutely immaterial and uncreated: it does not exists in space or time*» (Moevs's Italics). CHRISTIAN MOEVS, *The Metaphysics of Dante's Commedia*, p. 21. Of a different opinion was Attilio Mellone, who argued that Dante posited another heaven, the *coelum Trinitatis*, beyond the Empyrean. Only this heaven would coincide with God himself, and would be eternal and infinite. According to Moevs, Dante fused the pre-Scholastic *coelum Trinitatis* with the Empyrean. For a discussion on this point, and the medieval sources of the *coelum Trinitatis*, see CHRISTIAN MOEVS, *The Metaphysics of Dante's Commedia*, pp. 25-28 and n. 34 to p. 27; ATTILIO MELLONE, *Empireo*, in *ED*, pp. 668-671, and ID. *La dottrina di Dante Alighieri sulla prima crezione*, Nocera, Convento di Santa Maria degli Angeli, 1950, *passim*. Other scholars have argued that the Empyrean cannot be considered a space in material terms and have suggested that Dante's vision of the spiritual realm is close to the modern notion of hypersphere: a curved, finite, but boundless space. If however the Empyrean coincides with the uncreated, eternal, 'mind' of God, it is not possible that such a space could be finite «perché non è in luogo e non s'impola» («That sphere is not in space and has no poles», *Par.* XXII, 67). For a discussion of the notion of hypersphere as applied to Dante's image of the cosmos, see JOHN FRECCERO, *Dante's Cosmos*, Binghamton, Center for Medieval & Renaissance Studies, State University of New York at Binghamton, 1998 (Bernardo Lecture Series, 6), MARK PETERSON, *Dante and the 3-³PHERE*, «American Journal of Physics», XLVII, 1979, pp. 103-135; ROBERT OSSERMAN, *Poetry of the Universe*, New York, 1995.

moment), desire corresponds to a moment of contraction, the centripetal force that attracts the scattered fragments that compose reality back into unity. Desire is an almost erotic want of 'possession' realized through the process of imitation and participation that all beings experience, according to their individual nature and capacities. It involves angels, Blessed, and humankind alike.[53] Dante himself experiences this desire as he enters the heaven of the moon. «Desiderato», the passive voice of the verb «to desire», indicates God's condition of passivity as object of desire. Since longing of likeness to God activates all movements in the heavens, it is clear that God does not, and cannot, move anything. God is the final (and formal) cause of movement, but only by being the desired object of the intellectual substances, which – as instruments of the divine Providence – are its efficient causes.[54]

f. *Order and Ontology*. Paradiso II

Beatrice answers Dante's inquiry about the different areas of luminosity in the moon by setting his question within the broader issue of the relation between the One and the multiplicity, the eternal and the contingent. The varying degrees of luminosity visible on the moon and on the stars are not the effects of a variation in the material density of the celestial bodies. They are the effects of a qualitative rather than a quantitative cause, for they depend on the different formal principles the Intelligences pass on the celestial bodies. In the *Convivio* Dante had followed the Averroist quantitative theory, but in *Paradiso*, he refuted this idea.[55] The Averroists attributed the spots to the different degrees of rarefaction of matter in the moon. The quantitative explanation of the spots implied the existence of a single formal principle common to all

---

[53] The 'desire' of God is an important aspect of medieval mysticism. See for example the importance of desire in the theology of Gregory the Great in JEAN LECLERCQ, *L'amour des lettres et le désire de Dieu*, Paris, Cerf, 1990 (1958), pp. 36-37. For the poetics of desire in Dante, a theme widely debated among scholars, see at least LINO PERTILE, *La punta del disio: semantica del desiderio nella Commedia*, Firenze, Cadmo, 2005, and particularly on the semantic of desire in the *Commedia* (19-38). For a perspective on Dante's poetics of desire within the context of past lyric tradition, see TEODOLINDA BAROLINI, *The Origins of Italian Literary Culture*, New York, Fordham University Press, 2006, pp. 23-121.

[54] In Aristotle the prime cause did not know multiplicity, as it was self-subsisting thought in the act of thinking itself. The idea of its 'passivity' as mover derived from this definition of the prime mover, as Aristotle said in the book *Lambda* of his *Metaphysics*: «Thus the prime mover moves as object of love [loved], while all other things move by being moved». ARISTOTLE, *Meth.* Λ, 7, 1072b.

[55] See ENRICO PROTO, *La dottrina Dantesca delle macchie lunari*, in *Scritti vari di erudizione e critica in onore di Rodolfo Reiner*, Torino, Reiner, 1912, pp. 197-213, and BRUNO NARDI, *La dottrina delle macchie lunari nel secondo canto del Paradiso*, in *Saggi di filosofia Dantesca*, pp. 3-39, and CESARE VASOLI, *Il canto II del Paradiso*.

bodies. This, however, contradicted the fact that to different effects (the ones the stars operate on the inferior world) different causes must correspond. The Aristotelian principle of causality determined Dante's choice of a theory of diversified formal principles, which his revised angelology now allowed. In this respect, the ontological implications of the role the hierarchies play in originating celestial motion is generally underestimated, but it is crucial. Only a few among the ancient commentators seem to pay attention to the philosophical and the astrological implications of the above quoted tercets of *Paradiso* II, 40-42, which refer to Incarnation. A few, however, including Daniello and Pietro di Dante, quote Albert the Great's commentary on Aristotle's tract on *Meteors* and refer to the works of the Arabian astronomer Albumasar (Abū Ma'shar). Francesco da Buti engages in a sort of short treatise trying to explain how Dante might have reconciled the Aristotelian theory of hierarchical causes with the Christian idea of un-mediated creation. He attributed the diurnal movement to the natural properties of celestial bodies, and the contrary movement of the zodiac to the angels. Although the canto presents two distinguished moments, one expounding the movement of the spheres and the other attributing its cause to the angels, it does not seem possible to agree with Francesco da Buti's attempt to reconcile the apparent discrepancy existing in the two textual segments of *Paradiso* II, 112-123 and 127-132 discussed below.

Coherently with the criterion of unifying order, there cannot be two different causes of motion, one natural, and one supernatural. The planets do not move by their own will, as Beatrice clearly states, but are moved by the blessed movers. Planets and angels represent philosophical and theological instances that Dante reconciled in his definition of angelic ministries by 'Christianizing' the assumption of celestial causality. If the heavens were the instruments of the Holy Spirit in determining causality, it was then necessary to construct a theory of sidereal influence without assuming secondary formal causes in creation. Beatrice's cosmological discourse in *Paradiso* II should be considered as complementary to her exposition in cantos XXVIII and XXIX.

Following the well-known doctrines of his time, Dante had figured a universe formed by ten heavens composed by the seven spheres of planets, the sphere of fixed stars, the Primum Mobile, and the Empyrean.[56] Christian

---

[56] Besides the studies already cited, and those contained in the lectures and commentaries on cantos II, VIII, XXVIII, XXIX, specific contributions utilized for this work include: GIORGIO STABILE, *Cosmologia e teologia nella Commedia: la caduta di Lucifero e il rovesciamento del mondo*, «Letture classensi», XII, Ravenna, Longo, 1983, pp. 139-173; ALESSANDRO GHISALBERTI, *La cosmologia del Duecento e Dante*, «Letture Classensi», XIII, Ravenna, Longo, 1984, pp. 33-48; GIOVANNI FALLANI,

scholars had identified the Aristotelian immovable prime mover with the God of creation, situating Him in the Empyrean quiet heaven, which enclosed all inferior spheres.[57] In designing their Christianized cosmology, philosophers, astronomers, and theologians concentrated on the necessity of explaining the movements of the spheres, and their reciprocal relations, in order to give scientific foundation to the passage from the absolute quietness of the heaven of bliss to the movement of the heavens below. Beatrice illustrates the mechanics of the communication of movement from higher to lower spheres in the dense doctrinal tercets of *Paradiso* II, 112-120.

> Dentro dal ciel de la divina pace
> si gira un corpo ne la cui virtute
> l'esser di tutto suo contento giace.
> Lo ciel seguente, c'ha tante vedute,
> quell'esser parte per diverse essenze,
> da lui distratte e da lui contenute.
> Li altri giron per varie differenze
> le distinzion che dentro da sé hanno
> dispongono a lor fini e lor semenze (*Par.* II, 112-120).

> Within the heaven of the godly peace
> revolves a body in whose power lies
> the being of all things that it enfolds.
> The sphere that follows, where so much is shown,
> to varied essences bestows that being,
> to stars distinct and yet contained in it.
> The other spheres, in ways diverse, direct
> the diverse powers they possess, so that
> these forces can bear fruit, attain their aims.

Most commentators have interpreted these tercets as explaining the origin and the continuation of the movement of the spheres from the Primum Mobile down to the lowest heaven of the moon. Aristotelian physics provided an

---

Dante poeta teologo, Milano, Marzorati, 1965, pp. 245-259. See also, for the cosmological implications of Dante's *Quaestio de acqua et terra*, BRUNO NARDI, *La caduta di Lucifero e l'autenticità della "quaestio de acqua et terra"* in *"Lecturae" e altri studi danteschi*, R. Abardo ed., Firenze, Le Lettere, 1959, pp. 227-265; FRANCESCO MAZZONI, *Dante misuratore di mondi*, in *Dante e la Scienza*, pp. 25-53. See also GIOACHINO CHIARINI, *Dante e la simbologia classica dei sette pianeti*, in *Dante e la fabbrica della Commedia*, Ravenna, Longo, 2008 («Interventi classensi n. 22»), pp. 189-197.

[57] For the sources and the interpretations of Dante's doctrine of the Empyrean, see BRUNO NARDI, *La dottrina dell'Empireo e la sua genesi storica e nel pensiero dantesco*, and CHRISTIAN MOEVS, *The Methaphysics*, pp. 15-35. See also note 52 in this chapter.

explanation of the causes of physical movements, but did not allow for a theory of the production of the infinite array of forms coherent with Christian doctrine.[58] The above quoted tercets illustrate the formation of the universe from the point of view of the creation of beings, reconstructing the way forms are generated and passed through the heavens to mold the sublunar world. Dante seems to engage here in an ontological, rather than a physical dissertation. Reading it in terms of cosmological explanation, is indeed problematic if considered against the immediately following tercets in which Beatrice clearly attributes the origin of the movement to angels:[59]

> Lo moto e la virtù d'i santi giri,
> come dal fabbro l'arte del martello,
> da' beati motor convien che spiri; (*Par.* II, 127-129)
>
> The force and motion of the holy spheres
> must be inspired by the blessed movers,
> just as the smith imparts the hammer's art;

These lines clearly state that angelic hierarchies, the «beati motori», are the origin of the capacity of motion and influence («virtù») of the celestial

---

[58] «lo Stagirita, benché fosse del tutto convinto della dipendenza dei cieli e della natura inferiore da un unico principio, si era limitato a spiegare piuttosto il moto che la formazione delle cose, e non aveva compiutamente delineato quel processo che dall'unità del sommo principio produce la molteplicità sempre più distinta e diversa delle realtà particolari», [«Although the Stagirite was convinced that heavens and inferior nature depended upon a single principle, he explained motion rather than the formation of things. He did not completely delineate the process that from the unity of the highest principle produces the multiplicity of particular realities»]. C. VASOLI, *Il canto II del Paradiso*, in *I primi undici canti del Paradiso*, A. Mellone ed., Roma, Bulzoni, 1992 («Lectura Dantis Metelliana»), p. 44. The same kind of argument, which Vasoli recalls, is in BRUNO NARDI, *La dottrina delle macchie lunari*, p. 16. The inclination to see elements of Avicennian Neo-Platonism in Dante's theory of celestial motion seems to derive from the marginalization of the role of the hierarchies. Their combined functions of movers and mirrors of forms, seems to avoid the danger of emanatism. As Vasoli notes, scholastic theologians tended to utilize the *Liber de causis* to exclude necessary emanatism. The same approach seems to characterize Dante's attempt to correct the Neo-Platonic elements of Dionysian angelology with the Aristotelian theory of motion. It seems that his success would depend on the separation maintained between spiritual and material spheres, and on the role of angels in yoking the two.

[59] As Moevs effectively explains: the «blessed movers are angelic Intelligences; *virtù* is the causal-formative influence intrinsic to intellect/will. A particular *virtù* arises from intelligence as the "hammer's art" "breathes" from a smith: the intelligence is the smith, the stars are the hammer, the formative influence of the stars (ultimately displayed in sublunar matter) is the hammer's art». CHRISTIAN MOEVS, *The Metaphysics*, p. 118. Like the hammer, the stars operate only in consequence of angels's operation. It is therefore implicit that there cannot be any form of intrinsic natural principle of movement in the celestial bodies. In his commentary to *Convivio* II, v, 18 Cesare Vasoli observes that Dante's explanation of the action of the movers in *Par.* II, 127-144 is close to Albert the Great's exposition of the same argument in his commentary on Aristotle's *De coelo et mundo* (II, tr. 3, 14), in which recurs the same simile of the hammer. See *Convivio*, p. 167, n. 18.

bodies. The metaphor of the hammer, recurrent in astrological texts, illustrates the relationship linking Intelligences and planets: the angels are like smiths forging the world through the virtue they transfer to the stars.

If one accepts the hypothesis that *Paradiso* II, 112-120 illustrates a sort of animist theory of the sideral bodies, ascribing to them an immanent principle of movement, why then should Beatrice immediately afterword say that angels cause their motion? Furthermore, this hypothesis would also contradict Beatrice's exposition in *Paradiso* XXVIII, where she illustrates the inverse relation between hierarchies and heavens. She explains that the different speed of motion of each sphere is inversely related to the degree of perfection of each angelic mover, reflected by its vicinity to God.[60]

These tercets therefore seem to present Beatrice's explanation of how the divine mind generates and orders multiplicity through the movement originated by the hierarchies. The process that generates cosmos and nature originates in the ninth sphere:[61]

> Dentro dal ciel de la divina pace
> si gira un corpo ne la cui virtute
> l'esser di tutto suo contento giace (*Par.* II, 112-114).

> Within the heaven of the godly peace
> revolves a body in whose power lies
> the being of all things that it enfolds.

We may notice that «pace» (quiet) and «virtute», (power) are keywords introducing Dante's ontology. «Pace» indicates the condition of absence of movement characterizing the Empyrean. As movement is the expression of a potentiality with respect to space, it is impossible to ascribe it to God, who is pure act. More problematic is the word «virtue», which should be interpreted in its physical meaning of power, as it refers to the capacity of the celestial bodies to receive and maintain the impulse of movement they get from the angels. Francesco da Buti, commenting on this tercet, interprets virtue in terms of the capacity of the celestial spheres to preserve and communicate their «esse» (being), whose perfection is in their mo-

---

[60] «per che, se tu a la virtù circonde / la tua misura, non a la parvenza / de le sustanze che t'appaion tonde, / tu vederai mirabil consequenza / di maggio a più e di minore a meno, / in ciascun cielo, a süa intelligenza», [«so that, if you but draw your measure round / the power within-and not the semblance of- / the angels that appear to you as circles, / you will discern a wonderful accord / between each sphere and its Intelligence: / greater accords with more, smaller with less»]. *Par.* XXVIII, 73-78.

[61] Cfr. *Par.* II, 106-108.

tion.⁶² Virtue is the activating force of movement that descends from the blessed circles of the Empyrean to infuse forms on the planets and stars.

The general framework of angelological functions seems to contradict the opinion of many commentators that Dante considered the sidereal bodies to be organs of the animated heavens, endowed with an intrinsic capacity of motion. Animation and emanatism were two major philosophical temptations in the doctrinal debates on cosmology in Dante's time, and were often connected to the theory of secondary formal causes in creation. To attribute the origin of motion to external causes was not sufficient to avoid falling into the perilous waters of heterodox doctrines. Dante's reinterpretation of the ministries of the hierarchies and of the Dionysian anagogical process seems however to aim at avoiding them.

That the tercets in *Par.* II, 112-120 provide an account of creation from an ontological point of view seems to be confirmed by another passage in *Paradiso* XXX where Dante returns onto the same question in analogous terms:

> Fassi di raggio tutta sua parvenza
> reflesso al sommo del mobile primo,
> che prende quindi vivere e potenza (*Par.* XXX, 106-108).

> All that one sees of it derives from one
> light-ray reflected from the summit of
> the Primum Mobile, which from it draws.

Escorted by Beatrice to the Empyrean where the blessed and the angels reside, Dante observes the river of light flowing around the luminous point of origin and end of creation. In the highest sphere of the firmament, he seems to recall his first vision of the circles in *Paradiso* II. There, however, he was still looking at God from a point of observation situated in the material part of the universe. Now, he gains a new vision from above and sees the state of the creatures once they have completed their return journey to *Paradiso*, where the image of universe is finally restored to its original perspective.⁶³ Now he

---

⁶² This capacity, according to the commentator, corresponds to the movement of the stars from East to West.

⁶³ Giorgio Stabile argues that organicistic models influenced Dante's perception of space, so that in the *Commedia* terrestrial world resulted inverted with respect to cosmological space. To such physical inversion, a moral inversion corresponded: «gli emispheri non sono riducibili a semplici reciproci spaziali, ma rappresentano due livelli cosmici di un medesimo e compatto ordine morale», [«it is not possible to reduce the hemispheres to simple spatial reciprocals, for they represent two cosmic levels of the same compact moral order»]. GIORGIO STABILE, *Cosmologia, teologia e viaggio dantesco*, in *L'idea e l'immagine dell'universo nell'opera di Dante*, pp. 21-59; 52.

can see that the luminous point at the center of universe produces a ray reflected («reflesso») on the surface of the Primum Mobile. All commentators agree in interpreting the mechanics of the optical description: the divine ray touches the convex surface of the highest material sphere and from there it reflects back to enlighten the Empyrean.[64] In touching the ninth sphere, the ray activates its motion and its influential virtue. This interpretation, however, does not take into account that Dante explicitly says that blessed and angels both enjoy a direct vision of God. Moreover, it seems strange that the divine light of the Empyrean would come from a mere reflection on a sphere of lesser nobility than that of the spiritual creatures. If the ray bounces back, how could divine light penetrate the Primum Mobile and the other spheres? Finally, how is this interpretation compatible with the fact that God does not move anything directly? It seems possible to answer these questions by interpreting the term «reflesso» as 'the ray being reflected' by the intermediary mirroring action of the hierarchies. From the Empyrean, Dante can see the effects of the action of the highest order of the seraphim on the ninth sphere, whence nature and time originate.

Vellutello is the only commentator who reads «reflesso» as hinting at the mirroring function of the angels.[65] Unlike all other ancient and modern commentators, he argued that the light the seraphim reflect gives to the Primum Mobile its life («viver») and its power («potenza»). He identified «life» with «motion» (moto) and «power» with virtue, or influence. While in *Paradiso* II Dante discerned the structure of the heavens from the point of view of the creation of beings, in the Empyrean he acquires a perfect view of light's dynamism in enlivening the universe. Only here he can see the origin and causes

---

[64] Most commentators agree on the interpretation of the «reflesso» as a ray «coming back» to give light to the Empyrean, so as to make God visible to angels and Blessed. Among the ancient ones, only Francesco da Buti (1324?-1406) explicitly interpreted the above quoted tercet in this sense: «lo primo mobile in che percuote lo detto raggio: imperò che in esso percuote, e quinde ritorna insù; e debbesi intendere che si rifletta intorno intorno, per tutta la parte di sopra del detto Cielo», [«the said ray hits the first mobile, and as soon as it hits [the surface of the Primum Mobile, it bounces back. It should be understood that [the ray] reflects around on all the surface of the above said heaven»]. This opinion is common among modern commentators, among which are John S. Carroll; C.H. Grandgent; G.A. Scartazzini and G. Vandelli, Carlo Grabher, and Attilio Momigliano.

[65] In his commentary (1544), Alessandro Vellutello interprets 'reflesso' in these terms: «Fassi tutta sua parvenza, ciò è, tutto quello che pare di questo tal lume, si fa di raggio reflesso al sommo, ciò è, a la superficie, del primo mobile, perché prima si difonde ne l'ordine de' Serafini, e da questi per reflesso poi nel primo mobile, come di sopra vedemmo, che, il qual primo mobile, prende quindi da esso reflesso raggio viver e potenza, perché da lui prende 'l moto e la influentia, che participa poi con gli altri cieli, e quelli con gli elementi contenuti da lui». [«All its appearance, that is, all that appear of this light, becomes a ray reflected on the peak, that is, on the surface of the prime mobile, because it first reaches the order of Seraphim, and from them it is then reflected on the prime mobile, as we saw above»].

of life and understand them. Only here he obtains the perfect vision of the angelic choirs, their nature, their ministries, and their power in shaping the influence of the heavens.[66]

g. *The Angels and the Mechanics of the Spheres*

It is necessary to distinguish between the mechanics of the transmission of forms through reflected light, and the effects such enlightenment produces on the material spheres. The seraphim, closest to God in likeness, signify the unifying principle of love – caritas, which inflames and makes them 'inflaming'. By revolving the Primum Mobile, they attribute to it the properties of uniformity and unity of the first manifestation of being. Immediately below, in the heaven of the fixed stars, takes place the passage from this indistinct and uniform being into the multiplicity of beings:

> e 'l ciel cui tanti lumi fanno bello,
> de la mente profonda che lui volve
> prende l'image e fassene suggello (*Par.* II, 130-132).
>
> and so, from the deep Mind that makes it wheel,
> the sphere that many lights adorn receives
> that stamp of which it then becomes the seal.

In line 131 the expression «profound mind» («mente profonda») refers not to God, but to the angelic order that moves it, the cherubim. The stars become the «suggello», the mold (but the word «suggello» also derives from the Latin translation of the Greek term for idea) that reproduces the divine images the cherubim possess and pass onto them. Although remaining as one, the Intelligence's goodness («bontade», as fullness of knowledge) multiplies across the stars resolving into different kinds of virtues, just as the soul animates with different powers the different parts of the human body.[67] This

---

[66] Many commentators were influenced by the thought that Dante derived the influence of the Empyrean on celestial motion from Aquinas, who in his *Summa Theologiae* explained: «Caelum Empyreum habet influentiam super corpora quae moventur, licet ipsum non moveatur. Et propter hoc potest dici quod influit in primum caelum quod movetur, non aliquid transiens et adveniens per motum, sed aliquid fixum et stabile; puta virtutem continendi et causandi, vel aliquid huiusmodi ad dignitatem pertinens», [«For this reason it may be said that the influence of the empyrean upon that which is called the first heaven, and is moved, produces therein not something that comes and goes as a result of movement, but something of a fixed and stable nature, as the power of conservation or causation, or something of the kind pertaining to dignity»]. Unless 'Empyrean' here is interpreted as referring to the angels, once again Dante did not follow Aquinas' cosmology. AQUINAS, *S. Th.*, I^a, q. 66, a. 3.

[67] «E come l'alma dentro a vostra polve / per differenti membra e conformate / a diverse po-

process of «division» and multiplication takes place in the eighth sphere, the heaven of the innumerable stars («vedute») where Beatrice had explained to the admiring Dante that:

> Lo ciel seguente, c' ha tante vedute,
> quell'esser parte per diverse essenze,
> da lui distratte e da lui contenute (*Par.* II, 115-117).
>
> The sphere that follows, where so much is shown,
> to varied essences bestows that being,
> to stars distinct and yet contained in it.

To diversify and to unify, and to maintain harmony in the plurality, is the appropriate task of the cherubim, the order of wisdom. Their divine name, according to Dionysius, signifies *multitudo scientiae* (literally 'multiplicity of sciences', where science means knowledge of the immediate causes of things) and *fusio sapientiae* (literally 'uniformity of wisdom', where wisdom means the higher form of knowledge of the first causes of things).[68]

The «essences» in line 116 are the forms that, by uniting with the four elements, will originate the sublunar world of contingency. In combining with the different constellations, the cherubim's virtue forms different alloys («lega»), yet maintaining unity:

> Virtù diversa fa diversa lega
> col prezïoso corpo ch'ella avviva,
> nel qual, sì come vita in voi, si lega (*Par.* II, 139-141).
>
> It would be cause for wonder in you if,
> no longer hindered, you remained below,
> as if, on earth, a living flame stood still.

The cherubim's virtue is not simply a distinctive mark of their nature, but also derives, purified and perfected, from the Dionysian process of deification described in chapter II. The forms thus imprinted originate the influential

---

tenze si resolve», [«And as the soul within your dust is shared / by different organs, each most suited to / a different potency, so does that Mind»]. *Par.* II, 133-135. As commentators generally agree, the simile with the soul remains here metaphorical.

[68] See DIONYSIUS, *CH.* VII, 205b. The Latin text is Eriugena's translation of the Greek terms «*plethos gnoseos*» and «*chusin sofias*». In commenting on this passage of the *De Coelesti Hierarchia*, Albert the Great considered the attributes of seraphim and cherubim as referring to the unifying and attractive power of love-*caritas*, which in the seraphim is turned toward God as object of love, while in the cherubim is toward God as object of science. For Albert the Great, unlike Dante, the highest hierarchy does not have the task of disseminating but that of collecting and holding the fullness of science. See ALBERT THE GREAT, *Super Dionysium*, p. 93.

powers of the stars while remaining anchored to the unifying providential order governed by the angels. The task of the residual angelic orders is to direct the forms distinguished above (the «distinzioni») toward their proper and specific ends through the motion of their respective spheres:

> Li altri giron per varie differenze
> le distinzion che dentro da sé hanno
> dispongono a lor fini e lor semenze (*Par.* II, 118-120).

> The other spheres, in ways diverse, direct
> the diverse powers they possess, so that
> these forces can bear fruit, attain their aims.

The seven spheres of the planets direct the essences distinguished in the heaven of the fixed stars toward their ends («lor fini»), and ensure the continuous process of generation and corruption (indicated by the «semenze»). Each planet has different powers, which produce different effects on the sublunar world. The influence of the planets derives, as for the superior heavens, from their moving Intelligences, who also guarantee the process of communication from superior to inferior celestial bodies according to a certain proportion, or measure, and dictate the mathematical laws of harmony to space. The multiplicity of the formal principles ensured by the movers, explains the presence of the moon spots that Dante had believed to have been caused by a quantitative variation in the matter of the celestial bodies. As in the case of the different luminosity of stars, the moon spots are the effect of the diversification of the formal virtues in the celestial bodies.[69]

II. ASTROLOGICAL DETERMINISM AND FREE WILL

The absorption of science through the translations of the Greek-Arab astrological corpus involved a broader and complex assimilation of the cultures that had preserved and filtered the textual monuments of ancient astronomical wisdom.[70] Natural philosophy developed based on acceptance of the

---

[69] Both Jacopo della Lana (1324-28) and the Ottimo (Andrea Lancia, 1333-1340), in commenting the tercet that explains the origin of movement in the spheres (*Par.* II, 112-114) refer to ALBERT THE GREAT's *De Mineralibus*, II, 3,3.

[70] A consistent part of the scientific texts newly translated into Latin during the twelfth and the thirteenth centuries were astrological treatises, including Claudius Tolomeus' *Almagest* (*Mathematike Syntaxis*, or *Al-Magisti*); Albumasar's *Introductorium in Astronomiam*, Alfraganus' *Liber de aggregationibus scientiae stellarum et principiis coelestium motuum*; and Alpetragius's *Liber de motibus coelorum*,

principle of celestial causality, which constituted the cornerstone of Arab-Aristotelian physics and metaphysics. It provided a rationalization of myths and rituals connected to a superstitious cult of the stars. Although Isidore of Seville had already distinguished between astrology and astronomy, until the eleventh century the Latin world tended not to, and saw a strong connection between the mathematical science of star motion and the investigation of the astral influences it generated.[71] The images of the zodiac and the motion of stars were the visible signs of providential design. Deciphering them meant deciphering the universal and individual destiny of humankind. Theology and astrology shared the same interest in the «littera» of the universe, and both predicated on the human capacity of interpreting its language. The debate between theologians and astrologers involved similar issues, such as predestination and free will, prophetical knowledge, and sacred history.[72] The common standpoints of theological and astrological cultures – celestial causality and the assumption that planets and stars were providential instruments – led to a reinstatement of astrology in a position of preeminence among the liberal arts as it was deemed the science that unveiled the meaning of universal and individual destinies.[73] Theologians and astrologers however disagreed on

---

to mention some of Dante's most likely and important sources. For the *Almagest*, a *summa* of ancient astronomical wisdom written around the half of the second century A.D., see GERALD J. TOOMER, *Ptolemy's Almagest*, Princeton, Princeton University Press, 1998, and footnote 19 in this Chapter. According to Nardi, Dante knew Alpetragius, an Arabian-Hispanic astrologist of the twelfth century, probably through Albertus Magnus' commentary on ARISTOTLE's *De coelo*. See BRUNO NARDI, *Dante e Alpetragio*, in *Saggi*, pp. 139-166. For a complete list of the Latin translations of the Arabic treatises on astronomy in the twelfth century, see FRANCIS J. CARMODY, *Arabic Astronomical and Astrological Sciences in Latin Translation*, Berkeley, University of California, 1956. As Marie Thérèse d'Alverny observes, all translators were clerics, who sometimes dedicated their works to bishops, and whose orthodoxy was not in dispute. See MARIE-THÉRÈSE D'ALVERNY, *Astrologues et théologiens*, pp. 31-50.

[71] In his *Etymologiarum sive originum*, III, XXVII Isidore of Seville distinguished between astronomy and astrology. Astronomy was the science that studied the circular motion of the heavens, the constellations, and the origin of their names. Astrology was instead a natural discipline as far as it investigated the course of stars, but it became a form of superstition when astrologers presumed to infer predictions and horoscopes, or when they associated the zodiac with parts of the soul. See *The Etymologies of Isidore of Seville*, by S.A. Barney, W.J. Lewis, J.A. Beach, O. Berghof transls., Cambridge, Cambridge University Press, 2006.

[72] As Tullio Gregory explains: «andrà sottolineato come la causalità dei cieli costituisca la cornice fisica e metafisica entro la quale si iscrivono problemi cruciali per la riflessione teologica: l'inizio del tempo, l'esercizio della provvidenza, la libertà e la predestinazione, le forme della conoscenza profetica, la storia sacra, l'escatologia». [«It is worth underscoring that celestial causality constitutes the physical and metaphysical framework within which the issues crucial to theological reflection are situated: the beginning of time, the work of providence, freedom and predestination, the forms of prophetical knowledge, sacred history, and eschatology»]. According to Gregory, celestial causality also imposed the doctrine of moving Intelligences as philosophical necessity. TULLIO GREGORY, *I cieli, il tempo, la storia*, in *Speculum naturale*, p. 73.

[73] In the correspondence he established between heavens and disciplines in the *Convivio*,

the modes and limits of influences, and on the limits of human capacity to achieve a perfect, angelical, knowledge of future events.[74]

Those who attempted to give astrology religious dignity, such as Raymond of Marseille, Albert the Great, and Roger Bacon, believed that the heavens influenced only physical characteristics and natural inclinations, leaving untouched the rational capacity of human beings to choose and change their destiny even against their own inclinations. In this way, they not only allowed room for the exercise of free will but also geared it to a scientific doctrine of influence, although as a sort of exception. As ancient astrologers had argued, the uncouth and the uneducated were unable to use reason to resist their inclinations. Since these people represented the vast majority, astrology would be particularly strong in predicting the events of universal history.[75]

In the *Commedia*, and particularly in *Paradiso*, astrological elements are essential to the interpretation of the cantos dedicated to the themes of predes-

---

Dante placed astrology above all the other arts of the *trivium and quadrivium*, associating it with the heaven of Saturn: «E ancora: è altissima di tutte l'altre; però che, sì come dice Aristotile nel cominciamento dell'Anima, la scienza è alta di nobilitade per la nobilitade del suo subietto e per la sua certezza; e questa più che alcuna delle sopra dette è nobile e alta per nobile e alto subietto, ch'è dello movimento del cielo; e alta e nobile per la sua certezza, la quale è sanza ogni difetto, sì come quella che da perfettissimo e regolatissimo principio viene. E se difetto in lei si crede per alcuno, non è dalla sua parte, ma, sì come dice Tolomeo, è per la negligenza nostra, e a quella si dee imputare», [«Furthermore, it is far higher than all the others, since, as Aristotle says at the beginning of *On the Soul*, a science is high in nobility by virtue of the nobility of its subject and by virtue of its certainty; and this one, more than any of those mentioned above, is high and noble because of its high and noble subject, which regards the movement of the heaven, and high and noble because of its certainty, which is flawless, as coming from a most perfect and regular principle. And if anyone believes that there is a flaw in it, it does not pertain to the science, but as Ptolemy says, it results from our negligence, and so must be attributed to that»]. *Convivio* II, xiii, 30. In *Convivio* heavens and arts are associated as follows: (from below) grammar (moon), dialectics (Mercury), rhetoric (Venus), arithmetic (Sun), music (Mars), geometry (Jupiter), and astrology (Saturn).

74 Angels knew future events intellectually and immediately in their causes, because they could see them in God's mind. Human intellection could not achieve the same perfection and know events in their ultimate causes, but wise men literate in astrological science could achieve a similar type of knowledge as far as effects were concerned. Some, like Biagio Pelacani da Parma, believed that angels could directly pass their knowledge to human beings, or that with magical practices human beings could achieve cognition equal to that of the angels. For the ambiguous status of astrology as a discipline between mathematical science and magic, see GRAZIELLA FEDERICI VESCOVINI, *L'astrologia tra magia religione e scienza*, in *"Arti" e filosofia nel secolo XIV. Sulla tradizione aristotelica e i "moderni"*, pp. 171-193, and ID., *Astrologia e scienza. La crisi dell'aristotelismo sul cadere del Trecento e Biagio Pelacani da Parma*, Firenze, Nuovedizioni Enrico Vallecchi, 1979. For the debates on angelic intellectual knowledge see JAMES COLLINS, *The Thomistic Philosophy of the Angels* and TIZIANA SUAREZ-NANI, *Connaissance et langage des anges*. For Pietro d'Abano defense of astrology as science, and against its contamination with magic, see GRAZIELLA FEDERICI VESCOVINI, *Medioevo magico*, Torino, Utet, 2008, pp. 323-346.

75 On the general agreement on the validity of universal judicial astrology, see TULLIO GREGORY, *I cieli, il tempo, la storia*, in *Speculum naturale*, pp. 69-91.

tination, prescience, and angels.[76] In these cantos, Dante critically set the doctrine of human inclinations within the general scheme of astral influence. Although imbued with informative virtues, stars did not determine human destiny in a necessary way. The imperfection of matter and of human intellect limited the effects of astral influence, and this limitation was itself dictated by divine reason as salvation required freedom to choose the good. The material and the rational arguments for possible deviations from celestial influence indicated two different approaches to the issue of inclinations from two different, but complementary points of view. As instruments of angelic hierarchies, stars and planets exercised their influence as part of providential design. Their motion was a language speaking the eternal truths of love to the caducous contingent beings, and inclinations were part of the salvific plan. From an earthly perspective however, insofar as free will is an exercise of self-determination in view of salvation, inclinations become 'passions' human beings must curb to achieve virtue.

Dante anticipates his opinions on the determinism of astral influence in the proemial canto of *Paradiso* I, where Beatrice first illustrates the astrological doctrine of inclinations,

> Ne l'ordine ch'io dico sono accline
> tutte nature, per diverse sorti,
> più al principio loro e men vicine;
> onde si muovono a diversi porti
> per lo gran mar de l'essere, e ciascuna
> con istinto a lei dato che la porti (*Par.* I, 109-114).

> Within that order, every nature has
> its bent, according to a different station,
> nearer or less near to its origin.
> Therefore, these natures move to different ports
> across the mighty sea of being, each
> given the impulse that will bear it on.

Providential order operates in such a way that all creatures receive different «instincts» (according to their different degrees of proximity to the crea-

---

[76] Astrological observations open *Paradiso* VIII, where Carlo Martello introduces the doctrine of inclinations; *Paradiso* XX, on predestination, and *Paradiso* XXIX, the doctrinal canto on the nature and creation of angelic substances. Alison Cornish has shown the structural function of astrology in the *Commedia*, and its significance for Dante's reconciliation of faith and science. See ALISON CORNISH, *Planets and Angels in Paradiso XXIX: The First Moment*, «Dante Studies», CVIII, 1990, pp. 1-28, and ID., *Reading Dante's Stars*, Yale, Yale University Press, 2000. For a systematic study of astrology in *Paradiso*, see RICHARD KEY, *Dante's Christian Astrology*, Philadelphia, University of Pennsylvania Press, 1994.

tor) that lead them to fulfill their roles in the providential design by realizing their specific ends («diversi porti»). All intelligent creatures of different nature – meaning more or less close to God in their varying degrees of likeness – bend to the superior order of providence («accline» is an apax whose semantic resonance with «incline» stresses the link established between instinct and obedience). Inclinations guide all creatures to fulfill their individual goals and in so doing they move in the direction predisposed by providence. The word «instinct» in line 114 seems to indicate that providential order limits the action of sidereal influence to vegetative and sensitive faculties. Only rational human souls, which God creates immediately, are free from the influence of the planets. As Beatrice explains to Dante in *Paradiso* VII:

> Ciò che da essa sanza mezzo piove
> libero è tutto, perché non soggiace
> a la virtute de le cose nove (*Par.* VII, 70-72).

> Whatever rains from It immediately
> is fully free, for it is not constrained
> by any influence of other things.

Beatrice refers here to divine goodness («da essa») and the act of unmediated («sanza mezzo») creation of immortal immaterial substances («nove» because although immortal they were created), such as the human rational soul. They do not submit to the laws of celestial nature, but are free to choose the paths of likeness or dissimilitude from their creator, just as the rebel angels and Adam and Eve did in choosing to sin. Beatrice returns onto this idea few tercets later:

> L'anima d'ogne bruto e de le piante
> di complession potenzïata tira
> lo raggio e 'l moto de le luci sante;
>   ma vostra vita sanza mezzo spira
> la somma beninanza, e la innamora
> di sé sì che poi sempre la disira (*Par.* VII, 139-144).

> The rays and motion of the holy lights
> draw forth the soul of every animal
> and plant from matter able to take form;
>   but your life is breathed forth immediately
> by the Chief Good, who so enamors it
> of His own Self that it desires Him always.

The angels, through their action on the celestial bodies («luci sante»), «pull» upward the souls of animals and plants, namely the vegetative and sensitive souls. The vital principle of human life – the intellective soul – however, comes directly from God, as Statius had already explained to Dante in *Purgatorio* XXV.[77]

The doctrine of inclinations is of particular relevance not only for the cosmological structure of the *Commedia*, but also for its ethical and political implications. The role inclinations play in the universal design is parallel to the function they have in the organization of the earthly community. Plato argued in his *Republic* that in a well-regulated community citizens should assume different offices in consonance with their different inclinations.[78] This principle of labor division greatly influenced the literature devoted to the design of a well functioning political society. Cicero predicated the importance of inclinations in his *De officiis*. Ambrose, in his homonymous *De officiis* (a religious transposition of the work by the Roman politician), placed inclinations at the core of the orderly functioning of religious communities.[79] The different kinds of functions a community requires must correspond to the different natural capacities of its components.[80] Deviations from this principle inevitably

---

[77] Cfr. *Purg.* XXV, 67-75, where Statius explains the formation of the human soul as a three-stage process in which the vegetative soul transforms into sensitive until, when the brain of the embryo is perfect, the 'first cause' directly inflates a vital spirit into it. The rational soul thus evolves, maintaining its vital and sensitive characteristics: «che ciò che trova attivo quivi, tira / in sua sustanzia, e fassi un'alma sola,/ che vive e sente e sé in sé rigira». [«which draws all that is active in the fetus / into its substance and becomes one soul / that lives and feels and has self-consciousness»]. *Purg.* XXV, 73-75. See BRUNO NARDI, *Sull'origine dell'anima umana*, in *Dante e la cultura medievale*, pp. 207-224, for a discussion on Dante's philosophical sources on the origin of the soul. For the metaphysical implications of the embryology of *Purgatorio* XXV, see MANUELE GRAGNOLATI, *From Plurality of Forms to (Near) Unicity of Form: Embriology in Purgatorio 25*, in *Experiencing the Afterlife. Soul and Body in Dante and Medieval Culture*, Notre Dame, University of Notre Dame Press, 2005 («The William and Katherine Devers Series in Dante Studies», VII), pp. 67-70.

[78] In a passage of the *Republic*, Socrates explains to Adimantus that labor division is both efficient and natural. It is efficient because it allows for the satisfaction of everyone's needs. It is natural because it corresponds to the natural inclination of everyone participating in the community. PLATO, *Republic* 369, E-370 B.

[79] «Sic enim est faciendum, ut contra universam naturam nihil contendamus, ea tamen conservata propriam nostram sequamur, ut etiamsi sint alia graviora atque meliora, tamen nos studia nostra nostrae naturae regula metiamur», [«For we must act in such a way that we attempt nothing contrary to universal nature; but while conserving that, let us follow our own nature, so that even if other pursuits may be weightier and better, we should measure our own by the rule of our own nature»]. MARCUS TULLIUS CICERO, *On Duties (De officiis)*, M.T. Griffin and E.M. Atkins eds., Cambridge, Cambridge University Press, 1991, I, 110. *Duties of the Clergy*, in *Nicene and post-Nicene Fathers of the Christian Church*, II, P. Shaff and H. Wace eds., Grand Rapids, Michigan, W.B. Eerdmanns, 1955.

[80] For Dante's knowledge of Cicero's *De Officiis* see EDWARD MOORE, *Studies in Dante. First series. Scripture and classical authors in Dante*, Oxford, 1955, pp. 258ff, 353. For Dante's quotation of the *De officiis* in the *Monarchia*, see ID., *Studies in Dante. Second series, Miscellaneous Essays*, Oxford, 1968, p. 158.

involve inefficiency and injustice. In *Paradiso* VIII, Carlo Martello (Charles Martel) attributes the decadence of the Church and of Dante's contemporary society to these deviations:

> Sempre natura, se fortuna trova
> discorde a sé, com'ogne altra semente
> fuor di sua regïon, fa mala prova.
>   E se 'l mondo là giù ponesse mente
> al fondamento che natura pone,
> seguendo lui, avria buona la gente.
>   Ma voi torcete a la religïone
> tal che fia nato a cignersi la spada,
> e fate re di tal ch'è da sermone;
>   onde la traccia vostra è fuor di strada (*Par.* VIII, 139-147).

> Where Nature comes upon discrepant fortune,
> like any seed outside its proper region,
> Nature will always yield results awry.
>   But if the world below would set its mind
> on the foundation Nature lays as base
> to follow, it would have its people worthy.
>   But you twist to religion one whose birth
> made him more fit to gird a sword, and make
> a king of one more fit for sermoning,
>   so that the track you take is off the road.

Nature provides the foundation upon which human beings should build to develop their essence as rational creatures. It is a principle of individual happiness and at the same time, a principle of common good. The distortions caused by forcing someone destined to arms («tal che fia nato a cignersi la spada») to take religious vows («tal ch'è da sermone») – and vice versa – leads individuals and society astray from the right path («traccia») of virtue and justice. In this idea, individual and collective destinies interlace according to a vision that – as in Aristotle – harmonized ethics and politics.

In presenting the complex reasons of providential order, in *Paradiso* Dante stresses the negative consequences of deviating from natural inclinations, because they lead astray from the model of the ideal city.[81] The reason human beings often challenge their instincts resides in the imperfection of the matter upon which astral influences act:

---

[81] Cfr. «La carne d'i mortali è tanto blanda, / che giù non basta buon cominciamento / dal nascer de la quercia al far la ghianda», [«The flesh of mortals yields so easily- / on earth a good beginning does not run / from when the oak is born until the acorn». *Par.* XXII, 85-87.

> Vero è che, come forma non s'accorda
> molte fiate a l'intenzion de l'arte,
> perch'a risponder la materia è sorda,
>   così da questo corso si diparte
> talor la creatura, c' ha podere
> di piegar, così pinta, in altra parte; (*Par.* I, 127-132)
>
> Yet it is true that, even as a shape
> may, often, not accord with art's intent,
> since matter may be unresponsive, deaf,
>   so, from this course, the creature strays at times
> because he has the power, once impelled,
> to swerve elsewhere; as lightning from a cloud;

The simile of the art («arte») recalls the one in *Paradiso* II, 126, where it illustrated the instrumental relation between the superior immaterial angels and the stars. The imperfect matter of creatures can induce them to make the wrong choices and depart from the roles nature had assigned them. The rightful exercise of free will seems therefore in some way to go against preordered providential destiny, and project a negative shadow on the highest prerogative of rational beings.

To further explore the political implications of Dante's ideas on inclinations, it is worth looking at the specific heaven in which Carlo Martello speaks to the pilgrim. It is the third heaven of Venus, which is particularly relevant because of its implicit connection with the second treatise of *Convivio*, where Dante expounded a quite different view on angelic ordering and functions. In that treatise, Dante comments on his doctrinal *canzone* «Voi che 'ntendendo il terzo ciel movete». Venus's heaven is the heaven of love (associated in *Convivio* with the art of rhetoric) and the frontier of Saint Paul's ascension into the holy circles. In Dante's doctrinal treatise, the angelic order moving this heaven was that of the Thrones, through which, according to Gregory, «God accomplishes His judgments».[82] In *Paradiso* however, following Dionysius, the Thrones are the lowest order of the highest hierarchy and move the heaven of Saturn. The Thrones, according to Dionysius, have immediate knowledge of the 'types' of divine works necessary to creation.[83] In canto VIII of *Paradiso*, the movers of

---

[82] GREGORY THE GREAT, *Hom.* XXIV *in Evangelium*, in *Opere*, II, *Omelie sui Vangeli*, Roma, Città Nuova, 1994.

[83] On the different functions attributed to the Thrones in *Convivio* and in the *Commedia* and the influence of Albert the Great's commentary on *The Coelestial Hierarchy* on Dante, see DIEGO SBACCHI, *La presenza di Dionigi Areopagita*, pp. 1-20. Based on Dionysius's definition, Aquinas specifies that

Venus are the Principalities, whose name, again according to the Aeropagite, «refers to those who possess a godlike and princely hegemony, with a sacred order most suited to princely powers».[84] The Principalities oversee the lowest hierarchy, which governs over human affairs, and rule over archangels and angels. Venus is the planet of love, which is intended in a high philosophical sense as the harmony of friendship that allows for a peaceful and well-functioning community. The underlying angelic structure thus further stresses the connection between the doctrine of influences and its political implications, and evidences the structural necessity of the angelic ordering Dante adopted in *Paradiso*, and of the ministries allotted to each order.[85]

The political aspect of the governing functions attributed to angels is a key factor in Dante's architecture of the heavens, which appears as a mirror image, as well as the exemplar, of the ideal worldly city. In commenting on the governing functions of Dionysian hierarchies, Thomas Aquinas stressed their implications in terms of an analogical model of earthly and celestial cities, making explicit the political parallel that:

[...] una hierarchia est unus principatus, idest una multitudo ordinata uno modo sub principis gubernatione. Non autem esset multitudo ordinata, sed confusa, si in

---

«ordo thronorum habet excellentiam prae inferioribus ordinibus in hoc, quod immediate in Deo rationes divinorum operum cognoscere possunt», [«The order of the "Thrones" excels the inferior orders as having an immediate knowledge of the types of the Divine works»]. AQUINAS, *S. Th.* I^a, q. 108, a. 5.

[84] The full definition of Principalities is the following: «Manifestat enim ipsa quidem coelestium principatuum illud deformiter principale et ductivum cum ordine sacro et principalibus decentissimis virtutibus, et ad superprincipale principium eas universaliter converti, et alias hierarchice ducere ad illum ipsum, quantum possible, formari principium manifestareque seperessentialem eius ordinationem ornatumque principalium virtutum», [«The term "heavenly principalities" refers to those who possess a godlike and princely hegemony, with a sacred order most suited to princely powers, the ability to be returned completely toward that principle which is above all principles and to lead others to him like a prince, te power to receive to the full the mark of the Principle of principles and, by their harmounious exercise of princely powers, to make manifest this transcendent principle of all order»]. *CH.* IX 1, 257B.

[85] «Primo igitur videamus rationem assignationis Dionysii. In qua considerandum est quod, sicut supra dictum est, prima hierarchia accipit rationes rerum in ipso Deo; secunda vero in causis universalibus; tertia vero secundum determinationem ad speciales effectus. Et quia Deus est finis non solum angelicorum ministeriorum, sed etiam totius creaturae, ad primam hierarchiam pertinet consideratio finis; ad mediam vero dispositio universalis de agendis; ad ultimam autem applicatio dispositionis ad effectum, quae est operis executio; haec enim tria manifestum est in qualibet operatione inveniri», [«Let us then first examine the reason for the ordering of Dionysius, in which we see, that, as said above (*S.Th.* I, q. 108, Article 1), the highest hierarchy contemplates the ideas of things in God Himself; the second in the universal causes; and third in their application to particular effects. And because God is the end not only of the angelic ministrations, but also of the whole creation, it belongs to the first hierarchy to consider the end; to the middle one belongs the universal disposition of what is to be done; and to the last belongs the application of this disposition to the effect, which is the carrying out of the work; for it is clear that these three things exist in every kind of operation»]. AQUINAS, *S. Th.* I, q. 108, a. 6.

multitudine diversi ordines non essent. Ipsa ergo ratio hierarchiae requirit ordinum diversitatem. Quae quidem diversitas ordinum secundum diversa officia et actus consideratur. Sicut patet quod in una civitate sunt diversi ordines secundum diversos actus, nam alius est ordo iudicantium, alius pugnantium, alius laborantium in agris, et sic de aliis. Sed quamvis multi sint unius civitatis ordines, omnes tamen ad tres possunt reduci, secundum quod quaelibet multitudo perfecta habet principium, medium et finem. Unde et in civitatibus triplex ordo hominum invenitur, quidam enim sunt supremi, ut optimates; quidam autem sunt infimi, ut vilis populus; quidam autem sunt medii, ut populus honorabilis (*S. Th.* Ia, q. 108, a. 2).

[...] one hierarchy is one principality – that is, one multitude ordered in one way under the rule of a prince. Now such a multitude would not be ordered, but confused, if there were not in it different orders. So the nature of a hierarchy requires diversity of orders. This diversity of order arises from the diversity of offices and actions, as appears in one city where there are different orders according to the different actions; [...] But although one city thus comprises several orders, all may be reduced to three, when we consider that every multitude has a beginning, a middle, and an end. So in every city, a threefold order of men is to be seen, some of whom are supreme, as the nobles; others are the last, as the common people, while others hold a place between these, as the middle-class [«populus honorabilis»].[86]

While expressing the parallel between the celestial and earthly orders in political terms was common, placing a «scientifically» structured angelic ordering as the nexus linking the two was not.

The contextualization of the characters presented in this heaven further accentuates the political nuances of the doctrine of influences. Carlo Martello, a descendent of Carlo II of Anjou, is a political figure. Most of his conversation with Dante centers on political themes. He introduces himself as the heir of the crowns of Hungary, Provence, and Naples. The art of governing is at the center of his speech against his brother, Robert of Anjou, who was accused of avarice, one of the vices that compromises justice and good governance. In the same heaven Cunizza da Romano, an example of lust corrected by reason, evokes the figure of her sanguinary brother, the tyrant Ezzelino. Her speech, as Carlo's, mostly concerns political issues and explicitly introduces the theme of justice. At the end of her apostrophe, Cunizza invokes the order of the Thrones, thus again connecting the themes of earthly and divine justice.[87]

---

[86] See also Chapter II, p. 37.
[87] «Su sono specchi, voi dicete Troni, / Onde refulge a noi Dio giudicante; / Sí che quest parlar ne paion buoni», [«Above are mirrors-Thrones is what you call / them- / and from them God in judgment shines on us» (*Par.* IX, 61-63).

The mention of the Thrones illustrates the subordination of human to divine justice in the hierarchy of angelic orders. The influence of Principalities is directed to the well functioning of princedoms, according to the laws that bind the community. Alternation in fortune guarantees alternance in offices, and this is, as Carlo Martello explains, the purpose of the differences in human natures, as in fathers and sons. These differences are necessary to guarantee an efficient political community, as they are at the foundation of labor division and, consequently, of justice. In the heaven of Venus the function of the moving Intelligences is to direct individual erotic passion toward the supreme form of *caritas* that finds its highest realization in the political life of the city.

Although the prevalent opinion among theologians was that astronomical influence did not affect the rational soul, thus reconciling celestial causality and free will, a considerable number of astrologers believed in the determinism of astral influences. Their belief relied on the idea that the principle of causal influences resided not in the otherworld of spiritual substances, but immanently in the cosmos. Astrologers such as Michael Scot and Guido Bonatti, believed Providence bestowed on a select few the gift of divination so they could obtain perfect foresight of future events.[88] Augustine and Gregory the Great had been among the fiercest opponents of judicial astrology and had condemned its claim of knowing by divination what only faith could reveal. The vast corpus of Arabic scientific treatises such as those by Rasis, Alkindi, Albumasar, and Alhazen, contributed to the development of astrology as a mathematical science, and had a tremendous impact on the formation of

---

[88] Guido Bonatti (c. 1210- c. 1300), a mathematician from Forlì, was with John Holywood (Sacrobosco) one of the most famous astrologers of the thirteenth century. He was counselor to Frederick II, Ezzelino da Romano, Guido Novello da Polenta, and Guido da Montefeltro. His career testifies to the importance that astrologers had in influencing political decisions and is probably one of the reasons that made him and Scot, who was also counselor to Frederick II, even more blameworthy in Dante's views than Cecco d'Ascoli. Bonatti's best known work was the *Liber decem continens tractatus astronomiae*, written around 1277. Michael Scot (1175-1232?) was a Scottish mathematician, translator of Aristotle and astrologer whose major works dealt not only with astronomy but also with alchemy and occult sciences. According to Haskins, who identifies him with Michael of Cornwall, present at Chartres between 1252-54, the commentary on the *De sphaera* is attributed to him, but not with certainty. CHARLES H. HASKINS, *Studies in the History of Medieval Science*, p. 277. Scot's major astrological work was the *Liber introductorius*, where he denied that stars were the causes of events. They were nonetheless signs that, if correctly interpreted, allowed for predicting the future. For Dante's condemnation of Scot as a sustainer of the superiority of natural philosophy, and the relation between this figure and Nimrod relative to the claim of perfect astrological foresight, see MARCELLO CICCUTO, *Michele Scoto e la naturalis philosophia di Dante*, «Tenzone», I, 2000, pp. 31-41. On magic, religion and astrology, see also GRAZIELLA FEDERICI VESCOVINI, *Medioevo magico*, pp. 225-275, and in the same volume in particular on Michael Scot, pp. 47-69. See also PIERO MORPURGO, *Michele Scoto e Dante: una continuità di modelli culturali?*, in *Filosofia scienza e astrologia nel Trecento europeo*, Graziella Federici Vescovini and F. Barocelli eds., Padova, Il Poligrafo, 1992, pp. 79-94.

medieval astrology. The new scientific context of the thirteenth century led many to believe in the self-sufficiency of this discipline. In relying on deterministic causality, they further developed its connections with divinatory art. Indeed, although it assumed scientific dignity, astrology never completely lost its ancient relation to the sacred and to magic. In the treatises of astrologers such as Cecco d'Ascoli and Michael Scot, the boundaries between deterministic judicial astrology, magic, and alchemy were often blurred, thus eliciting the diffidence of the Church, and provoking great debates among the theologians who defended legitimate astrology.[89]

Dante attacked astrological determinism in at least two main occurrences, in *Inferno* XX, and in *Purgatorio* XVI. In *Inferno* XX, Dante and Virgil meet the diviners in the fourth ditch of the eighth circle of hell. Looking at the bodies that have lost their natural identity, with their heads turned toward their backs, Dante feels compassion and cries. He is so weakened that he needs to hold on one of the rocks surrounding the circle. Virgil rebukes him for his feelings, since compassion is not admissible before divine justice. However, as in the canto of Francesca, Dante's weakness is perhaps the sign of a past, repressed, temptation of believing in the possibility that natural philosophy could be sufficient to penetrate the truths of the origin and causes of the world. This was indeed the claim of the two astrologers he meets in the ditch, Michael Scot and Guido Bonatti. They believed astrological science was a gift similar to prophecy, and that those who could decipher the message contained in the celestial signs would be able to know the past and the future. Through these two figures, Dante condemns astrology as an esoteric superior form of intellectual hubris that presumed to unveil divine misteries and challenged the superiority of theology.[90]

The viewpoints that astrology was a 'sacred' science superior to all other disciplines, and that was contaminated by magic, were perhaps the reasons we find Michael Scot «che veramente de le magiche frode seppe 'l gioco» (*Inf.* XX, 117) and Guido Bonatti in canto XX. Similarly to Ulysses, these diviners

---

[89] A text of capital importance for the sacralization of astrology in the Middle Ages was *Picatrix*. See GRAZIELLA FEDERICI VESCOVINI, *L'astrologia tra magia, religione e scienza*, in "*Arti*" *e filosofia nel secolo XIV. Studi sulla tradizione aristotelica e i "moderni"*, Firenze, Nuovedizioni Enrico Vallecchi, 1983, pp. 171-193. For the importance of the debate on astrology in the XVth century, and on the influence of Picatrix and the Hermetic corpus, see EUGENIO GARIN, *Lo zodiaco della vita. La polemica sull'astrologia dal Trecento al Cinquecento*, Bari-Roma, Laterza, 1996.

[90] In the proemium to his commentary on Sacrobosco's *De sphaera*, Cecco d'Ascoli wrote that "this science [astrology] makes man divine" (haec est enim scientia quae humanum facit esse divinum). The quotation is from TULLIO GREGORY, *Astrologia e teologia nella cultura medievale*, in *Mundana Sapientia*, p. 307.

failed in their claim of true knowledge due to lack of faith. Thus, their immoderate desire to see ahead was turned into a perennial sight of what was behind. A third famous astrologer, paradoxically evoked by his very absence in the ditch of the diviners, was another Dante contemporary, Cecco d'Ascoli (Francesco degli Stabili), who harshly attacked the spiritual cosmology of the *Commedia* in his *Acerba etas*.[91]

Representatives of heretical, superstitious pride such as Bonatti, Cecco, and Scot, claimed the superiority of natural philosophy over theology, and believed that astrology, associated with magical practices, could unveil the secrets of providential destiny. More than anything else, these ideas revealed the diviners' belief that natural virtue was sufficient to comprehend divine truths. Their immoderate desire of knowledge was in a way similar to Ulysses' *cupiditas sciendi*, for it did not rely upon the necessary *auxilium* of grace. These extreme positions emerged in the commentaries that Scot and d'Ascoli wrote on John of Holywood (Sacrobosco)'s treatise on *De sphaera* (*On Sphere*), where they also introduced elements of necromancy and demonology.[92] While Dante had attempted to reconcile astrological science with religious spirituality, attributing some miraculous events such as the eclipses at Jesus' death to divine intervention, Scot and Cecco interpreted it in the light of astrological determinism. Even more dangerous than their impiety was their claim that astrology was a prophetical science, superior to metaphysics and theology, and the only one able to deify humankind.

An indirect response to Scot and Bonatti is in Marco Lombardo's warning to Dante in *Purgatorio* XVI. Here Lombardo states that to believe that the heavens necessarily cause all events on earth is a mistake and an illusion:[93]

---

[91] Cecco d'Ascoli (Francesco degli Stabili, 1257-1327), astrologer and poet, engaged in famous disputes with Dante over poetry and cosmology. In his *Acerba* (*Acerba etas*), an unfinished poem of five books, he openly polemized with Dante on the question of astral influences, and argued for the superiority of natural instincts over education. In commenting on Sacrobosco's *De sphaera* Cecco introduced elements of demonology and necromancy. Cecco d'Ascoli's heretical opinions led to his condemnation to the stake in 1327. For a discussion of the magical implications of Cecco d'Ascoli's astrology, see GRAZIELLA FEDERICI-VESCOVINI, *Medioevo magico*, pp. 277-311.

[92] John Holywood (known as Sacrobosco, 1195-c. 1256) was an English mathematician and astrologer, and the author of one of the most famous medieval treatises on cosmology, *Tractatus de sphaera* (1230). This treatise was taught in Universities in Dante's time. See Lynn Thorndike's introduction to his *The Sphere of Sacrobosco and Its Commentators*, Chicago, The University of Chicago Press, 1949, pp. 1-58.

[93] Commentators observe a similar condemnation of determinism in Thomas Aquinas. The *Doctor Angelicus* touched upon the question of astrology in many of his works. Of particular relevance to the above quoted tercets is a passage from the proemium of his *De iudiciis astrorum*: «Hoc autem omnino tenere oportet, quod voluntas hominis non est subiecta necessitate astrorum; alioquin periret liberum arbitrium, quo sublato non deputarentur homini neque bona opera ad meritum, neque mala ad culpam». [«It is necessary to maintain that human will it be not subject to the

> Voi che vivete ogne cagion recate
> pur suso al cielo, pur come se tutto
> movesse seco di necessitate.
>   Se così fosse, in voi fora distrutto
> libero arbitrio, e non fora giustizia
> per ben letizia, e per male aver lutto.
>   Lo cielo i vostri movimenti inizia;
> non dico tutti, ma, posto ch'i' 'l dica,
> lume v'è dato a bene e a malizia,
>   e libero voler; che, se fatica
> ne le prime battaglie col ciel dura,
> poi vince tutto, se ben si notrica (*Purg.* XVI, 67-78).
>
>   You living ones continue to assign
> to heaven every cause, as if it were
> the necessary source of every motion.
>   If this were so, then your free will would be
> destroyed, and there would be no equity
> in joy for doing good, in grief for evil.
>   The heavens set your appetites in motion-
> not all your appetites, but even if
> that were the case, you have received both light
>   on good and evil, and free will, which though
> it struggle in its first wars with the heavens,
> then conquers all, if it has been well nurtured.

The stars are not the necessary causes of everything that happens on earth. Their influence does not extend to the sphere of moral action, and human beings must take responsibility of their choices. Lombardo clearly states that by determining inclinations, the heavens only initiate and cannot determine the course of human actions. The capacity to discerning good is the result of wisdom. Men must cultivate their intellectual virtues to be able to choose what to avoid and what to pursue, so that they can freely desire what God desires. Without this freedom to choose between good and evil, there would be no moral responsibility, and no justice could exist («lume v'è dato a bene e a malizia», *Purg.* XVI, 75).

Marco Lombardo offers a different point of view from which to look at inclinations in their effects and in their relation to free will, as distinct but complementary factors. On moral ground, what really counts is the use of rea-

---

stars by necessity; for otherwise free will would be destroyed. By denying free will, it would be impossible either to attribute merit for good actions, or responsibility for the bad ones»].

son in choosing actions. Reason leads and educates instincts so that inclinations and free will concur in realizing the grand design of providential order. In *Paradiso*, Dante discussed inclinations as aspects of our nature that we must know and use in the active life. To go against them meant to deviate from the natural course of the universal destiny of salvation. In *Purgatorio*, however, Dante seems interested in discussing celestial influences from a moral point of view, so that inclinations are likened to passions, and must be educated in virtue. Thus Lombardo says that free will in the beginning must engage in «battles» against the heavens, and only when nurtured with wisdom («se ben si notrica», *Purg.* XVI, 78) it is able to win over inclinations:

> A maggior forza e a miglior natura
> liberi soggiacete; e quella cria
> la mente in voi, che 'l ciel non ha in sua cura (*Purg.* XVI, 79-81).
>
> On greater power and a better nature
> you, who are free, depend; that Force engenders
> the mind in you, outside the heavens' sway.

Only the educated mind will be free to choose good. In the famous oxymoron of line 80, «liberi soggiacete», the ancient belief that only wisdom makes one truly free resonates and anticipates the Paradisiacal harmony where the desire of the blessed and the desire of God coincide. Beatitude consists in this ethical harmony. The expression «liberi soggiacete» in line 80 establishes a distant, but strong link with Martello's speech in *Paradiso* VII, where the same terms «libero», and «soggiace» appear – as in line 80 – in the same line (71) to place greater stress on the dialectic between astrological influence and free will in the superior design of Providence. If Marco Lombardo in *Purgatorio* XVI denies determinism and the direct influence of astrology on human destiny, in *Paradiso*, Beatrice then shows how free will and astrological influence harmonize in the ethical order of the universe and in that of the *Commedia*.

CONCLUSIONS

This chapter has shown how angelology casts new light on the interpretation of the cantos dedicated to creation and cosmology. The analysis of angelic cosmological operations has evidenced that the dialects of knowledge and desire is at the core of the participative angelology of the *Commedia*. The cognitive moment represented by the angelic contemplative-illuminative operations precedes and causes the angels' desire to unify with God, thus giving origin to the process that ultimately transforms the knowledge so acquired into active participation in the government of the cosmos. The most important operation that this contemplative-active process triggers consists in the movement of the celestial spheres. We have seen that Dante configured a structure in which each order presides over the motion of a corresponding heaven in reversed order of angelic nobility and magnitude of the spheres. The illumination of angels actualizes motion and activates influential virtue in the orbs. By acting as mirrors, they do not create but merely pass formal principles onto the stars. By extending to hierarchies both moving and intellectual functions, and by applying the laws of *perspettiva*, Dante thus provided a solution to the issue of the transmission of formal principles by avoiding the assumption of direct divine intervention. His original solution relied on his particular interpretation of angelic hierarchies as efficient causes in the dialectics between the One and multiplicity. This contemplative-operative angelology forms the kernel of his cosmology, and distinguishes it from contemporary representations of the cosmos.

This chapter also demonstrated that the structure of the *Commedia* reflects Dante's need to reconcile a scientific theory of celestial movements with the principles of the Christian doctrine of creation. The physical explanation of terrestrial phenomena is grafted onto metaphysical ground so as to yoke the demonstrative truths of philosophy to the intuitive truths of theology. In incorporating the functions of Aristotelian Intelligences into the Dionysian angelology, and extending them analogically to the human sphere, Dante builds a system that harmonizes grace and free will, and places the mystical experience of contemplation within an ethical perspective.[94] Angelology, cosmology, and astrology combine in the *Commedia* to offer a general hypothesis of how creation works, from the primary source of divine light to the actions of human beings. Dante's system relies on the original connection he established between the angelic operations of light transmission and their action as movers of the celestial spheres. The simultaneous combination of these angelic operations determines the astrological influence of the stars. In the image of the cosmos that the *Commedia* contains and represents, the astrological influence of angels is the result of a supreme synthesis of contemplation and operation. Metaphorically, this synthesis indicates that speculative thought must apply to practical life to make it morally relevant. Against this backdrop, we have seen that Dante's doctrine of inclinations explains how the cosmological

---

[94] See *Par.* II, 127-148.

order extends its effects to the moral and political spheres in the human world.[95] In *Paradiso*, questions about creation, angelic operations, cosmological motion, astrological influences, free will, and predestination interlace to illustrate the scientific foundation of Dante's theological poetry.

---

[95] Cfr. *Purg.* XVI, 73-78 and the doctrine of astrological influence in *Par.* VIII, 97-105 and 123-135. «Lo ben che tutto il regno che tu scandi / volge e contenta, fa esser virtute / sua provedenza in questi corpi grandi», [«The Good that moves and makes content the realm / through which you now ascend, makes providence / act as a force in these great heavens' bodies»]. *Par.* VIII, 97-99.

Chapter four

OTHER ANGELIC OPERATIONS: ANGELS IN *PURGATORIO*, THE HEAVENLY MESSENGER, AND BEATRICE

Chapter III has shown that Dante subordinated the speculative function of angels to their intervention in the providential government of the cosmos and of universal history. Enlightening and moving the celestial spheres, however, are not the sole operations angels perform in the *Commedia*. The separate substances animating the Empyrean are not creatures locked in the eternal circular motion around the origin of light and life. With the levity that distinguishes all paradoxes, angels freely move from their circles and, as laborious bees, disperse the pollen of their luminous wisdom on the petals of the blessed rose and the inferior spheres.

The amalgamation of Biblical, Aristotelian, and Neo-Platonic traditions, filtered through the Scholastic synthesis of thought that operates in the *Commedia*, is visible in the variety of functions Dante attributed to the angels. Besides their main tasks as mirrors of divine light and chronocrators, angels intervene in the material space to assist human souls in finding their way toward salvation.[1] From *Inferno* to *Paradiso*, there is a pervasive presence of angelic figures: the *messo celeste* (heavenly messenger) of *Inferno* IX; the neutral angels mingled with the *ignavi* in eternal wait at the shores of hell; the fallen an-

---

[1] The Fathers of the Church generally agreed that there were different kinds of angels, some intent on contemplation, and others active in the material world. As we have seen in chapters I and II, Dante unified these functions so that all angels contemplate and intervene in the sublunar world. Opinions on which orders could be sent on 'mission' varied among the theologians, but in Dante it seems that all angels can be equally commanded to special tasks both in the world and in the otherworld. Franciscan Scholasticism, more than the Dominican one, absorbed the Gregorian interest in the angelic active intervention in the human sphere. For the Franciscans, and particularly Bonaventure, the angels were models for their mission of wandering to deliver Christ's word everywhere. The presence of angels performing various 'menial' tasks in the *Commedia* might be a sign of their influence. On Franciscan and Bonaventure's angelology, see David Keck, *Angels and Angelology*, pp. 129-154. For the opinions of theologians on the missions of angelic orders, see Attilio Mellone, *Angelo*, in *ED*, p. 270.

gels that oversee divine justice in hell; the anthropomorphic angels that guard the terraces of *Purgatorio*; the incorporeal Intelligences that move the heavens in *Paradiso*; and, in analogy with angels, Dante's luminous guide, Beatrice. With the exception of the neutral angels, who do not play any active role in the cosmic order and yet belong to it as exemplarily punished figures of ignavy, all these characters perform various tasks.[2] They are messengers, assistants, or revealing signs of divine power, interfering wherever necessary with the sphere below the heaven of the Moon. The variety of operations that angels perform in the *Commedia* finds its ultimate justification in the fact that angels act as instruments of divine Providence, and govern sublunar reality by performing different functions, according to their capacities and their hierarchies.

This chapter analyzes three specific cases of angelic operations, different but complementary to the motion of the celestial bodies. These are the operations performed by the *Purgatorial* angels, the heavenly messenger in *Inferno* IX, and Beatrice. All represent special instances of Dante's interpretation of how angels can intervene to assist the souls, and the poet, in the narrative journey through the canticles.

I. THE ANGELS AND THE BLESSED. A PREMISE

In the Empyrean, where the pilgrim finally sees the «general form of Paradise» («la forma general di Paradiso», *Par.* XXXI, 52), Dante appears astonished by the vision of the angels continually moving not only around the radiant 'point' but also back and forth from the petals of the candid rose, like a swarm of bees around a flower:

> sì come schiera d'ape che s'infiora
> una fiata e una si ritorna
> là dove suo laboro s'insapora,
>   nel gran fior discendeva che s'addorna
> di tante foglie, e quindi risaliva
> là dove 'l süo amor sempre soggiorna (*Par.* XXXI, 7-12).

---

[2] According to John Freccero, the neutral angels are those who chose to be «per sé» («by themselves», *Inf.* III, 39). Their position outside Hell (but still among the punished) does not reflect their initial lack of acceptance of the Good – which caused their damnation together with the rebel angels- but their stillness before action: «The second, privation of action, won for the *per sé* angels complete isolation in Dante's cosmos». JOHN FRECCERO, *The Neutral Angels*, in *Dante and the Poetics of Conversion*, R. Jacoff ed., Cambridge, Harvard University Press, 1986, p. 116.

> just like a swarm of bees that, at one moment,
> enters the flowers and, at another, turns
> back to that labor which yields such sweet savor,
> descended into that vast flower graced
> with many petals, then again rose up
> to the eternal dwelling of its love.

Dante expands the mystic metaphor of the angels – bees, incessantly circling around God and producing the honey of charity, to include a vertical motion to and from the candid rose («discendeva [...] e quindi risaliva»). The spatially extended metaphor intensifies the image of 'laboriousness' of the angels and underscores their interaction with the blessed circles. This image evidences the condition of mobility of the angels not only within but also outside their ordered circles, showing that they are not fixed in their allotted places in *Paradiso*, but move according to necessity, continuously interacting with the Blessed.

Various passages of *Paradiso* suggest a relation of similitude between blessed souls and angels. Such affinity constitutes the ground upon which Dante attributed to the Blessed functions similar to those of the angels, leading to his invention of 'angelic characters'.[3] Like the angels, the souls differ in degrees of Beatitude according to their different capacity of receiving and cooperating with grace. Like the angels, the Blessed are arranged in different ranks, represented by the seats they occupy in the mystic rose. Their vicinity to the Virgin Mary varies with their varying degrees of merit (as for the angels, where merit is intended as their capacity of vision and therefore of cooperation with grace): «puoi tu veder cosi' di soglia in soglia / giù digradar» («these you can see /from rank to rank as I» *Par.* XXXII, 13-14). In *Paradiso* III Piccarda Donati explains to Dante that although Paradise is «everywhere in heaven», grace «does not rain equally» over all its creatures («d'un modo non vi piove» *Par.* III, 90). The hierarchical ranking of the Blessed is visible in the order in which the evanescent shadows appear to Dante in the nine spheres.[4] Like the angels, the souls have a direct vision of the first essence, although only through angelic enlightenment do they acquire a capacity of vision that

---

[3] The analogy between angels and Blessed had its theological foundation in the *New Testament*, in the words of Luke (20:36), for whom «the children of the resurrection will be equal to angels» [«aequales enim angelis sunt et filii sunt Dei»], Matthew (22:30), who said that the at resurrection all «are like the angels of God» [«sunt sicut angeli in caelo»], and Paul, who in his first Letter to the Corinthians (1 Corinthians, 6:3) warns: «Do you not realize that we shall be the judges of angels?» [«Nescitis quoniam angelos iudicabimus»].

[4] Cfr. *Par.* IV, 28-42.

allow them to bear the vision of divine light. As Peter Damian explains in *Paradiso* XXI:

> Luce divina sopra me s'appunta,
> penetrando per questa in ch'io m'inventro,
>   la cui virtù, col mio veder congiunta,
> mi leva sopra me tanto, ch'i' veggio
> la somma essenza de la quale è munta (*Par.* XXI, 83-87).
>
> Light from the Deity descends on me;
> it penetrates the light that enwombs me;
>   its power, as it joins my power of sight,
> lifts me so far beyond myself that I
> see the High Source from which that light derives.

In line 83 «luce divina» clearly indicates the light that angels reflect, which, in turn, increases the blessed soul's capacity of seeing, a phenomenon stressed by the intensification of the terms indicating light and sight: «luce», «veder», «veggio», so that Damian may look at the supreme essence («somma essenza»). The above quoted lines insist on the circular effect of the ray that hits the souls from above («s'appunta»), penetrating and sharpening Damian's sight until he can perceive the very origin of light.

Other elements reveal the parallel Dante established between angels and Blessed, which he extended to humankind as a modular structure characterizing all creatures, and which adapted to their different natures. An important element shared by the two celestial families of angels and Blessed is the distinction of the act of seeing in the two subsequent moments of knowing and loving, as Salomon reveals in *Paradiso* XIV when he unveils to Dante the doctrine of the glorious bodies.[5]

Although the blessed souls occupy an inferior place in the ladder of beings, Dante continually stresses the analogy between them and the angels, often applying to the former typical attributes belonging to the latter. In so doing, Dante makes the two heavenly *militiae* appear similar, intentionally overlapping their relative imagery so that angels can even be mistaken for blessed souls, and vice versa. Thus, when Salomon speaks, Dante does not see who is answering his question about the glorious body. The first thing

---

[5] *Par.* XIV, 40-42: «La sua chiarezza séguita l'ardore; / l'ardor la visïone, e quella è tanta, / quant' ha di grazia sovra suo valore.» [«Its brightness takes its measure from our ardor, / our ardor from our vision, which is measured / by what grace each receives beyond his merit.»] Cfr. Also *Par.* V, 8-9: «l'etterna luce, / che, vista, sola e sempre amore accende» [«once seen, / that light, alone and always, kindles love»].

he perceives is the sound of a humble voice, which leads him to think it may be the voice of the archangel Gabriel.

The affinity between the two sacred families also emerges in the spatial organization of Dante's Empyrean. The hierarchization of angelic and Blessed choirs in the immaterial 'whereless' space of the Empyrean suggests the influence of the debate about the position the Blessed should occupy in the celestial Jerusalem. As admitted by theologians such as Aquinas and Bonaventure, who followed in the steps of Gregory the Great, the elect formed a sort of tenth hierarchy, similar to those of the angels. Because of this similarity, the elect would be able to perform typical angelic functions aimed at assisting human souls in their spiritual movement toward God.

In the theological debate, the question of the affinity between Blessed and angels developed within the context of the exegesis of creation. Ever since the early centuries of the Christian interpretation of *Genesis*, the hypothesis emerged that God had created human beings as a consequence of the fall of Lucifer and the rebel angels. Chased from heaven, they had left the ranks of angelic choirs incomplete. Human souls in a perfect state of grace would eventually sit beside the angels (for they would be similar to them) to restore the lost perfection of the choirs.

Gregory the Great spoke of the tenth order (*decimus ordo*) of the elect in his *Homily* XXXIV, 6 where he interpreted the parable of the ten drachms (Luke 15:8-11) as symbolizing the addition of the order of the Blessed to the nine orders of the angels.[6] Scholastic thought absorbed this 'theory of Restoration', but with differences among theologians. The Gregorian theory of the tenth order tended to privilege the centrality of angelic substances in creation, confining the saved souls to a secondary restoring function. Although maintaining the idea of assimilation of the blessed souls to angels, Aquinas later stressed the difference existing between their natures, indirectly reaffirming the necessity of an independent reason for the creation of human beings. In his *Summa Theologiae* he argued that although human beings cannot be assumed into the angelic orders by nature, they could however become equal to the angels by gift of grace:

Respondeo dicendum quod, sicut supra dictum est, ordines Angelorum distinguuntur et secundum conditionem naturae, et secundum dona gratiae. Si ergo considerentur Angelorum ordines solum quantum ad gradum naturae, sic homines nullo modo assumi possunt ad ordines Angelorum, quia semper remanebit naturarum distinctio. Quam quidam considerantes, posuerunt quod nullo modo homines transferri

---

[6] The first author to mention the so-called 'Theory of Restoration' and to speak of the tenth order was probably Salonius of Geneve in his *Expositio mistica in Ecclesiasten* (died after 450 A.D.). According to Bruno Nardi the idea derived from Origen.

possunt ad aequalitatem Angelorum. Quod est erroneum, repugnat enim promissioni Christi, dicentis, Lucae XX, quod filii resurrectionis erunt aequales Angelis in caelis. Illud enim quod est ex parte naturae, se habet ut materiale in ratione ordinis, completivum vero est quod est ex dono gratiae, quae dependet ex liberalitate Dei, non ex ordine naturae. *Et ideo per donum gratiae homines mereri possunt tantam gloriam, ut Angelis aequentur secundum singulos Angelorum gradus.* Quod est homines ad ordines Angelorum assumi. Quidam tamen dicunt quod ad ordines Angelorum non assumuntur omnes qui salvantur, sed soli virgines vel perfecti; alii vero suum ordinem constituent, quasi condivisum toti societati Angelorum. Sed hoc est contra Augustinum, qui dicit XII de Civ. Dei, quod non erunt duae societates hominum et Angelorum, sed una, quia omnium beatitudo est adhaerere uni Deo (*S. Th.* I, q. 108, a. 8).

I answer that, as above explained (Articles 4,7), the orders of the angels are distinguished according to the conditions of nature and according to the gifts of grace. Considered only as regards the grade of nature, men can in no way be assumed into the angelic orders; for the natural distinction will always remain. In view of this distinction, some asserted that men can in no way be transferred to an equality with the angels; but this is erroneous, contradicting as it does the promise of Christ saying that the children of the resurrection will be equal to the angels in heaven (Lk. 20:36). For whatever belongs to nature is the material part of an order; whilst that which perfects is from grace which depends on the liberality of God, and not on the order of nature. *Therefore by the gift of grace men can merit glory in such a degree as to be equal to the angels, in each of the angelic grades; and this implies that men are taken up into the orders of the angels.* Some, however, say that not all who are saved are assumed into the angelic orders, but only virgins or the perfect; and that the other will constitute their own order, as it were, corresponding to the whole society of the angels. But this is against what Augustine says (*De Civ. Dei* xii, 9), that "there will not be two societies of men and angels, but only one; because the Beatitude of all is to cleave to God alone".[7]

Although Aquinas here advocates for the unification of Blessed and angels in a same society, he also maintains an ontological distinction between the two. It is interesting to notice that the Testamentary sources he quotes refer to the 'restoration' of angels only after the Last Judgment. Dante was aware of this distinction, as a passage in *Convivio* II V and his exposition of the doctrine of the glorious body in *Paradiso* XIV clearly indicates.[8] In the *Commedia*, he does not seem to assimilate Blessed and angels before the *parusia*, for the Blessed cannot achieve perfection of Beatitude before resurrection, lest the supreme effect of Incarnation be diminished.[9] The spiritual world Dante de-

---

[7] My Italics.
[8] *Convivio* II, V, 12.
[9] In *Convivio* II, v, 12 Dante followed the theory of restoration: «Dico che di tutti questi ordini

scribes is the world before Apocalypse, in which the two societies are distinct, and we cannot infer what the order of the Empyrean will be afterwards. The picture we contemplate in the *Commedia* shows empty stalls in the ranks of the candid rose, but we do not know anything about the angelic choirs. It is not clear whether at the end of time, the elect sitting in the rose will all become angels. We see a form of paradise that is not eternal but immersed in providential time; the time of angels. In this temporal configuration, Beatrice seats in the third row below the Virgin Mary, indicating that she belongs to the society of the Blessed and not to that of the angels.[10] Upon order by Mary, however, she gains the gift of grace that makes her able to operate as an angel.[11] If Dante still followed the Gregorian theory of restoration in the *Commedia*, then the rose would exist only prior the Last Judgment, for eventually only nine angelic choirs would remain. In this case, Beatrice's attributes would hint at her future

---

si perderono alquanti tosto che furono creati, forse in numero della decima parte: alla quale restaurare fue l'umana natura poi creata», [«I say that of all these orders a certain number were lost as soon as they were created, perhaps one-tenth in number, for the restoration of which part human nature was afterwards created»]. As Nardi suggests, a passage from GREGORY THE GREAT's *Homily in Evangelium XXXIV*, 6 seems to be Dante's source. Nardi traced the origin of the doctrine of the tenth order back to Origen, who argued that the creation of the human species and of the world was a consequence of the fall of the rebel angels. Augustine condemned this idea in his *De civitate Dei* XI, 23. According to Nardi, Dante's formulation shows evidence of some influence of Origen's thought. See BRUNO NARDI, *Tutto il frutto ricolto*, pp. 253-255, and CESARE VASOLI' commentary on *Convivio* II, v, 12, p. 164.

[10] «Ne l'ordine che fanno i terzi sedi, / siede Rachel di sotto da costei / con Bëatrice, sì come tu vedi». [«Below her, in the seats of the third rank, / Rachel and Beatrice, as you see, sit.»]. *Par.* XXXII, 7-9. See also *Inf.* II, 102 and *Par.* XXXI, 67-69. Most commentators agree on the allegorical interpretation of Beatrice as Theology, sitting aside Contemplative Life (Rachel), who represents a necessary condition to the study of theology. Others interpret Beatrice as *preveniens* grace, complementary to the illuminating grace represented by Lucia. These allegorical interpretations tend to limit the meaning of the function of Beatrice. If we take into account the literal Beatrice, she appears here as one of the figures analogically associated to the figure of Mary, as Rachel and the other women mentioned in the Rose.

[11] Other textual occurrences referring to Beatrice's position among the Blessed and the angels are: *VN* 19, 1, [XXVIII, 1] «lo segnore della giustizia chiamoe questa *gentilissima* a gloriare sotto la insegna di quella regina benedetta virgo Maria» [«the God of Justice called this most gracious one to glory under the banner of that blessèd Queen, the Virgin Mary»], where Mary is the Queen of angels; 20, 10, [XXXI, 10]: «Ita n'è Beatrice in l'alto cielo, / nel reame ove gli angeli hanno pace, / e sta con loro [...]», [«Beatrice has gone home to highest Heaven, / into the peaceful realm where angels live; / she is with them;»]; 22, 8, [XXXIII, 8], 21-16: «divenne spiritual bellezza grande, / che per lo cielo spande / luce d'amor, che li angeli saluta, / e l'intelletto loro alto, sottile / face maravigliar, sì v'è gentile», [«became transformed to beauty of the soul, / diffusing through the heavens / a light of love that greets the angels there, / moving their subtle, lofty intellects / to marvel at this miracle of grace»], vv . Furthermore, Beatrice is claimed by the angels in *VN* 10, 18, [XIX, 7], 15-18. In *Paradiso* XXXI, 70-72: «Sanza risponder, li occhi su levai, / e vidi lei che si facea corona / riflettendo da sé li etterni rai», [«I, without answering, then looked on high / and saw that round her now a crown took shape / as she reflected the eternal rays»]. The allusion to the «reflection of the eternal rays» cannot be overlooked as the main characteristic of angels, as illustrated in *Purg.* XV, 16-24.

angelic destiny. If not, and a tenth order is destined to remain even after the Last Judgment, then these attributes would testify to the presence of an analogical relation between angels and Blessed. The likely influence of Aquinas and Bonaventure on this question seems to support the last hypothesis.

Aquinas' articulated views on the necessity to maintain a distinction between separate substances was shared, and taught – although with differences – by other Scholastic theologians, particularly by Bonaventure. Bonaventure, whose vision of creation was more anthropocentric than angelocentric, argued that the creation of human beings did not occur only to replace the fallen angels, but it had its proper place in the divine design of salvation, for otherwise the importance of Incarnation as supreme providential event in sacred history would be diminished. For Bonaventure, the Blessed who did not achieve a degree of perfection (depending on the gifts of grace) sufficient to be integrated in the angelic hierarchies, formed the tenth choir (the tenth drachma).[12] The idea of different degrees of perfection emerges in Dante's hierarchical arrangement of the Blessed in the rose, visually represented by the appearance of the holy souls in the various circles of heaven. As Piccarda Donati explains in *Paradiso* III, the Blessed achieve perfection relatively to their different capacities to receive grace. Dante's spatial arrangement in the Empyrean seems

---

[12] «Et ideo tertia positio est adhuc, quod supra novem ordines angelorum addetur ordo decimus ex his, qui in vita ista non pervenerunt ad tantam meritorum excellentiam, ut exaltentur ad ordines angelorum; sed meritis Christi salvati, decimum tenent gradum sicut pro eorum salute sol iustitiae Christus per decem gradus descendit, quod signatum fuit in miraculo facto Ezechiae, quarti Regum vigesimo. Et ista positio videtur esse satis probabilis, tum propter imperfectionem meritorum, quam habent multi in via, tum etiam propter perfectionem electorum quantum ad numerum, quae erit in Ierusalem superna. Perfectio autem maxima in denario consistit, sicut infra dicetur, quando agetur de decalogo, vel de decimatione. Et ideo positionem istam possumus sustinere, quia eam probabilem reddit decimationis significatio et decimae drachmae inventio et solis per decem gradus descensio et ipsius denarii perfectio». [«And for that reason there is yet a third position», that above the nine Orders of the Angels there is added a tenth Order out of those, who in this life did not arrive at so great an excellence of merits, as to be exalted to the Orders of the Angels; but having been saved by the merits of Christ, hold the tenth step, just as for their salvation Christ, the Sun of Justice, descended through ten steps, which was signified in the miracle worked for Hezekiah, in the twentieth (chapter) of the Fourth (Book) of Kings. And this position seems to be sufficiently probable, both on account of the imperfection of the merits, which many have in the way, and also on account of the perfection of the Elect as much as regards the number, which will be in the Supernal Jerusalem. Moreover the greatest perfection consists in the group of ten, just as will be said below, when the Decalogue, and/or tithing, will be dealt with. And for that reason we can sustain this position of theirs, because the signification of tithing and the finding of the tenth drachma and the descent of the Sun through ten steps and the perfection of the group of ten itself render it probable»]. BONAVENTURE, *Commentaria in quatuor libros sententiarum (Commentaries on the Four Books of Sentences)* II, d. 9, art. un., q. 7, resp. (tome II, page 254 in Quaracchi edition). English translation by Alexis Bugnolo, based on Quaracchi edition, at http://www.franciscan-archive.org/bonaventura/index.html. See also BONAVENTURE, *Collationes in Hexaemeron*, Conference XXI, 16, where the saint talks about the 'hierarchic spirits', and GREGORY THE GREAT, *Homily 24*. Cfr. *Par.* XXX, 43-45, and *Par.* XIV, 40-51.

to reflect Bonaventure's idea of an eternal tenth order, although with a more radical distinction between the heavenly dispositions before and after the Last Judgment. Sitting in the candid rose are not only the Blessed but also the saints, who otherwise should be with the angels. It is possible that Dante assumed that the assimilation of the 'perfect' Blessed would occur only at the Last Judgment, after they recovered unity with their glorious body. Salomon's exposition of the doctrine of the glorious body in *Paradiso* XIV seems to support this hypothesis.[13] Before the end of time, however, by gift of grace, the souls of the rose might perform, as needed, angelic functions.

II. The role of Angels in *Purgatorio*. Beatitudes and the Gifts of the Holy Spirit

An important case in which we find angels operating outside the celestial realm in a structured way is in *Purgatorio*. Generally Dante critics consider them as a sort of 'guardians', performing different functions of governance. The angels however share these functions with other characters performing similar roles. The first among these is Cato, the soul Dante and Virgil encounter as they set foot on the shores of the mountain of *Purgatorio*.[14] Still enfolded in the infernal haze, they discern – old and solitary – the Roman politician who chose suicide as an extreme form of freedom from Caesar's tyranny. Whereas in Inferno all guardians were demons or mythological figures, the first character symbolizing the order of divine justice in *Purgatorio* is a historical controversial figure. According to most commentators, he is the guardian of *Purgatorio* (or of lower *Purgatorio*). His appearance anticipates the figuration of the bright ministers governing the intermediate realm of atonement. Although not an angel, Cato seems to belong to the saved, placed at the liminal space of *Purgatorio*, where the souls wait to begin their ascent. The long stay, the steep cliff, and the rarely opened door that leads to *Purgatorio* proper (*Purg.* IX), are elements that indicate the difficulty of achieving awareness of sin and willingness to atone. Unlike the damned, the souls of *Purgatorio* are

---

[13] *Par.* XIV, 37-60.

[14] For the figure of Cato and its interpretations, see Mario Fubini, *Catone*, in *ED* I, pp. 876-882. Ezio Raimondi sees in Cato a providential figure, and suggests his possible identification with one of the souls Christ freed when harrowed Hell. See Ezio Raimondi, *Ritual and Story*, in *Purgatorio*, A. Mandelbaum, A. Oldcorn, C. Ross eds., Berkeley, University of California Press, 2008, pp. 1-10. For an interpretation of Cato as exemplary figure representing the highest achievement in terms of moral freedom in pre-Christian antiquity see Romano Manescalchi, *Una nuova interpretazione del Catone dantesco*, Critica Letteraria, XXXVI, n. 140, fasc. 3, 2008, pp. 419-446.

not passive subjects of divine punishment administered by inflexible guardians. They freely choose to actively cooperate with grace by voluntarily submitting to the torments of the seven circles. It is a process that encompasses willingness to expiate and determination to perfect virtue in view of Beatitude.

Besides Beatrice in *Inferno* II, Cato is the first figure of the Blessed performing offices outside the Rose. Dante describes him by hinting at angelic attributes. When he speaks, his white hair and white beard move like «honorable plumes» («oneste piume», *Purg.* I, 42), and four holy lights («quattro luci sante») enlighten his face:

> Li raggi de le quattro luci sante
> fregiavan sì la sua faccia di lume,
> ch'i' 'l vedea come 'l sol fosse davante (*Purg.* I, 37-39).
>
> The rays of the four holy stars so framed
> his face with light that in my sight he seemed
> like one who is confronted by the sun.

The description of Cato insists on accumulation of terms underlying the startling luminosity of the ancient hero, here allegorically associated with the vision of the sun: «rays» («raggi»), «lights» («luci», «lume»), «sun» («sol»). Dante's sense of sight is almost blinded, as happens when looking straight into the sun. Similarly to Beatrice, with his luminosity Cato reveals the presence of divine. His ministry and his angelic attributes are significant, for they anticipate the figure of Beatrice in *Purgatorio*, and give us a glimpse of Dante's modes of invention of his angelic characters. Cato is an example of a soul cooperating with the angels in the administration of divine justice. At the higher limit of *Purgatorio*, in Eden, Matelda performs a similar role, one that is complementary and analogous to that of the *Purgatorial* angels.[15]

The ancient Roman introduces Dante and Virgil to the first angel the two poets see in *Purgatorio*. Approaching from the sea surrounding the austral mount, the helmsangel («angelo nocchiero») appears as the anti-type of Char-

---

[15] The interpretation of Matelda is still object of debate among critics. The opinion that Matelda represents the Edenic, preternatural condition, seems to prevail. See FIORENZO FORTI, *Matelda*, in *ED* III, pp. 854-860. For an interpretation of Matelda as representing human prelapsarian state see also CHARLES SINGLETON, *Journey to Beatrice*, Baltimore, The Johns Hopkins University Press, 1977, pp. 184-221. For Matelda's song as «teodia», testifying to Dante's suggestion that lyric and the arts are forms of worship that may contribute to re-acquire a condition similar to that lost with the fall, see PETER S. HAWKINS, *Dante's Testament. Essays in Scriptural Imagination*, Stanford, Stanford University Press, 1999, pp. 159-179. Many scholars have suggested Guido Cavalcanti's *pastorella* as a subtext of Dante's encounter with Matelda. Among these, see TEODOLINDA BAROLINI, *Dante's Poets*, Princeton, Princeton University Press, 1984, pp. 148-152.

on, and sails the souls from the estuary of Tiber to the shores of the sacred mountain. He is the «first minister» of *Purgatorio*, says Cato, alluding to the governing functions of the angels in the realm of atonement (*Purg.* I, 98-99). His wings move the boat, and his brightness, as Cato's, is unbearable to Dante's sight.[16] Cato and the angelic sailor precede the other figures of angels that populate the ordered system of *Purgatorio* at each stage of the pilgrims' ascent.[17] As Virgil and Dante move toward the high walls leading to *Purgatorio* proper, they walk through a little valley where the princes who did not rule wisely await. Accompanied by Sordello, the two poets assist to a sort of sacred representation in which two angels descended from Mary's womb come to chase away the serpent with their swords of fire (*Purg.* VIII, 25-39). In *Purgatorio* IX an angel guards the door at the entrance to the terraces where the souls initiate their atonement, and is symmetrically situated with respect to *Inferno* IX, where another angelic figure opens the gates of the infernal city of Dis. The porter holds the silver and golden keys of the door and a sword with which he draws seven letters 'P' (indicating the seven virtues acquired in the circles) on Dante's forehead, and directs the exceptional visitors as they become spiritually ready to move ahead to the next circle of atonement.[18] As

---

[16] Cato stresses the quality of the angels of *Purgatorio* by referring to them as «ministries» and «officers». Cfr. *Purg.* II, 29-39.

[17] A brief excursus of angelic descriptions in *Purgatorio* illustrates the centrality of the characteristics of luminosity and of the metaphor of the mirror for the act of transmission of vision: «poi, come più e più verso noi venne / l'uccel divino, più chiaro appariva; / perché l'occhio da presso nol sostenne» [«Then he-that bird divine-as he drew closer / and closer to us, seemed to gain in brightness, / so that my eyes could not endure his nearness»], *Purg.* II, 37-39; «ben discernea in lor la testa bionda; / ma nella faccia l'occhio si smarria, / come virtù ch'a troppo si confonda» [«My eyes made out their blond heads clearly, but / my sight was dazzled by their faces-just / like any sense bewildered by excess»], *Purg.* VIII, 35-37; «tal nella faccia ch'io nol soffersi; / e una spada nuda avea in mano / che refletteva i raggi sí ver noi, / ch'io dirizzava spesso il viso invano» [«his face so radiant, I could not bear it; / and in his hand he held a naked sword, / which so reflected rays toward us that I, / time and again, tried to sustain that sight in vain»], *Purg.* IX, 81-84; *Purg.* XII, 88-90; «che è quel, dolce padre, a che non posso / schermar lo viso tanto che mi vaglia?» [«Kind father, what is that against which I / have tried in vain, I said, to screen my eyes?»], *Purg.* XV, 25-26; «Drizzai la testa per veder chi fossi; / e già mai non si videro in fornace vetri o metalli sí lucenti e rossi» [«I raised my head to see who it might be; / no glass or metal ever seen within / a furnace was so glowing or so red as one I saw»], *Purg.* XXIV, 136-138 (Seraphim are also red, see for example the pious fires «fuochi pii» of *Paradiso* IX, 77); «tal che mi vinse e guardar nol potei» [«I could not look at such intensity»]. *Purg.* XXVII, 60. The color white characterizes also the angels of the Rose of Paradise: «Le facce tutte avean di fiamma viva, / e l'ali d'oro, e l'altro tanto bianco, / che nulla neve a quel termine arriva» [«Their faces were all living flame; their wings / were gold; and for the rest, their white was so / intense, no snow can match the white they showed»]. *Par.* XXXI, 13-15. For the connection between whiteness and the notion of light corporeality in Dante, see Silvia Finazzi, *La metafora scientifica*, in *La metafora in Dante*, pp. 171-184.

[18] Commentators have variously interpreted the symbols of the sword, the keys, and the vest of the angel. There is general agreement on the fact that the porter angel performs priestly offices in

the pilgrims approach the summit of *Purgatorio*, angels gradually become brighter. They are luminous presences standing at the liminal space of each terrace, unexpectedly appearing to indicate the way up to the pilgrims. In contrast to the angels of the *anti-Purgatorio*, awesome and terrible figures reminding of the creatures of the Old Testament, the angels of the terraces are evangelical figures who enlighten, sing, and speak.[19]

Angels finally appear in Beatrice's train in Eden. There, they accompany the Blessed in triumph, partly linking this «real» image of Beatrice to the vision described in the *Vita Nuova*, and partly establishing a relation between Beatrice and the angels that reminds us of the figure of the Virgin as Queen of angels. Through the analogy imposed by Beatrice's angelic train, her image indirectly reminds the reader of the very principle and cause of Dante's journey, and sets the encounter within the salvific perspective of the *Commedia*. Angels have here the poetical purpose of making evident the link between Beatrice, who serves Lucy's command, and the Virgin, who first sent Lucy. However, this is not the only function angels have in this crucial episode. They also directly assist the scorned pilgrim by interceding with Beatrice in his favor, and try to mediate between the poet and his ancient flame:[20]

> così fui sanza lagrime e sospiri
> anzi 'l cantar di quei che notan sempre
> dietro a le note de li etterni giri;
> ma poi che 'ntesi ne le dolci tempre

---

administering the sacrament of confession. For an analysis of the symbols of *Purgatorio* IX, see CHARLES ROSS, *The Ritual Keys*, in *Purgatorio*, pp. 85-94.

[19] Romano Guardini notices that in the *Commedia* Dante reproduces the characteristics of the angels of both the Old and the New Testament, differentiating them according to the task and the place in which the angels appear: «non vengono di loro iniziativa, ma in essi viene e agisce Dio. Gli angeli sono messaggeri nel senso tremendo che in certo modo portano Colui stesso che li invia.[...] Nel Nuovo Testamento si attenua la natura tremenda – talvolta, starei per dire, selvaggia – degli angeli [...] nella poesia di Dante [...] essi sono pienamente cristiani. Sono gli "araldi celesti," gli eserciti del Dio vivente, le prime creature del Signore del mondo che si sono decise per la santità [...] sono vicini a quelli dell'amico Giotto. Hanno abbandonato la fissità ieratica, accorrono ed agiscono, ma con l'autorevolezza di creature celesti. Sono belli e beati, ma nella purezza della serietà Cristiana». [«They do not come by their initiative, for in them God comes and acts. The angels are messengers in the terrible sense that in a way they bring the One who sends them [...] In the New Testament the angels' terrible nature – and I would say almost wild- becomes more gentle [...] in Dante's poetry they are fully Christian. They are the "celestial heralds", the host of the living God, His first creatures of the world, who chose sanctity [...] they are close to those by his friend Giotto. They have abandoned their hieratic stiffness, they intervene and act, but with the authority of celestial creatures. They are beautiful and blessed, but with the purity of Christian severity»]. ROMANO GUARDINI, *L'angelo nella Divina Commedia*, in *Studi su Dante*, Brescia, Morcelliana, 1967, pp. 32-41.

[20] *Purg.* XXX, 91-99. After Beatrice's first rebuke (*Purg.* XXX, 73-75), the angels sing «In te, Domine, speravi» *Purg.* XXX, 83.

> lor compartire a me, par che se detto
> avesser: 'Donna, perché sì lo stempre?',
>   lo gel che m'era intorno al cor ristretto,
> spirito e acqua fessi, e con angoscia
> de la bocca e de li occhi uscì del petto (*Purg.* XXX, 91-99).
>
> so I, before I'd heard the song of those
> whose notes always accompany the notes
> of the eternal spheres, was without tears
>   and sighs; but when I heard the sympathy
> for me within their gentle harmonies,
> as if they'd said: «Lady, why shame him so?» –
>   then did the ice that had restrained my heart
> become water and breath; and from my breast
> and through my lips and eyes they issued-anguished.

The song and music of the angels, in which the celestial harmony of the spheres resounds, touches Dante and dissolves his anguish in sighs and tears. He interprets their chant as a prayer to Beatrice to be merciful, and feels reassured. Nevertheless, Beatrice's long reply to the angels reminds them that neither can they intercede, nor can they interfere with her mission. Their ministry is to enlighten the souls so they may find the necessary moral strength to proceed along their path of atonement; no more.

Besides the tasks of these angelic figures of *Purgatorio*, I argue that angels play a structural function in the design of the sacred mountain. The preceding chapter, showed the centrality of the Holy Spirit in Dante's cosmological structure, and the existence of an instrumental relation between the Holy Spirit and the angels, as privileged instruments in carrying out providential dispositions. In *Purgatorio*, by reciting the Beatitudes – the ideal Christian horizon of perfection inspired by the Holy Spirit – angels connect each step of the atonement process to the gifts of the third Trinitarian Person.[21] The il-

---

[21] «Beati pauperes spiritu quoniam ipsorum est regnum caelorum. Beati mites quoniam ipsi possidebunt terram. Beati qui lugent quoniam ipsi consolabuntur. Beati qui esuriunt et sitiunt iustitiam quoniam ipsi saturabuntur. Beati misericordes quia ipsi misericordiam consequentur. Beati mundo corde quoniam ipsi Deum videbunt. Beati pacifici quoniam filii Dei vocabuntur. Beati qui persecutionem patiuntur propter iustitiam quoniam ipsorum est regnum caelorum», [«Blessed are the poor in spirit: for theirs is the kingdom of heaven. Blessed are the meek: for they shall possess the land. Blessed are they that mourn: for they shall be comforted. "The poor in spirit"... That is, the humble; and they whose spirit is not set upon riches. Blessed are they that hunger and thirst after justice: for they shall have their fill. Blessed are the merciful: for they shall obtain mercy. Blessed are the clean of heart: for they shall see God. Blessed are the peacemakers: for they shall be called children of God. Blessed are they that suffer persecution for justice' sake: for theirs is the kingdom of heaven»]. *Matth.*, 5:3-10.

luminating presence of the angels thus reflects the association Dante established in *Paradiso* between angels as ministries of Providence and the Holy Spirit, making them the heralds of the inspiring Love that moves and orders the universe. The path to salvation that angels enlighten passes through the grades of purification represented by the seven terraces of *Purgatorio*. Along this path, the souls must achieve perfection in virtue. Such perfection, however, is beyond human reach. It can be achieved only through the gifts of the Holy Spirit.[22] The science of the gifts is aimed at salvation; it is knowledge that produces perfection in virtuous actions such as those represented by the examples in the terraces.

In the circle of the Proud (*Purg.* XII), Dante introduces the first angelic guide who invites the two characters to follow him to the steps leading to the upper terrace. This «beautiful bird» speaks, but in this canto it is the choir of souls that sings the Beatitude *Beati pauperes spiritu*. The circle of the envious in *Purgatorio* XV stages a sharp contrast between the blindness and obscurity of envy and the brightness of the angel. At this level of ascent, the angel's clarity is still unbearable to Dante, but he is now able to understand how its light propagates. He thus describes the angel's enlightening by engaging in a scientific digression on the optical laws of light reflection.[23] As in the circles of Empyrean, the winged creatures on the terraces are reflecting mirrors of divine light to enlighten the souls and assist them in their progress toward perfection of virtue.

---

[22] The gifts of the Holy Spirit (Wisdom, Understanding, Counsel, Fortitude, Knowledge, Piety, and Fear of the Lord) are connected to grace and the theological virtues. Sanctifying grace is God's supernatural gift the human soul receives with the sacrament of baptism. With sanctifying grace the soul participates in the Holy Spirit, and this participation consists of the first divine gift of Love. Grace evolves in the theological virtues of faith, hope, and love (*caritas*), which, together with the cardinal virtues (prudence, justice, fortitude, temperance) induce in human beings the desire to perfect their souls in the likeness of God. The gifts are the flowering of the highest of theological virtues, *caritas*, and stimulate the pursuit of perfect virtue. The doctrine of the gifts of the Holy Spirit was based on *Isaiah* 11:2-3, where the prophet enumerated them: «spiritus sapientiae et intellectus, spiritu consilii et fortitudinis, spiritus scientiae et pietatis et replebit eum spiritus timoris Domini» [«the spirit of wisdom, and of understanding, the spirit of counsel, and of fortitude, the spirit of knowledge, and of godliness. And he shall be filled with the spirit of the fear of the Lord»]. These gifts are of two types: of intellect and of will. Both types are the sources of supernatural and moral operations as inspired by the Holy Spirit. For a description of the gifts, their role, and distinction from the cardinal virtues, see A. GARDEIL, *Dons du Saint-Esprit*, in *Dictionnaire de Theologie Catholique*, pp. 1728-1781. SAINT PAUL, in I *Ad Romanos* 5:5 stresses the relation between the Holy Spirit and the virtue of caritas: «caritas Dei diffusa est in cordibus nostris per Spiritum Sanctum qui datus est nobis» [«the charity of God is poured forth in our hearts, by the Holy Ghost, who is given to us»]. See also AQUINAS, *S. Th*. II, II^a, q. 23, a. 2. In *Paradiso*, where the souls already possess Beatitude, each within the limits of their different natures, the gifts are no longer needed.

[23] *Purg*. XV, 142-145. See Chapter II, pp. 81-83.

The journey to purification, inspired by the ideal horizon dictated by the Beatitudes, reveals an essential ethical aspect, evidenced by the dialects of sins and virtues visible in the examples the souls see or hear on the terraces. The examples of virtue, taken from the life of Mary and from classical and biblical histories, represent the natural fruits of achieved perfection.[24] They illustrate the effects of the capacity to operate in harmony with the inspiration of the Holy Spirit; a capacity produced by the gifts. Beatitudes and gifts are thus intimately related, and both connected to the angels, revealing their structural importance in *Purgatorio*.

Reading the second canticle of the *Commedia* strictly in terms of classical ethics would not lead to a thorough comprehension of its spiritual structure. While in *Inferno* space is organized according to the Aristotelian division of vices, *Purgatorio* is a spiritual space in which the souls progress along the mystical path of *conversio mori* to regain a condition of prelapsarian innocence. Its topography reflects the Christian doctrine of the capital sins and is modeled upon a path to purification that relies upon cooperation with grace to achieve perfection in virtue.

Anna Maria Chiavacci Leonardi has observed that the Beatitudes in *Purgatorio* are not a marginal aesthetic element added to strengthen the poetic structure of the canticle; rather they express the very essence of Dante's conception of *Purgatorio*. The relation between Beatitudes and Marian virtues, which the scholar suggested Dante could have derived from Hugh and Richard of Saint Victor, is crucial.[25] Chiavacci Leonardi's analysis was an important step forward in the critical assessment of Purgatorial structure. However, it did not solve the issue of the function of the angels in the terraces and their textual relation with the Beatitudes. Similarly to the evangelic values of the

---

[24] The first visual example in each terrace is taken by the life of Mary, who is a model of perfection in that particular virtue. Anna Maria Chiavacci Leonardi has shown the connection between Beatitudes and Marian virtues in ANNA MARIA CHIAVACCI LEONARDI, *Le Beatitudini e la struttura poetica del Purgatorio*, «Giornale Storico della Letteratura Italiana», CLXI, fasc. 513, 1984, pp. 1-29.

[25] The scholar interprets the second canticle as illustrating the passage from the law of the Old Testament – from a God of Justice – to the new law of love of which the *Sermon of the Mount* is the *Magna Charta* (Matthew 5:3-10). The Beatitudes thus reflect the revolutionary moment in the spiritual history that Incarnation had made possible, and constitute the structural backbone of the canticle. The author suggests Hugh of Saint Victor's *De quinque septenis* and Richard of Saint Victor's *Liber exceptionum* as possible sources for Dante. For a comment on the tradition of these texts see ANNA MARIA CHIAVACCI LEONARDI, *Le Beatitudini*, p. 24. Other possible sources of Dante's connection between capital vices and the process of purgation are the manuals of confessors used in Dante's time. See ALISON MORGAN, *Dante and the Medieval Other World*, Cambridge University Press, 1990, pp. 132 ff. See also SERGIO CRISTALDI, *Dalle Beatitudini all'Apocalisse: il Nuovo Testamento nella Commedia*, «Letture Classensi», XVII, pp. 23-67.

Sermon of the Mount, angels are not an aesthetic motive integrating the general poetic framework of the *Purgatorio*, but are a crucial element in Dante's conception of the poetical and geographical space intermediate between *Inferno* and *Paradiso*.

To understand the role of angels and their relation to the Beatitudes it is necessary to analyze the relation between the Beatitudes and the gifts of the Holy Spirit.[26] In the light of the functions of the angels as operators of the Holy Spirit discussed in Chapter III, the relation linking angels and Beatitudes through the gifts testifies to the central role the third Trinitarian person plays both in the purgatorial realm and in the cosmological order of the *Commedia*.

To illustrate this point, I should explore two influential sources of the theological doctrine of the gifts, namely Augustine and Aquinas, who – whether through Victorine re-elaborations or not – are the most likely sources of Dante's inspiration of purgatorial ordering. Augustine connected the spiritual virtues of the Beatitudes to the gifts, and established a tradition that continued in medieval theology, present in the authors of the school of Chartres and in Aquinas. Augustine associated Beatitudes and gifts in several works, particularly in his commentary on the *Sermon of the Mount* (*De sermone domini in monte*), the *Exposition on Psalm* 11, and the fragment of *Letter* 171/A to Massimo. In these texts, the bishop of Hippo connected the seven gifts to the seven Beatitudes, and identified them with the seven steps of purification of the heart.[27]

---

[26] For Thomas Aquinas the gifts of the Holy Spirit derived from the theologian virtues and allow for the perfection of spiritual and ethical virtue, *S. Th.* I-II, q. 68, a. 2-4. For the connection between the Beatitudes and gifts of the Holy Spirit, see also Susanna Barsella, *The Role and Function of Angels in Dante Alighieri's Purgatorio*, paper presented at the Graduate Colloquium on French, Hispanic, German and Italian Literature, Culture and Language, The Catholic University of America, Washington, March 1996. On the same topic, see Nicola Fosca, *Beatitudini e processo di purgazione*, Electronic Bulletin of the Dante Society of America, 2002, http://www.princeton.edu/~dante/ebdsa.

[27] Augustine based his idea of the seven steps of purification to Beatitude on *Isaiah* 11:2-3 and *Matthew* 5:3-10. See also *Luke* 6:20-22. Augustine reduced the original eight Beatitudes present in *Matthew* 5:3-10 to seven in order to match them with the seven gifts. He interpreted the last Beatitude («blessed are they that suffer persecution for justice's sake», *Matth.* 5:10) as a synthesis of the other seven, which he associated to the seven steps of purification: «We may realize that that entire long sermon relies on these seven sentences. For the eighth, which says "blessed are they that suffer persecution for justice's sake", signifies the fire itself, with which the silver is purified seven times». [«de quibus sententiis septem totum illum sermonem prolixum dictum esse animadverti potest; nam octava, ubi dictum est, Beati qui persecutionem patiuntur propter iustitiam, ipsum ignem significat, quo septempliciter probatur argentum»]. Augustine, *Exposition on Psalm XI, 7 (XII)*. The influence of Augustine's doctrine of the gifts is visible in the seven candles opening the Purgatorial procession in *Purgatorio* XIX. For the interpretation of the seven candles as representing the seven gifts based

In explaining his doctrine of the gifts, and commenting on the Beatitudes, Aquinas recalled Augustine's postulated relation between the *Sermon of the Mount* and the spiritual gifts. In the *quaestio* 69 of his *Summa Theologiae* II^a-II^{ae}, he specified the sense of this relation, and affirmed that the Beatitudes represent an ideal state that human beings approach through their capacity to operate in harmony with the inspiration of the Holy Spirit. The gifts are *habiti* that induce such capacity.[28] This ideal state of perfection in virtue represents the ultimate goal of human life; a superhuman condition that reason alone – as guide to virtue – cannot achieve without the gifts. They produce a capacity to discern and operate the good that is essential to complete the spiritual journey leading to celestial Beatitude.[29] The examples of the terraces, together with angelic enlightenment, lead the atoning souls through their purification process, which restores the lost prelapsarian innocence of the Edenic state and opens to them the doors of *Paradiso*.[30] The role Dante assigned to angels as privileged instruments of the Holy Spirit justifies the textual relation between the heavenly ministers and the Beatitudes pronounced in the terraces, and testifies to the importance of the science of gifts in the purgatorial process.

Although Augustine seems to be an important source for the ascending structure of the sacred mountain, Dante based his arrangement of the terraces

---

on AUGUSTINE's *De doctrina Christiana*, see SIMONE MARCHESI, *Dante, Virgilio (e Agostino) di fronte ai sette candelabri: Purgatorio 29.43-57*, Electronic Bulletin of the Dante Society, 2002.

[28] See also AQUINAS, *S. Th.* I^a-II^{ae}, q. 68.

[29] Thus Aquinas effectively explains the Beatitudes in a philosophical key: «beatitudo est ultimus finis humanae vitae. [...] Spes autem de fine consequendo insurgit ex hoc quod aliquis convenienter movetur ad finem, et appropinquat ad ipsum, quod quidem fit per aliquam actionem. Ad finem autem beatitudinis movetur aliquis et appropinquat per operationes virtutum; et praecipue per operationes donorum, si loquamur de beatitudine aeterna, ad quam ratio non sufficit, sed in eam inducit spiritus sanctus, ad cuius obedientiam et sequelam per dona perficimur. Et ideo beatitudines distinguuntur quidem a virtutibus et donis, non sicut habitus ab eis distincti, sed sicut actus distinguuntur ab habitibus». [«happiness is the last end of human life. [...] Again, we hope to obtain an end, because we are suitably moved towards that end, and approach thereto; and this implies some action. And a man is moved towards, and approaches the happy end by works of virtue, and above all by the works of the gifts, if we speak of eternal happiness, for which our reason is not sufficient, since we need to be moved by the Holy Ghost, and to be perfected with His gifts that we may obey and follow him. Consequently the Beatitudes differ from the virtues and gifts, not as habit, but as act from habit»]. AQUINAS, *S. Th.* I^a-II^{ae}, q. 69, a. 1r.

[30] More than properly virtues, the examples from Mary's life illustrate the operations of the gifts. As Aquinas explains referring to meekness, justice, and mercy: «mititas accipitur pro actu mansuetudinis, et similiter dicendum est de iustitia et de misericordia. Et quamvis haec videantur esse virtutes, attribuuntur tamen donis, quia etiam dona perficiunt hominem circa omnia circa quae perficiunt virtutes, ut dictum est». [«Meekness is to be taken as denoting the act of meekness: and the same applies to justice and mercy. And though these might seem to be virtues, they are nevertheless ascribed to gifts, because the gifts perfect man in all matters wherein the virtues perfect him, as stated above (Question 68, Article 2)»]. AQUINAS, *S. Th.* I^a-II^{ae}, q. 69, a. 1.

on the criterion of Love, which is the source of the gifts and of the ordering principle of the cosmos. Love in *Purgatorio* is always imperfect: either directed to the wrong object (Proud, Envy, Wrath), insufficient (Sloth), or directed to the right object but in an excessive way (Gluttony, Lust).[31] Love, and therefore the Holy Spirit and its intervention through the gifts, is the criterion that governs the logic of purgatorial ascent.

Although following Augustine and Aquinas, Dante introduced two variations in the scheme that associates Beatitudes and gifts. He eliminated the Beatitude of the meek, and split into two the Beatitude of justice («Beati qui esuriunt et sitiunt iustitiam, quoniam ipsi saturabuntur», [«Blessed are they that hunger and thirst after justice», *Matth.* V, 6] referring thirst to the avaricious and prodigals, and hunger to the gluttons.[32] Both Dantean Beatitudes refer to justice, but in different ways. The first («*beati qui sitiunt*») is directed to those either too much or too little attached to earthly possessions, and invokes meekness as a renunciation to fighting and instigating fights for the love of deciduous goods. This Beatitude substitutes that of the meek, which Dante reinterpreted in terms of justice. In so doing, Dante likely relied on Augustine, who associated the meek with the holy gift of Piety:

> Rixentur ergo immites et dimicent pro terrenis et temporalibus rebus: *Beati autem mites, quoniam ipsi haereditate possidebunt terram*, de qua pelli non possunt (*De sermone domini*, I, 2).
>
> Therefore, let those who are not meek struggle and contend for earthly and temporal things; but *blessed are the meek, for they shall possess the land by inheritance* from which they cannot be expelled.[33]

Augustine thus places meekness in a political context of contending factions fighting for possession of material goods, with an implicit recall to distributive justice. Those disinherited on earth will inherit in the kingdom of God, for divine justice reverts and corrects mundane iniquity. Following Augustine, Aquinas made the connection with justice more explicit by extending the relation between gifts and Beatitudes to include the seven virtues (cardinal and theological). In his scheme, the gift of Piety is associated with the cardinal virtue of Justice.[34]

---

[31] Virgil's exposition of the doctrine of love spans over two central cantos, *Purg.* XVII, 91, XVIII, 75.

[32] *Purg.* XXII, 6 and *Purg.* XXIV, 151-154 respectively.

[33] AUGUSTINE, *De sermone domini in monte*, I, 2, in AURELII AUGUSTINI, *Opera Omnia* (www.augustinus.it). English translation by J.J. Jepson, AUGUSTINE, *The Lord's Sermon on the Mount*, Ancient Christian Writers, New York, The Newman Press, 1948.

[34] AQUINAS, *S. Th.* II^a-II^{ae}, q. 141, a. 1. Also Augustine had associated the Beatitudes with the

The second half of the evangelical Beatitude (*Purg.* XXIV) refers to the gluttons: «Blessed are they that hunger after justice». Almost as if to justify his reinterpretation, instead of an exact quotation of the verset, Dante explains to his readers its inner meaning, and places it in a very privileged position at the end of the long canto XXIV, introduced by the gentle, perfumed breeze that the angel's wings provoke when removing another 'P' from Dante's forehead:

> E senti' dir: «Beati cui alluma
> tanto di grazia, che l'amor del gusto
> nel petto lor troppo disir non fuma,
> esurïendo sempre quanto è giusto!» (*Purg.* XXIV, 151-154).
>
> And then I heard: «Blessed are those whom grace
> illumines so, that, in their breasts, the love
> of taste does not awake too much desire –
> whose hungering is always in just measure.»

The term «esuriendo», a Latinism referring to hunger, hints at the Beatitude of justice («Blessed are those who hunger and thirst justice») interpreted as the strength needed to dominate passions so as to desire only what is just: the «food» only God provides («quanto è giusto»). By appealing to this strength in virtue, Augustine associated the Beatitude to the gift of Fortitude.

By rearranging the Beatitudes, and insisting on their explicit connections with the gifts that Augustine and Aquinas had authoritatively established, Dante designed the atonement process as consisting of three steps, reflecting the tripartite anagogical process that characterizes angelic hierarchies. As the angels, the souls in *Purgatorio* undergo voluntary punishment (purgation), acquire virtue through the examples (enlightenment), and unify with the divine via the perfection they achieve with the gifts of the Holy Spirit (participation). The angels, celestial instruments of the Holy Spirit, supervise this process, and represent the enlightenment on the science of the gifts the souls receive, thus securing their capacity to achieve that perfection in virtue that is necessary to become citizens of the celestial Jerusalem. The textual link between angels and Beatitudes, which they sing or pronounce, reflects the structural importance of angelic presence in *Purgatorio*. This presence is coherent with the angels' function as ministries of the Holy Spirit, which finds its origin in the providential structure of Dante's cosmos.

---

cardinal and theologal virtues, see for example Augustine, *Letter* 171/A, 2. Cfr. also AUGUSTINE, *De sermone* II, 87.

III. The Heavenly Messenger (Inferno IX, 79-103)

Inferno is the canticle of obscurity and clamor; the antithesis of harmonious and luminous *Paradiso*. Yet, *Inferno* is ordered chaos, obeying in its structure to a division dictated by divine justice, in which even the ingnavious angels find precise collocation.³⁵ By condemning all forms of vicious and malicious deviations from the good, infernal justice is anchored in the providential principle that moves and governs the cosmos. In the allegorical metaspace of *Inferno*, divine justice applies equally to the tormentors and the tormenters, for it compels demons and mythological monsters to enforce the law they ignored or against which they rebelled. Two lights illuminate the oppressive darkness of Lucifer's pit: the fire of the great spirits in the castle of wisdom in the first circle; and the much brighter light of the figure sent from heaven (the *messo*) to help Virgil and Dante enter the city of Dis (*Inf.* IX). This figure, which represents external divine intervention in the exceptional circumstance of Dante's voyage, resolves one of the most frightening impasses Virgil and Dante face in their descent.

The sequence of events anticipating the arrival of the heavenly messenger illustrates the interlacing pagan and Christian symbolisms that characterizes this figure, and reflects the ideological texture of the episode. Frightened by the Erynnes's invocation of Medusa, Virgil intimates Dante to cover his eyes, and presses his hands on Dante's, lest Medusa's petrifying glance would keep him captive in hell forever, held by the enchantment of the ancient divinity, punished for looking directly at what one cannot see without the mediation of Perseus's mirror. In the salvific logic of the *Commedia* this mirror assumes the sembiance of the angel, abating the malignant effect of Medusa's glance. As the protagonist plunges in total blindness, the author interrupts the narrative continuity of the episode to introduce an hermeneutical address to the readers (*Inf.* IX, 61-63), which precedes the arrival of the messenger where the juxtaposition of pagan and Christian elements intesifies. In blindness Dante perceives a loud sound, wind, and tremor preannouncing the arrival of the *messo*, and suggesting to the reader a parallel with the earthquake that accompanied Christ's Harrowing.³⁶ Only at this point Virgil allows Dante to look at the figure approaching the walls of Dis. The simile of the

---

³⁵ On Dante and the neutral angels, see JOHN FRECCERO, *The Neutral Angels*, in *Dante*, pp. 110-118.
³⁶ *Inf.* VIII, 124-127. Cfr. *Inf.* XII, 34-45. Similar effects are also described in *Vita Nuova* XXIII, when the poet models the vision of Beatrice's death on the description of Christ's death in the Gosples of Matthew and Luke, and in John's *Apocalypse*.

frogs derived from Ovid, immediately precedes the entrance on stage of the *messo*.[37]

> Come le rane innanzi a la nimica
> biscia per l'acqua si dileguan tutte,
> fin ch'a la terra ciascuna s'abbica,
>   vid'io più di mille anime distrutte
> fuggir così dinanzi ad un ch'al passo
> passava Stige con le piante asciutte (*Inf.* IX, 77-82).
>
> As frogs confronted by their enemy,
> the snake, will scatter underwater till
> each hunches in a heap along the bottom,
>   so did the thousand ruined souls I saw
> take flight before a figure crossing Styx
> who walked as if on land and with dry soles.

The *messo* («un») walks over the muddy waters of Styx, again recalling the figure of Christ, before opening the gates with a light touch of the *verghetta* (little wand) he holds in his right hand.[38] These actions, and the light the *messo* emanates, help Dante – and the reader – to immediately realize that he belongs to the celestial society of the Empyrean. Yet, the deliberate figural ambiguity that surrounds him suggests Dante's intent to prevent a straitforward identification of this figure with an angel. He needs to be both, retaining the meanings that his association with the pagan Mercury-Hermes imply, and are so crucial to the interpretation of the episode.

The pagan and Christian elements interlacing in the episode that span two cantos (*Inf.* VIII and IX), have led commentators to identify the *messo* either with Mercury or with an angel. The former hypothesis originated with Pietro di Dante (1340), the latter with Giovanni Boccaccio (1374). Among the early commentators, only Bernardino Daniello (1568) interpreted the messo as a synthesis of both, stressing the importance of the co-presence of both mercurial and angelic attributes.[39] Most commentators after the sixteenth century

---

[37] *Metam.* VI, 370-381.

[38] *Matth.* 14, 26: «Et videntes eum super mare ambulantem, turbati sunt, dicentes: Quia phantasma est. Et prae timore clamaverunt», [«And they seeing him walk upon the sea, were troubled, saying: It is an apparition. And they cried out for fear»]. Cfr. *Mk* 6,45-52; *Joh.* 6, 16-21. These Christian references induce the reader to associate the *messo*, analogically, to the figure of Christ. Like the Redemptor, the messenger submits the pagan world to higher Intelligence, and defeats the evil forces obstructing Dante's way.

[39] This hypothesis had little success in the exegetical history of the *messo*, for the majority of

shared the angelic interpretation, while the hypothesis of Mercury prevailed among thirteenth and fourteenth-century interpreters such as Benvenuto (1375) and Serravalle (1416).[40] Other hypotheses have included Hercules, Aeneas, Henry VII, the Veltro, and Christ, but none of these readings has been satisfactorily defended. The attributes and function of the *messo* suggest Dante's intentional combination of elements pertaining to both angelic Intelligences and Mercury, the chthonic messenger. Heroes such as Aeneas and Hercules, nonetheless, belong to the mythological texture of cantos VIII and IX, and frame their literary meaning.[41]

The *messo*'s intervention in aid of the two poets is coherent with the providential design of Dante's angelology, and reflects the importance of angelic functions in the narrative pattern of the *Commedia*. I maintain that this polysemic character, charged with a plurality of symbolic references and object of much debate among Dante scholars, finds a coherent explanation if considered in the light of Dante's general design of angelic ministries as instruments

---

commentators tended to read this figure in isolation rather than in connection with Dante's angelology. The first to suggest that the *messo* was an angel with the appearance of Mercury was Bernardino Daniello in his commentary to *Inferno* IX, 88-89: «Né è dubbio, che il Poeta in questo luogo hà voluto imitar Virgilio, si come quasi in tutti suol fare, quando egli nel quarto libro finge, che Mercurio scenda in terra mandato da Giove ad Enea, per rimuoverlo dall'amor di Didone, & essortarlo à venir in Italia; ma il nostro Poeta come Cristiano fà, che l'angelo tegna la persona di Mercurio, mandato da Dio con la verga, ch'è l'auttorità ad aprir la porta della città di Dite». [«There is no doubt that the poet wanted here to imitate Virgil – as he almost always does – when in the fourth book [of the Aeneid] he invents that Mercury descends on earth sent by Jupiter to Aeneas to dissuade him from loving Dido, and to exhort him to go to Italy. Our poet, however, being Christian, makes the angel assumes the identity of Mercury, sent by God with the wand, who has the authority to open the gates of the city of Dis»]. Daniello introduced two elements crucial to the interpretation of the episode of Dis: a principle of *inventio* (the imitation of Virgil), and a criterion of mythological exegesis (*integumentum*). His interpretation, however, was generally overlooked. The perception of a dual nature in the figure of the *messo* reappeared in Nicola Zingarelli, Paul Renucci – who interpreted the messo in the light of Dante's approach to mythology –, and Stephen Bemrose, who sees in the Mercurial shell of the *messo* an assimilation of pagan motives in Dante's angels. See NICOLA ZINGARELLI, *I tempi, la vita e le opere di Dante. Storia letteraria d'Italia*, Milano, Vallardi, 1939, p. 924; PAUL RENUCCI, *Dante disciple et juge du monde gréco-latin*, Paris, Les Belles Lettres, 1958, pp. 214-217, and STEPHEN BEMROSE, *Dante's Angelic Intelligences*, pp. 140-142; 171-178.

[40] Serravalle rejected the identification of the *messo* with an angelic creature because «hoc alienum est ab intentione auctoris, quia ipse auctor non ponit angelos in *Inferno*», [«this was not in the intention of the author, who does not place angels in Hell»]. For the ancient commentator, the *messo* would not represent grace but only heavenly help. The belief that angels cannot descend into Hell led also Giovanni Pascoli to identify the *messo* with Aeneas.

[41] The term syncretism is used here in a broad sense, meaning the juxtaposition of different religions and philosophies in a Christian perspective. Dante's use of mythology can be seen in the context of the medieval reinterpretation of pagan culture aiming at retrieving and integrating the classical heritage into Christian thought. For the syncretic character of the heavenly messenger see SUSANNA BARSELLA, *The Mercurial Integumentum of the Heavenly Messenger (Inferno IX, 79-103)*, «Letteratura Italiana Antica», IV, 2003, pp. 371-395.

of divine order. His assistance in *Inferno*, which Beatrice guaranteed Virgil, not only testifies to the providential meaning of Dante's descent into hell, but also is crucial to define Dante's criteria of literary invention and hermeneutics.[42] His Mercurial appearance is not only a figurative mark that makes the *messo* homogenous with the infernal imagery displayed in hell, but also is critical to the meaning of the episode within literary and hermeneutic perspectives. While it is functional to characterize Dante's descent as 'virtuous', contextually the messenger offers the key to its interpretation.

The *messo* is a sign of the divine disposition that allowed Dante to enter the circular path from a state of moral confusion to a renovated spiritual condition through the visionary experience of the divine. Beyond this, however, Dante's sapient amalgamation of mythological and Christian elements in this hybrid figure serves to figuratively connect the poets' entrance into Dis to the literary topos of infernal descents. The angel-Mercury represents Dante's dialogue with Christian and pagan traditions in an episode that stages the unprecedented descent of a living being (and author) into the depths of Tartarus.

Rhetorical reasons also induced Dante to deliberately combine angelic and pagan elements in the figure of the messenger. Intervening to restore order in hell, he must interact with the language, the rhetoric, and the symbology that belonged to Dante's mytho-Christian universe; one dominated by classical pagan imagery. The *messo*'s Mercurial appearance and language are a rhetorical artifice necessary to make his Christian message effective and understandable within a pagan context. Dante's *contaminatio* of pagan and Christian motives appears – almost paradoxically – to respond to a principle of realism that transforms the *messo* into a literary specimen of what Panofsky called the medieval «principle of disjunction», whereby artists clad Christian contents in classical forms, and vice versa.[43] When this happened, the symbols and the imageries of Christian and pagan worlds became interchangeable, and widened the range of rhetorical devices available to medieval authors.[44] Dante could thus utilize all the elements per-

---

[42] Most likely Virgil refers to her in *Inferno* IX, 8 when he reassures Dante that someone had guaranteed assistance to them – «Tal se n'offerse».

[43] «Wherever in the high and later Middle Ages a work of art borrows its form from a classical model, this form is almost invariably invested with a non-classical, normally Christian, significance; wherever in the high and later Middle Ages a work of art borrows its theme from classical poetry, legend, history or mythology, this theme is quite invariably presented in a non classical, normally contemporary, form». ERWIN PANOFSKY, *Renaissance and Renascences in Western Art*, Copenhagen, Russak & Company Ltd., 1960, p. 84.

[44] Christian content in pagan form emerged in many fields, mostly in figurative art. The artists from the eleventh to the thirteenth centuries «subjected classical originals to an *interpretatio* Christi-

taining to the two cultural universes and make them converge into a new meaning.

a. *The Messo's Veil*

Medieval exegesis established a relation of continuity between classical and Christian cultures, and made them interact. Their imageries were seen in a figural relation, where the pagan era prefigured the Christian one. This intellectual attitude was founded on the belief that God manifested Himself to men even before Revelation, so that pagan divinities were but the signs of God's intervention in the world.

The Christianization of mythical materials was the result of a complex and controversial process that lasted centuries, and the so called 'pagan humanism' of medieval schools, such as those of Chartres and Saint Victor, greatly contributed to make it an essential part of the medieval cultural *koiné*. Critical to enlighten the presence of this humanistic attitude toward the mythological past is Dante's hermeneutical tercet in *Inferno* IX, 61-63:[45]

> O voi ch'avete li 'ntelletti sani,
> mirate la dottrina che s'asconde
> sotto 'l velame de li versi strani (*Inf.* IX, 61-63).
>
> O you possessed of sturdy intellects,
> observe the teaching that is hidden here
> beneath the veil of verses so obscure.[46]

---

ana, the term Christiana here meant to include, in addition to that which can be found in Scripture or in hagiology, all kinds of concepts that come under the heading of Christian philosophy», ERWIN PANOFSKY, *Renaissance*, p. 83. A famous example of this tendency is the Nicola Pisano's transformation of Hercules into the cardinal virtue of Fortitude in the pulpit of Pisa Cathedral (c. 1260).

[45] As in *Purgatorio* IX, the address precedes the description of the angel at the door of *Purgatorio* proper. As John Freccero argues, the tercet relates to the theological allegory of Medusa, and signals not just narrative but a hermeneutical crisis: «The address to the reader is not thus a stage direction, but an exhortation to conversion, a command to await the celestial messenger so that we, like the pilgrim, may "trapassare dentro"». JOHN FRECCERO, *Dante: The Poetics of Conversion*, Harvard, Harvard University Press, 1986, pp. 134-135. According to Franke, *Inferno* IX represents a hermeneutical rite of passage: «Hermeneutics comes forward as the only means by which both Dante's journey as protagonist and his poem can make genuine progress, though the effect of hermeneutics is precisely to collapse these two together, revealing the one as the meaning behind the other». Amilcare Iannucci read the episode as a narratological impasse, where Virgil's rhetorical *auctoritas* revealed itself insufficient for proceeding into a region he had never described. See WILLIAM FRANKE, *Dante's Hermeneutic Rite of Passage, Inferno IX*, «Religion and Literature», XXVI, 2, 1994, pp. 1-26; see also of the same author, *Dante's Interpretive Journey*, Chicago, University of Chicago Press, 1996; AMILCARE A. IANNUCCI *Virgil's Erichthean Descent and the Crisis of Intertextuality*, «Forum Italicum», XXXIII, n. 1, 1999, pp. 13-26.

[46] Dante here paraphrases his own definition of the allegory of poets given in *Convivio* II, I, 2-3: «[...] le scritture si possono intendere e deonsi esponere massimamente per quattro sensi. L'uno si

The term «veil» (*velame*), used to indicate the rhetorical covering of the doctrine this tercet hides, echoes the definition of *integumentum* (cover, veil), which is an exegetical criterion for the interpretation of pagan mythology that medieval theologians such as Bernardus Silvestris developed in the Middle Ages.[47] The definition of *integumentum* can be found in Silvestris's Commentary to the book VI of the *Aeneid*, and is associated with the interpretation of Aeneas' descent into Avernus accompanied by the Sybil.[48]

Integumentum vero est *genus demonstrationis* sub fabulosa narratione *veritatis* involvens *intellectum*, unde et *involucrum* dicitur (*In Aeneid, Proemium*, 44-46).

---

chiama litterale e questo è quello che non va oltre a ciò che suona la parola fittizia, si come né le favole dei poeti. L'altro si chiama allegorico e questo è quello che si nasconde sotto 'l manto di queste favole, ed è una veritate ascosa sotto bella menzogna». [«[...] it should be explained that texts can be interpreted, and must therefore be elucidated, principally in four senses. The first is called literal: this is the sense conveyed simply by the overt meaning of the words of a fictitious story, as, for example, in the case of fables told by poets. The second is called allegorical: this is the sense concealed under the cloak of these fables, and consists of a truth hidden under a beautiful lie»].

[47] J. Freccero suggests Paul's *Corinthians* II, 3:12-16 as another relevant source for the interpretative meaning of the Biblical motive of the veil in the quoted tercet. JHON FRECCERO, *Dante*, p. 122.

[48] BERNARDUS SILVESTRIS, *Commentum Bernardi Silvestris super sex libros Eneidos Virgilii*, W. Riedel ed., Gryphiswaldae (Greifswald), Abel, 1924, p. 2, English translation: *The Commentary on the First Six Books of the Aeneid of Vergil Commonly Attributed to Bernardus Silvestris*, J.W. Jones and E.F. Jones eds., Lincoln, University of Nebraska Press, 1977. The School of Chartres was one of the twelfth – century centers of Neo-Platonism with proto-humanistic characteristics. This school significantly contributed to the recovery of the pagan heritage by elaborating criteria for an allegorized Christian interpretation of mythology and classical texts. Bernardus Silvestris was connected to this school. The technical term *integumentum* alludes to the narration of religious beliefs in mythological form and originally indicated Plato's use of narratives to expound philosophical truths. William of Conches used the word *integumentum* to describe mythology as the fabulous cover of philosophical, moral, or theological truths. Bernardus Silvestris distinguished two methods, one applied to mythology and another to the Scriptures. He described integumental exegesis in both his main commentaries, to Virgil's *Aeneid* and to Capella's *De Nuptiis*. Marie-Dominique Chenu observed that the use of the term *integumentum* reveals the influence of Platonism on twelfth – century theology. For a critical discussion on the medieval concept of *integumentum* and its philosophical and theological implications, see MARIE-DOMINIQUE CHENU, *Involucrum. Le mythe selon les théologiens médiévaux*, «Archives d'Histoire Doctrinale et Litteraire du Moyen Âge», XXII, 1955, pp. 75-79. Robert Hollander has stressed the importance of *integumentum* especially in reference to medieval interpretations of Virgil. See ROBERT HOLLANDER, *Allegory in Dante's Commedia*, Princeton, Princeton University Press, 1969, pp. 101-103. Jean Pépin also observed that Dante probably had in mind Bernardus' definition of *integumentum* in *Convivio* II, I, 3, where he defines the allegory of the poets with words close to Silvester's. See JEAN PÉPIN, *Dante et la tradition de l'allégorie*, Paris, Librairie J. Vrin, 1970, pp. 66-67. These scholars, however, did not notice the relation between Bernardus Silvestris and *Inf.* IX, 61-63. See also GIUSEPPE MAZZOTTA, *Dante, Poet of the Desert: History and Allegory in the Divine Commedia*, Princeton, Princeton University Press, 1979, pp. 227-274. For the notion of medieval allegory see also ERNST R. CURTIUS, *European Literature and the Latin Middle Ages*, Princeton, Princeton University Press, 1990 (1948), pp. 108-113, and ERWIN PANOFSKY, *Renaissance*, pp. 71-73. For the influence of the School of Chartres see WINTHROP WETHERBEE, *Platonism and Poetry in the Twelfth Century: The Literary Influence of the School of Chartres*, Princeton, Princeton University Press, 1972.

Integumentum is indeed *an explanation of a truth* whose meaning is wrapped in a fabulous narration, and for this reason is also called *covering*.

The italics in the quotation above stress the parallel between Silvestris' definition and Dante's tercet. The term «dottrina» (doctrine) indicates a body of teaching concerning the truth, («genus demonstrationis [...] veritatis»), while the word «velame» seems to almost literally translate Bernard's «involucrum». The principles of *integumentum* and disjunction seem to provide the hermeneutical keys for evaluating the angelic and mythical figure that opens the gates of Dis. Together they effectively capture the ideological framework and the cultural context in which Dante conceived of this figure.

b. *The Poetic Impasse*

The entrance to Dis marks a crucial moment in the diegetical pattern of the *Commedia*, for it represents not only the author's entrance in a literary unexplored world, but it also defines the ethical character of Dante's fictional descent. The syncretic figure of the *messo* is fundamental to individuate the system of relations Dante establishes between his own experience and the Christianized heroic traditions of classical descents.

The critical moment triggering the dynamics of canto IX occurs at the end of canto VIII (115-17) when the demons and monsters guarding the entrance to the lower part of hell shut the doors in Virgil's face. Virgil's dismay at the demons' tracotance throws Dante into deep anguish, worried about his fate. The Latin poet reassures him by anticipating the arrival of someone («tal») who will force their way into the lower infernal circles. He assures Dante that the demons must yield to divine authority, as they had already done when Christ harrowed hell.[49] Moreover, Virgil guarantees Dante of his competence as his guide by mentioning his previous descent into Tartarus, when the sorceress Erichtho sent him to summon the soul of one of Pompey's soldiers.[50] In

---

[49] Virgil mentions his own previous descent into Tartarus in *Inf.* IX, 22-27, and Christ's harrowing of Hell in *Inf.* VIII, 124-126: «Questa lor tracotanza non è nova; / che già l'usar a men secreta porta, / la qual senza serrame ancor si trova». [«This insolence of theirs is nothing new; / they used it once before and at a gate / less secret – it is still without its bolts»]. Andrea Lancia, the author of the *Ottimo Commento* (1333), already noticed the relation between *Inf.* VIII, 124-26 and *Inf.* IX, 22-27. See Ottimo's commentary to *Inf.* VIII, 124-126.

[50] «Ver è ch'altra fïata qua giù fui, / congiurato da quella Eritón cruda / che richiamava l'ombre a' corpi sui. / Di poco era di me la carne nuda, / ch'ella mi fece intrar dentr'a quel muro, / per trarne un spirto del cerchio di Giuda». [«But I, in truth, have been here once before: / that savage witch Erichton, she who called / the shades back to their bodies, summoned me. / My flesh had not been long stripped off when she / had me descend through all the rings of Hell, / to draw a spirit back

reassuring the frightened Dante, Virgil thus indirectly connects Christian and pagan descents, introducing the main underlying theme of the episode that culminates with the *messo*'s intervention. These descents have a common trait in the rebellion of infernal creatures, vanquished by a superior supernatural power.[51] The figural link between sacred and profane descents hinges on the modes of facing and defeating the rebellion of the infernal guardians.[52] This is the main operation the *messo* performs in his mission to restore the order threatened by the demons.

As seen in the previous paragraph, before the iron walls of Dis the text intensifies cross references to mythological and Christian elements, thus preparing the readers to perceive and decode the double nature of the messenger. Demons «fallen from Heaven» and mythological monsters, such as the Erinnyes and the Gorgon, crowd the battlements. They inveigh against those who attempt to violate the regions of Tartarus, and try to reassess their dominion over the inner part of hell.[53]

The Erinnyes, the ancient goddesses of vengeance who in pagan mythology intervened whenever someone overstepped the limits imposed by cosmic order (as Virgil and Dante intend to do), cast an angry apostrophe against the

---

from Judas' circle»] *Inf.* IX, 22-27. The reference is to Book VI of Lucan's *Pharsalia*, where Pompey's son asks the sorceress Erichtho to divine the outcome of the battle of Pharsalus. See LUCAN, *Pharsalia* VI, 640-645. Ottimo's extensive commentary on *Inf.* IX, 22-24 also illustrates the similarities between Dante and Lucan's descriptions of the entrances to the underworld.

[51] The topos of defeated rebellion in hell recurs in Virgil, Lucan, and Statius, as well as in the apocryphal gospel of Nicodemus, a major source of Christ's harrowing of hell. In *Aen.* VI, 388-407 and 417-423, Charon and Cerberus try to oppose Aeneas' descent. In *Phar.* VI ,625-751 and *Theb.* IV, 443-518 respectively, Erichtho and Tiresias invoke the shadows of Tartarus and at first they do not receive answer. They need to rebuke Hades' divinities to make them comply with their wishes.

[52] There is no notice of the demons' resistance at Christ's Harrowing in the *New Testament* (Acts 2:27; 2:31; Peter's *Epistle* 3:19-20; 4:6), which only indirectly refers to the event. This 'rebellion pattern', typical of pagan descents, is however present in other religious texts. Iannucci suggests the apocryphal *Gospel of Nicodemus*, the Franciscan *lauda*, used by Perugian Disciplinati as *Devozione* for Holy Saturday morning – the same time as Dante's *statione* before the city of Dis – and Vincent of Beauvais' paraphrase of this apocryphal Gospel in his *Speculum maius*. Other possible sources of the harrowing are medieval mystery plays. See AMILCARE A. IANNUCCI, *The Harrowing of Dante from upper Hell*, in *Inferno. Lectura Dantis*, A. Mandelbaum, A. Oldcorn, C. Ross, eds., Berkeley, University of California Press, 1998, pp. 123-135.

[53] Medusa and Tesiphon are also at the entrance of *Orcus* in *Aeneid* VI, 273-89, together with other mythological monsters. Iannucci noted the presence of these classic sources at the crucial moment of the entrance in lower hell. See AMILCARE A. IANNUCCI, *Virgil's Erichthean Descent and the Crisis of Intertextuality*, pp. 13-26. Book VI of the *Aeneid* is the main subtext of the entrance into Dis, not only for the strong presence of Virgilian materials but also because Lucan and Statius partly modeled their descents into Tartarus on Aeneas' journey into Avernus. For a detailed list of Dante's quotations from *Aeneid* in *Inf.* IX, see ROBERT HOLLANDER, *Le opere di Virgilio nella Commedia*, in *Dante e la «Bella scola» della poesia. Autorità e sfida poetica*, Ravenna, Longo (Studi Danteschi), 1993, pp. 271-274.

two poets, and curse those who dare to descend into Tartarus before them.⁵⁴ They refer to Theseus and Pirithous, held captive in Hades for their attempt to kidnap Persephone. The Erinnyes' main target, however, is Hercules, who freed Theseus and Pirithous and infringed the laws of fate that even Jupiter was forced to obey. The Erynnes allude to Hercules' twelfth labor, when he not only freed the captive heroes, but also captured Cerberus, the three-headed guardian of Hades.⁵⁵ The reference to Hercules' twelfth labor is particularly significant for it reappears in the *messo*'s rebuke to the demons in *Inf.* IX, 91-99, and therefore encloses the entire episode, casting an indirect figural relation between ancient and modern chthonic heroes.⁵⁶

Hercules, Theseus, and Pirithous are connected to another important model of virtuous descent: that of Aeneas. In *Aeneid* VI, Aeneas recalls these heroes when he prays to have access to the cave of Averno: «si fratrem Pollux alterna morte redemit / itque reditque uiam totiens. quid Thesea, magnum / quid memorem Alciden ...?» [«time and again traversing the same road up and down; / if Theseus, mighty Hercules – must I mention them?»] (*Aeneid*, VI 121-123).⁵⁷ Charon at the crossing of Styx cast an invective against the same heroes similar to that of the Erynnes:

> nec uero Alciden me sum laetatus euntem
> accepisse lacu, nec Thesea Pirithoumque,

---

⁵⁴ «"Vegna Medusa: sí 'l farem di smalto" / dicevan tutte riguardando in giuso: / "mal non vengiammo in Teseo l'assalto"», [«"Just let Medusa come; then we shall turn / him into stone," they all cried, looking down; / "we should have punished Theseus' assault"»], *Inf.* IX, 52. The most ancient commentaries to the Erinnyes' verses witness the version of the myth known in Dante's time. According to this medieval version, Theseus and Pirithous together make an oath to marry only daughters of Zeus. They descend to Tartarus but fail to kidnap Proserpina as a spouse for Pirithous. As a punishment, they are put in chains and left under the Erinnyes' surveillance. According to the prevalent variant of the myth, Pirithous escaped and asked for Hercules's aid. The hero then descended to Tartarus to free Theseus. On his way he beat and enchained Cerberus, who had tried to stop him. When he returned to the upperworld with Theseus he brought the infernal guardian with him.

⁵⁵ This version of the myth, where Hercules goes to Hell to rescue his friend of his own will and captures Cerberus, was more compatible with medieval allegorical reading of the pagan descent to the underworld.

⁵⁶ In Neo-Platonic exegesis, Theseus and Pirithous represented wisdom and eloquence respectively, while Hercules represented virtuous strength. According to this interpretation, and in particular to Bernardus Silvestris', the descent to hell represented the itinerary the soul followed through the temporal world in order to return fortified to God. Strength, wisdom, and eloquence (the ability to retell the undergone experience) are necessary to complete this itinerary. See Bernardus Silvestris' commentary on *Aen.* VI, 392 and 393. Hercules and Theseus were strictly associated and their figures sometimes even confused. In particular, «In Bernardus, Orpheus, Theseus, and Hercules as types of Aeneas descending virtuously also master the underworld». JANE CHANCE, *Medieval Mythography. Volume 1*: From Roman North Africa to the School of Chartres. AD 433-1177, Gainesville, University Press of Florida, 1994, p. 458. See also GIORGIO PADOAN, *Ercole, ED* II, pp. 717-718 and ID., *Teseo*, in *ED* V, pp. 595-596.

⁵⁷ English translations are by Robert Fagles, *The Aeneid*, New York, Penguin, 2006.

> dis quamquam geniti atque inuicti uiribus essent.
> Tartareum ille manu custodem in uincla petiuit
> ipsius a solio regis traxitque trementem;
> hi dominam Ditis thalamo deducere adorti.
>
> *(Aeneid*, VI, 392-97)

> The law forbids me to carry living bodies across
> In my Stygian boat. I'd little joy, believe me,
> When Hercules came and I sailed the hero over,
> Or Theseus, Pirithous, sons of gods as they were
> With their high and mighty power. Hercules stole
> Our watchdog – chained him, the poor trembling creature,
> Dragged him away from our king's very throne! The others
> Tried to snatch our queen from the bridal bed of Death!

Here, as done by the Erynnes in the *Commedia*, Charon directs his rage against the heroes' act of violence against the laws of Hades, for their violation affirms the superiority of the power of the solar gods over the princes of Erebus. Like Aeneas, the boatman also insists on the heroes' capacity not only to descend but also to return from Hades; a trait that ties these heroes and makes them victorious figures of virtue. Their capacity to come back from the realm of the dead is a sign of their superior qualities, but also of the existence of a more powerful force that allows them to infringe the laws of infernal divinities. In the version of the myth underlying Virgil's narrative in book VI, the heroes indeed do need the assistance of a guide and the protection of the herald of Jupiter, Mercury. The power assisting Dante is however superior to that of the ancient gods, for he will be able to penetrate the deepest parts of the infernal abyss, while the classical heroes never descended into the depths of Tartarus.

The classical topos of the descent in the underworld was of particular relevance in the Christian reinterpretation of pagan mythology.[58] Neo-Platonic

---

[58] In Christian Neo-Platonism the descent to Hades was seen as an *exitus a Deo*, a necessary passage through the earthly world of false good in order to ascend to the contemplation of the true Good (*reditus in Deum*). «Descensus autem ad inferos quadrifarius est: est autem unus naturae, alius virtutis, tertius vitii, quartus artificii. Naturalis est nativitas hominis [...] sed iste omnium communis est. – Est autem alius virtutis qui fit dum sapiens aliquis ad mundana per considerationem descendit, non ut in eis intentionem ponat sed ut eorum cognita fragilitate eis abiectis ad invisibilia penitus se convertat et creaturarum cognitione creatorem evidentius agnoscat. Secundum hoc Orpheus et Hercules qui sapientes dicti sunt descenderunt. – Est vero tertius vitii, qui vulgaris est, quo ad temporalia devenit⟨ur⟩ itaque intentio tota in eis ponitur eisque tota mente servitur nec ab eis animus amplius dimovetur. Taliter Euridicem legimus descendisse». [«There are four types of descents into the underworld: natural, virtuous, vicious, and artificial. The natural one is the human birth [...] and this is

exegetes interpreted the heroes' descents as prefigurations of Christ's Harrowing of Hell,[59] and as allegories of the virtuous soul's itinerary to God through earthly experience (the Augustinian *regio dissimilitudinis*), a step in the anagogical ascent to heaven.[60]

c. *The Virtuous Descent*

The above-mentioned heroes, and their literary sources, indicate the models with which Dante dialogued and delimited the ideal space of these authoritative texts.[61] These semi-gods (with the exception of Orpheus) were all exemplary figures of virtuous descents, particularly in the *Aeneid* commentary tradition. In his comment to *Aeneid* VI Bernardus Silvestris associated Mer-

---

common to all. There is however another virtuous descent a wise man undergoes in order to know the earthly goods. Not because he desires them, but so that, after having known their fragility and rejected them, he may completely dedicate himself to the heavenly goods and, by knowing the creatures, he may improve his knowledge of their creator. Orpheus and Hercules, who are considered wise men, descended in this way. There is indeed a third, vicious, descent, which is the ordinary one, whereby one attains at the secular goods and places in them all his desires. His whole mind is a slave to them, and his soul cannot be detached from them. We know that Eurydice descended in this way»]. BERNARDUS SILVESTRIS, *Commentum Bernardi Silvestris super Sex Libros Eneidos Virgilii*, VI 50 r 1. The fourth descent is the one by «nigromantico artificio».

[59] Both Virgil and Apollodorus (*Bibliotheca* II, 5, 11-12) refer to the same source for this version of Hercules's descent. Dante knew Hercules's stories as he quoted them in *Convivio*. *Convivio* III, 3, 7 and he knew the version Virgil seems to follow in *Aeneid* VI, with Mercury psychopomp substituted by the Sibyl (SERVIUS's *Commentary* on *Aeneid* VI, 392). The principle that violence is useless against the shadows characterizes this story and also appears in the Sibyl's admonishment to Aeneas not to use his sword against them (*Aeneid* VI, 290). It is uncertain whether Dante had direct knowledge of Apollodorus's *Bibliotheca*. The book had a medieval transmission as authors such as Servius, Hyginus and Fulgentius show some acquaintance with it and it is quoted among Vatican Mythographers' sources. These authors, however, do not report this version of the myth. In a previous study, I proposed an additional conjectural source, already signaled by Silvio Pasquazi but only in reference to the Mercurial characteristics of the *messo*: Horace's *Ode* III, 11, where the poet prizes Mercury for having won Cerberus with his soothing art. See SUSANNA BARSELLA, *The Mercurial Integumentum*, pp. 383-387.

[60] Augustine used this expression to describe his condition prior to conversion in his *Confessiones* (*Confessions*), VII, 10, 16). The term *dissimilitudo* (unlikeness) indicated a state of remoteness from God, in which the likeness in which He created human beings is lost, so that they become 'dissimilar' to Him and to what is essentially human.

[61] Orpheus is enigmatically absent in Dante's text, although the Neo-Platonic commentators associated him with the chthonic heroes. A reason for the exclusion of the myth of Orpheus and Eurydice may be that commentators interpreted Eurydice as a figure of concupiscence representing the vicious descent into the secular world. In Boethius' interpretation Orpheus symbolizes the failure of the virtuous descent. See SEVERINUS BOETHIUS, *Consolatio philosophiae* III, 12, 50-58. As Guglielmo Gorni suggests, Orpheus is a silenced anti-type of Dante. Guglielmo Gorni interpreted the lack of Dante's references to Eurydice as a case of negation (*Verneinung*), suggesting that the story of the poet who failed to save his woman be the anti-model of Dante's own story of a poet successfully saved by his beloved. See GUGLIELMO GORNI, *La Beatrice di Dante, dal tempo all'eterno*, in DANTE ALIGHIERI, *Vita Nova*, L.C. Rossi ed., Milano, Mondadori, 1999.

cury to Aeneas's descent, thus introducing the figure of the god to represent the divine assistance necessary to successfully complete the hero's journey.[62] The presence of Mercury accompanying Aeneas as a sort of interpreter resolving moments of impasse was already attested to in the different versions of the myth of Hercules' twelfth labor that inspired Virgil's narrative in *Aeneid* VI. The same function of guide-interpreter characterizes Dante's *messo*.

The mythological nucleus that Virgil developed and others, including Dante, followed, is patterned after a version of the myth reported both in Servius' commentary on the *Aeneid* and in Apollodorus' *Bibliotheca*. According to this version of the myth, Hercules descends into Tartarus escorted by Mercury psychopomp. His presence signals to the inhabitants of Hades that the gods protect Hercules' incursion in their world, but Mercury-Hermes has also the task of making chthonic laws comprehensible to the hero. After crossing the Styx, and facing the ghosts of Meleager and the Gorgon Medusa, Hercules' first reaction is to use his sword against them. Mercury however warns him that arms are useless against shadows.[63] Following the pattern of the myth, Virgil transferred the function of the psychopomp to the Sibyl, who similarly prevented the Trojan hero from using force to overcome Charon's reluctance to ferry him over to the other shore of the Acheron. Aeneas eventually gains his entrance into Avernus by showing the golden bough he hides under his mantle, a sign of authority over the dead, strictly related to the special rod of Mercury the psychopomp, and which reappears as the *messo*'s little bough («verghetta», *Inf.* IX, 90).[64]

The chthonic functions of the pagan god remained a privileged trait in the medieval Christianization of Mercury, which reached its peak in the allegori-

---

[62] «Bernardus's commentary was the most important and influential medieval commentary on Virgil's work, although its approach and specific glosses can be traced back to the earlier treatments by Servius, Macrobius, Fulgentius, Remigius, and William of Conches [...]». JANE CHANCE, *Medieval Mythography*, p. 458. For the influence of Bernardus's commentary on the *Commedia* see DAVID THOMPSON, *Dante and Bernard Silvestris*, «Viator», I, 1, pp. 201-206, and GIORGIO PADOAN, *Traduzione e fortuna del commento all'Eneide di Bernardo Silvestre*, «Italia Medievale e Umanistica», III, 1960, pp. 227-240.

[63] Similarly the Sibyl's admonishes Aeneas not to use his sword against the creatures of Avernus: «Corripit hic subita trepidus formidine ferrum / Aeneas strictamque aciem venientibus offert, / et ni docta comes tenuis sine corpore vita / admoneat volitare cava sub imagine formae, / inruat et frustra ferro diverberet umbras», [«instantly struck with terror, Aeneas grips his sword / and offers its naked edge against them as they come, / and if his experienced comrade had not warned him / they are mere disembodied creatures, flimsy / will – o' – the whisps that flit like living forms, / he would have rushed them all, slashed through empty phantoms with his blade»] (*Aeneid* VI, 290-294).

[64] In *Aeneid* VI 405-410 Virgil calls the golden bough «fatalis virga». Aeneas's golden bough is connected to orphic initiation (Aeneas needs to be purified before he can enter the sacred wood to find the bough. *Aeneid* VI, 136-148) and to the shepherd's golden bough Mercury received from Apollo to guide the souls into the underworld. See SERVIUS, *In Aeneid* IV, 242.

CHAPTER FOUR

cal exegesis of Martianus Capella's *De Nuptiis Philologiae et Mercurii*.⁶⁵ The commentators of Capella – and in particular Bernardus Silvestris – linked again Mercury and Aeneas's descents, and in Dante's text we find an echo of their interpretation.⁶⁶ As the pagan figures representing the virtuous conquest of materiality had needed divine intermediation and protection, so does the Christian hero. In analogy with the classical explorers of the underworld, the pilgrim undertakes an allegorical journey in which he learns about vices and virtues. His experience, like that of the heroes in their allegorical interpretations, leads him to knowledge, and from knowledge to desire of purification, to which he will be initiated in *Purgatorio* to attain finally the longed for unification with the divine in the conclusive canto of *Paradiso*.

The direct and indirect references to the above mentioned version of the myth of Hercules (the only one testifying to the presence of Mercury and to Cerberus' loss of his 'beard') and to its later allegorical interpretations, allow for the setting of Dante's entrance into lower Hell within a well-established, allegorical tradition, and elevates the wayfarer to the Olympus of virtuous chthonic heroes. Dante, however, does not simply go back along the same path as his predecessors, but spirals through a circular pattern that will lead

---

⁶⁵ This fifth-century poem had considerable fortune in the Middle Ages and was the object of numerous commentaries. In Capella's allegorical narrative, Mercury crosses the underworld regions to re-ascend the heavens in search of a suitable bride. Virtue escorts him to Apollo's celestial palace from which the two gods reach Zeus's temple. There, a council of gods consents to the marriage of Mercury and Philology. Capella's *De nuptiis* was a crucial source for the medieval allegorical tradition of the descent-ascent motive. Various authors commented on both Capella's *De nuptiis* and Virgil's *Aeneid*: William of Conches and Bernardus Silvestris of the School of Chartres, and Alexander Neckam, who had contact with this school. «The mythographic tradition culminates in the twelfth century in the system of descents schematized by William of Conches and reinterpreted by Bernardus Silvestris [...] Virgil and Martianus are accorded a "Boethian" re-reading to single out the virtuous descent as their common denominator». JANE CHANCE, *Medieval Mythography*, p. 445. See also in the same volume, pp. 52-54; 445-492. For the importance and diffusion of Martianus' work and its commentaries, see also, *Texts and Transmission. A Survey of the Latin Classics*, L.D. Reynolds, R.J. Tarrant, M. Winterbottom, N.G. Wilson, P.K. Marshall eds., Oxford, Oxford University Press, 1984, pp. 245-46 and, for the commentaries on Capella, see also *The Berlin Commentary on Martianus Capella's De nuptiis Philologiae et Mercurii*, Jan H. Westra ed., Leiden, E.J. Brill, 1994, pp. IX-XXXIV.

⁶⁶ In Neo-Platonic allegorical exegesis both Theseus and Hercules were also *figurae Dei*. Padoan observes that Hercules, as well as Theseus, «fu ripetutamente accostato al Sansone biblico, di cui pareva costituire il *pendant* greco-latino: e come in Sansone si vide una prefigurazione di Cristo, cosí anche di Ercole si ebbe un'interpretazione figurale, assai diffusa [...], che in alcuni casi diviene dichiaratamente cristiana», [«was repeatedly associated with the Biblical Sanson. He constituted a sort of Sanson's Greek-Latin *pendant*. As Sanson was seen as Christ's prefiguration, so Hercules was widely interpreted in figural sense [...] in some cases in an openly Christian one»]. GIORGIO PADOAN, *Ercole*, in *ED*, p. 718. For Anthony Cassell, in the *messo*'s speech to the demons, Hercules is a prefiguration of Christ, ANTHONY K. CASSELL, *Dante's Fearful Art of Justice*, Toronto, University of Toronto Press, 1984, p. 165, n. 4.

him back to earth after experiencing the celestial regions. Dante's Christian descent, a truly successful path through knowledge of vice and punishment, provides the basis for a contextualized interpretation of the messenger as the Christianized figure of Mercury.[67]

d. *The Angelic Mercury: Eloquence and Wands*

The function of Mercury as a divine instrument the gods used to communicate with mortals (alive or dead) remained a distinct trait of the Christian reinterpretation of the Cyllean God as angelic Intelligence, and his functions as psychopomp and herald of the gods shape the figure of the *messo* sent from heaven to open the doors that fate would otherwise hold shut.[68]

Two key elements allow for identifying the *messo*'s *integumentum* with Mercury: his wand (*virga*), and his eloquence, where the *virga* is also the symbol of Mercury's eloquence.[69] The emblem of the *verghetta* to signify eloquence appears in Ovid, Statius, Horace, and Apuleius, and was perhaps the strongest element that induced ancient commentators to recognize Mercury in

---

[67] There is little doubt that the messenger is sent from heaven on this particular occasion. The word «messo» appears with the same meaning to indicate the angels in *Purgatorio*, who also assist the two poets by indicating to them their way through the various terraces. «"Non ti maravigliar s'ancor t'abbaglia / la famiglia del cielo", a me rispuose: / "messo è che viene ad invitar ch'om saglia"» (*Purg.* XV, 28-30).

[68] Iconographic examples of Mercury's *Christiana interpretatio* are the «Mercury in Exaltation», an illumination of Abu Ma'šar's *Liber astrologiae*, ca. 1403, New York, Pierpont Morgan Library, MS. 785, in ERWIN. PANOFSKY, *Renaissance* (fig. 90). Mercury, the planet, is dressed within a monk's frock, reading a book (knowledge) and holding a musical instrument while sitting on an ark (power over the dead). Underneath, a feminine Angel is represented holding three ears of wheat, symbols of Persephone, also linked to orphic mysteries, and a variation of Mercury's bough. Another illumination in Scotus's *Liber introductorious*, Munich, Staatsbibliothek, clm. 10268, fol. 85, represents Mercury as a bishop. See ID., *Renaissance*, fig. 89 and 90. For the iconography of Michael, see also LOUIS RÉAU, *Iconographie de l'art Chrétien*, Paris, Presses Universitaire de France, 1956, pp. 30-55.

[69] Ovid uses the term «virga» for Circe's wand in *Metam.* XIV, 278, 295, 300, 413, and Virgil employs it for Aeneas' bough in *Aen.* VII, 190. Ovid uses again «virga» to indicate Mercury's staff in *Metam.* I, 671, 675, 716; II, 736, 819; IX, 307. The term is present also in Lucan's *Thebaid*, II, 1-6; 11. Among the many examples in Ovid and Virgil and most classical authors, an interesting one is from Apuleius, who, indicating the caduceus as the typical attribute of Mercury, calls it «virgula», i.e. literally «verghetta»: «et virgula Mercurium indicabat», [«and even the rod indicated that he was Mercury»]. LUCIUS APULEIUS, *Le metamorfosi* (*Metamorphoseon*), Milano, Rizzoli, 1989, X, 30. The *verghetta*, like the golden bough, was connected with the rites of initiation and magic, and in particular with the cult of Persephone, as Virgil explicitly mentions in *Aeneid*, VI, 142. See FERDINAND J.M. DE WAELE, *The Magic Staff or Rod in Graeco-Italian Antiquity*, Ghent, Drukkery Erasmus, 1927, p. 82. For a study on Mercury as the god of language and oral communication see MAURIZIO BETTINI, *Le orecchie di Hermes. Studi di antropologia e letterature classiche*, Torino, Einaudi, 2000, pp. 5-51.

the figure of the *messo*.⁷⁰ Both in antiquity and during the Middle Ages the *virga* was associated with Mercury's chthonic powers, and its symbolism included authority over the dead and even magic.⁷¹ In parallel with Mercury and Aeneas's boughs, the *messo*'s *verghetta* is a sign of supernatural power, of authority over Hell and, symbolically, over its mythological and classic representations.

The association of Mercury and eloquence, rooted in mythology and attested to in classical literature, remained a fundamental trait of the god in the Middle Ages and later in the Renaissance.⁷² In Martianus Capella's allegorical text *De nuptiis Philologiae et Mercurii*, Mercury symbolizes Eloquence, the necessary complement of Wisdom. Referring to Capella, Fulgentius in his *Mythologiae* pointed to the deceitful aspect of Mercury's eloquence. This interpretation derived from Augustine's identification of the god with the eloquence of merchants and traders. Augustine stressed the symbolism of Mercury as god of language and interpretation, which he derived from his Greek name 'Hermes'.⁷³ The god's *facunditas* was also associated with the planet

---

⁷⁰ In Statius' *Thebaid* II, 1-6, Mercury appears as psychopomp to escort Laius's anabasis. Pietro di Dante suggested that Mercury's eloquence was the main key to Dante's allegorization of the *messo*. The messenger uses his wand to open the gates and his speech to tame the demons. The commentator parallels this episode with that of Mercury's rescue of Io from Argo in Ovid, *Metam.* I, 715-716. Another relevant Ovidian textual source of the virga is found in his *Metamorphoses* II, 814-819, in which Mercury opens Herse's door with a touch of his wand.

⁷¹ Kerény, tracing back the tradition to its most ancient sources (*Homeric Hymn to Hermes* 1-543 and APOLLODORUS, *Bibliotheca* iii.10.2 among the most important ones), mentions two of Mercury's different staffs: one is a shepherd's staff given to the young god by his brother Apollo in exchange for the lira (*virga* or *rabdos*); the other one was given to him by Zeus, when he chose Mercury as his messenger (*caduceus* or *karukeion*). The first is connected to Mercury's function as psychopomp; the second is associated to his function as Zeus's messenger and is a sign of authority. Both Zeus and Hades chose Mercury as their herald because of his eloquence. See OVID, *Metam.* II, 676-707, and ROBERT GRAVES, I, p. 65. Because of his power over the dead, the chthonic Mercury and his bough were also associated with magic. According to Hollander, Dante's choice of the word «verga» for Tiresias's wand in *Inf.* XX, connects it to the «verghetta» of *Inf.* IX and suggests a contraposition between the bad black magic associated with magicians' wands, and the good white magic associated with the *messo*'s wand. See ROBERT HOLLANDER, *The Tragedy of Divination in Inferno XX*, in *Studies in Dante*, Ravenna, Longo, 1980, p. 180. Giorgio Brugnoli suggests that these two wands are meant to signal Virgil as magician with an implicit negative connotation. See GIORGIO BRUGNOLI, *Nomen Omen. Due nomi parlanti in Dante*, in *I nomi da Dante ai contemporanei. Atti del IV Convegno Internazionale di Onomastica e Letteratura. Pisa, 27-28 febbraio 1998*, Viareggio, Baroni, 1999, pp. 35-45. Ferdinando M. De Waele notices that while in the Greek world Mercury's *karukeion* did not play any magic role, in Italian necromancy it was connected with the Orphic-Pythagorean cult related to reincarnation, F.J.M. DE WAELE, *The Magic Staff*, pp. 36-39; 56-59. For angelic magic, or *ars notoria* in the Middle Ages, see GRAZIELLA FEDERICI VESCOVINI, *Medioevo magico*, pp. 115-147.

⁷² The association of eloquence and Mercury remained constant throughout the centuries. In the seventeenth century, following the traditional interpretations of Augustine and Fulgentius, Cesare Ripa identified Mercury with humble eloquence, and also associated the god with fame. CESARE RIPA, *Iconologia*, New York, Georg Olms Verlag Hildesheim, 1970 (Roma, 1593), p. 143.

⁷³ Augustine, in *De civitate Dei*, interpreted the god as the personification of the cunning elo-

named after him. Albert the Great, in his *De rebus metallicis*, wrote that the engraved image of Mercury on a given stone would confer to its owner the power of eloquence and business ability.[74] Dante himself, in the *Convivio*, connected Dialectic with the virtues of the planet Mercury.[75] Indeed, the *messo* is the heavenly creature that delivers the longest speech in the *Commedia*, after the angel porter of *Purgatorio*.[76] While his wand opens the gates with no effort, his speech silences the demons, and reaffirms the power of the word over the sword, as in the original myths that form the subtexts of canto IX.

Mythological references, literary sources, and the association with eloquence, suggest that the mercurial characteristics of the *messo* do not play a merely symbolical role but refer, literally, to Mercury's attributes as psychopomp. This function, together with the elements characterizing it (the magic wand and eloquence), was absorbed into the cult of the archangel Michael as a result of the early syncretism of Mercury and Michael (as soothers of souls).[77] Although most ancient commentators were inclined to interpret

---

quence of traders and merchants: «Nam ideo Mercurius quasi medius currens dicitur appellatus, quod sermo currat inter homines medius; ideo Ἑρμῆς Graece, quod sermo, vel interpretatio, quae ad sermonem utique pertinet, ἑρμήνεια dicitur; ideo et mercibus praeesse, quia inter vendentes et ementes sermo fit medius; alas eius in capite et pedibus significare volucrem ferri per aera sermonem; nuntium dictum, quoniam per sermonem omnia cogitata enuntiantur». [«Now it may be said that it is language itself that is Mercury. This is suggested by the interpretation they give of him; for they derive the name Mercury from *medio currens*, "running in between", because speech "runs between" men. His name in Greek is Hermes, because speech, or rather, interpretation – which is clearly connected with speech – is called *hermeneia*. The reason why Mercury presides over commerce is that speech is the means of communications between sellers and buyers. The wings on his head and feet symbolize the swift flight of speech through the air; he is called a messenger because it is through speech that thoughts are conveyed»]. AUGUSTINE, *De civitate dei* 7, 14. English translation by H. Bettenson, *The City of God*, transl., London, Penguin, 1984 (1972). Augustine's interpretation became canonical, and Isidore of Seville repeated it in his *Etymologiae* VIII, 45-49.

[74] ALBERT THE GREAT, *De mineralibus et rebus metallicis*, English transl. by D. Wyckoff, *The Book of Minerals*, Oxford, Clarendon Press, 1967, p. 143.

[75] *Convivio* II, XIII, 11-13.

[76] *Purg.* IX, 64-145. The pattern of this Canto is symmetrical to that of *Inf.* IX, where the two poets need to pass through the door of the *Purgatorio* proper. Dante is scared and then reassured after Saint Lucy helps him. As in *Inf.* IX, the author addresses his readers before introducing the heavenly guardian (vv. 70-72). The speech of the angel at the door of *Purgatorio* takes 9 lines (*Purg.* IX, 121-129), exactly the same length of the *messo*'s speech (*Inf.* IX, 91-99).

[77] The cult of Michael, says Réau, «a remplacé celui des divinités païennes: du dieu égyptien Anubis et particulièrement de Mercure, l'Hermès psycopompe, qui jouait dans la mythologie un rôle analogue. Une colline de Vendée s'appelle encore aujourd'hui *Saint-Michael-Mont-Mercure*», [«has replaced that of pagan divinities: of the Egyptian god Anubis, and particularly that of Mercury, the Hermes psychopomp, who played an analogous role in mythology»]. LOUIS RÉAU, *Iconographie*, p. 44. The syncretism was still present in the fifteenth century, when Luca Signorelli painted the Archangel Michael with a caduceus engraved on the shell of his armor. Pasquazi showed that the syncretism of Michael and Mercury was already in existence in the fourth century A.D. See also GIOVANNI BUSNELLI, *Il messo del cielo alle porte di Dite*, Roma, Civiltà Cattolica, 1910. Christian icono-

the *messo* as a generic angelic figure, several modern interpreters, such as Silvio Pasquazi, identify him with Saint Michael, as the guardian angel of Limbo.[78] Michael, however, was mainly a figure of power: he was the prince of the heavenly Host who defeated the rebel angels, the protector of the Virgin and the Church, and the vanquisher of the dragon in the Apocalypse of John.[79] According to his different functions he was represented as an archangel, a principality, or a seraphim. Also the *messo* comes as a conqueror of a defensive stronghold, but in canto IX there is no real fight such as that against Lucifer or the dragon: on the contrary, many elements indicate Dante's willingness to depart from a knightly figure. The Mercurial angel does not carry a sword but a frail bough, and his main weapon is his intimidating speech.[80] In Dante's hell the epic battle against the rebel angels has been won and the demons subjugated: they are now infernal officers, coerced to enforce the laws of divine Justice.

All the elements composing the complex structure of this ritual canto indicate the strategic importance of Mercurial attributes in identifying the «tal» commanded from heaven. This figure not only restores the order of justice by

---

graphy also represented Mercury with an open book, sign of wisdom and knowledge. This image derived from a contamination of the Greek myth of Hermes psychopomp, the Egyptian God Thot – who taught the souls to achieve knowledge of divine things – and the Babylonian God Nebo, the writer-God. The images of Mercury dressed as a Bishop correspond to the Christian representation of these attributes. See JEAN SEZNEC, *The Survival of the Pagan Gods*, Princeton, Princeton University Press, 1972, pp. 158 and 254. In medieval iconographical tradition, Michael psychopomp is represented as an Archangel holding a staff and a scale to weigh the souls at the Last Judgment. The function of psychopomp attributed to Michael originates from the fact that Michael contended for Moses' soul with the devil. See SILVIO PASQUAZI, *Il messo celeste angelo del Limbo*, in *All'eterno dal tempo. Studi danteschi*, Roma, Bulzoni, 1985 (1966), p. 920. Ave Appiano shows that the attested existence of a relation between Mercury psychopomp with the sacrament of baptism is another example of the Christian reinterpretation of the god. See AVE APPIANO, *Le forme dell'immateriale*, Torino, Società Editrice Internazionale, 1996, p. 78. See also footnote 68.

[78] S. Pasquazi suggested that the *messo* could be the guardian angel of Limbo based on the tradition of *megalopsicoi* angels. Others have proposed that the messo could be the guardian of hell. It seems problematic, however, to assume the existence of an angelic figure permanently residing in hell. See Pasquazi's detailed argumentation in favor of an angel guardian of Limbo in SILVIO PASQUAZI, *Il messo celeste*, pp. 52-76. See also S. PASQUAZI, *Messo celeste*, in *ED* III, pp. 919-921.

[79] See FRANCISCO SUÁREZ, *Opera Omnia*, Paris, Ludovicum Vives, 1856, pp. 691-694.

[80] «L'angelo che irrompe impetuoso davanti alle porte di Dite per mettere in fuga le «furie infernal di sangue tinte», non lotta, in realtà, contro lo stesso principio demonico, ma si limita a ricollocarlo nei suoi giusti termini, per liberare a Dante la strada. La furia non è negata, bensì ri-collocata in seno all'"orribil arte" della divina Giustizia. Addirittura, questa divina Giustizia è chiamata, di fronte all'antico demone, col nome di Fato». [«The angel that impetuously irrupts before the gates of Dis to chase away the "furie infernal di sangue tinte", in reality does not fight against demonic rule. He just re-establishes it with its right limits to free Dante's way. The fury is not denied, but re-placed within the «orribil arte» of divine justice. Moreover, before the ancient demon this justice is called Fate»]. MASSIMO CACCIARI, *L'angelo necessario*, Milano, Adelphi, 1986, p. 111. For the conception of Justice in hell, ANTHONY K. CASSELL, *Dante's Fearful Art of Justice*, 1984.

curbing the demons' mutinous attitude, but also determines the ethical character of Dante's entrance into Dis through its allegorical link with the tradition of virtuous and heroic descents. The syncretism of the *messo* appears justified by these two main functions. If his role as psychopomp coincides with the function later attributed to Michael, his relation to heroic descents is however typical only of Mercury. It seems therefore coherent to suppose that Dante transformed the ancient psychopomp, the god of trivial speech and power over the dead, into the angelic intelligence that operates as infernal herald. Dante operated the same metamorphosis on the goddess Fortune (*Inf.* VII, 71-84), who, like Mercury, becomes an angelic intelligence governing human alterning fate.[81]

A further compelling element leads to interpret the *messo* as Mercury transformed into angelic Intelligence. Mercury was the Latin name of the Greek god of language and interpretation, Hermes, from whom the science of interpretation, hermeneutics, derives. Coherently with the meaning of his name, the *messo*'s function, literally and figuratively, is to overcome the impasse that prevents the pilgrim to enter Dis; to see beyond the walls as beyond what is known; to untangle the knots of interpretation. It is not by chance that the hermeneutical tercet in *Inferno* IX, 61-63 discussed above is set between the threat of the pagan, petryfying glance of Medusa, allegorically preventing insight, and the arrival of the «tal» who opens the text to a new understanding.

The sudden, violent irruption of the angel-Hermes at the gate of Hell forces us to re-think the question of textual impasse so often debated by scholars. After the initial impediment, when the demons at the gate deny access to the pilgrim, the fortified wall is no longer an obstacle. The angel 'opens up' the barrier. The action resumes, and the angel appears as the 'angel of interpretation'. He regains all the attributes of the mythical Hermes, god of the crossroads, psychopomp, and the very embodiment of language.

The vehement speech the *messo* addresses to the demons in front of the gates of Dis supports this hypothesis. Not only is it an example of exquisite Mercurial eloquence, but it also focuses on the heroic theme of Hercules's twelfth labor that underlies the whole episode, so relevant to the meaning of Dante's enterprise:

---

[81] *Inf.* VII, 73-96. Dante transforms the ancient goddess *Fortuna* into an angelic intelligence, of the same type and importance as the other *prime creature* (v. 95), that is the angels. Fortune is – like all angels – a divine instrument who «provede, giudica, e persegue / suo regno come il loro gli altri dei» [«She foresees, she judges, she maintains her reign, i as do the other heavenly powers.»] (vv. 86-87). One can see here the superimposition of the theological concept of Angel on the pagan idea of god, for Fortune retains the forms of a goddess although it is clear that her function is at the service of the true revealed God. The same principle applies to Mercury, who also becomes an angelic Intelligence.

> "O cacciati del ciel, gente dispetta",
> cominciò elli in su l'orribil soglia,
> "ond'esta oltracotanza in voi s'alletta?
>   Perché recalcitrate a quella voglia
> a cui non puote il fin mai esser mozzo,
> e che più volte v'ha cresciuta doglia?
>   Che giova nelle fata dar di cozzo?
> Cerbero vostro, se ben vi ricorda,
> ne porta ancor pelato il mento e 'l gozzo". (*Inf.* IX, 91-99)
>
> "O you cast out of Heaven, hated crowd,"
> were his first words upon that horrid threshold,
> "why do you harbor this presumptuousness?
>   Why are you so reluctant to endure
> that Will whose aim can never be cut short,
> and which so often added to your hurts?
>   What good is it to thrust against the fates?
> Your Cerberus, if you remember well,
> for that, had both his throat and chin stripped clean".

Delivered in a colloquial tone, and with words taken from common vernacular discourse, this speech is an example of *sermo communis*, the eloquence Augustine associated with Mercury. The *messo* directs his apostrophe to the rebel angels («Cacciati dal ciel») rather than to the whole community of monstrous creatures populating the battlements. The allusion to the banishment from heaven introduces the first reference to events of Christian history, and indirectly evokes the figure of Michael, who defeated Lucifer's army. In the following lines, the events interlace with the mythological narratives that prefigure Christ's harrowing of hell. The term "oltracotanza" in line 93 harks back, intensifying it, to the «tracotanza» of *Inf.* VIII, 124, the word Virgil used to recall the demons' defeated rebellion when Christ forced the doors at the entrance of hell. The episode signifies the subversion of the old law and the instauration of the new order of providential love. The dramatic sequence of rhetorical questions in tercet 94-96 reminds the demons of the necessity to obey the decrees of divine will («quella voglia / a cui non puote il fin mai esser mozzo»), and yield to providential order («fata») that governs universe, whereby the *messo* derives his power and authority. The mention of Cerberus in line 98 seals the messo's rebuke by recalling Hercules's twelfth labor. The infernal dog – the *messo* warns the demons – still bears the signs of its beating by Hercules. These «parole sante» conclude the episode with the image of the pagan hero, who emerges from the context as a prefiguration of the poet as Christian hero called to subvert the ancient poetic 'order' of those who explored the infernal abyss before him.[82]

IV. BEATRICE'S ANGELIC LIGHT

The pivotal importance of Dante's angelology in the *Commedia* leads to a reconsideration of the role of Beatrice in the light of his general design of angelic functions. Her heavenly operations – connected to those of the historical protagonist of Dante's juvenile book (*libello*), the *Vita Nuova* – appear in a different light when examined within the context of the *Commedia's* angelology. The comparison between the *Vita nuova* and the *Commedia* testifies to the evolution of the literary Beatrice from metaphor to analogical model of angel. Although in the libello she no longer was the metaphorical woman-angel of the lyric tradition, but a sort of incarnation of the miracle, only in the *Commedia* does her literary figure developed into an angelic character performing angelic operations in rescuing, initiating, and – most importantly – enlightening Dante in his itinerary to the contemplation of the Trinity.[83]

Although commentators have widely discussed the angelic aspect of Beatrice in the *Vita Nuova* and in the *Commedia*, the parallel between the «most

---

[82] The *messo*'s verses in *Inf.* IX, 98-99 («Cerbero vostro, se ben vi ricorda, / ne porta ancor pelato il mento e 'l gozzo») refer to the lesson Hercules gave to Cerberus. The sources of this version of the myth are in *Vatican Mythographers* I and II (I, 48, 57; II, 156) as well as in other mythological compilations and collections such as those by Hyginus (*Fable* XXXI). See Chance for the importance, composition, and diffusion of these compilations, pp. 158-204 and 300-346. As seen above, Dante likely knew other important mythological sources he could derive from the commentaries to classical texts and in particular to the *Aeneid* (Bernardus Silvestris, Servius, Macrobius), the *Thebaid* (Lactantius, Fulgentius), the *Pharsalia*, and the *De Nuptiis* by Martianus Capella (Bernardus Silvestris). For the influence and diffusion of these commentaries see again JANE CHANCE, *Medieval Mythographers*, pp. 445-492; *Texts and Transmission: A Survey of the Latin Classics*, pp. 245-246; ERNST R. CURTIUS, *European Literature*, pp. 354-355. For the medieval versions of Hercules's descent to Hell in the *Divina Commedia* see in particular GIOVANNI BOCCACCIO, *Esposizioni sopra la Comedia*, IX (i), 33-40. Boccaccio reports the different versions of the myth of Theseus and Pirithous also in *Genealogie*, IX, XXXIII, and X, XLIX, where the *Hercules Furens* by Seneca is indicated as a source for the detailed narration of the fight with Cerberus. For ancient versions of the myth and their sources see ROBERT GRAVES, *The Greek Myths*, London, Penguin, 1990 (1955), vol. 2, pp. 152-158; KÁROLY KERÉNY, *Gli Dei e gli Eroi della Grecia*, Milano, Garzanti, 1978 (1958), vol. 2, pp. 191-195.

[83] «The sacramental and Christological dimensions of the *Vita Nuova*'s Beatrice, the fact that she has come from heaven to earth as a manifest miracle, that the portents of her death are the portents of Christ's death, that she *is* the incarnate number nine, take Guinizzelli's solutions an enormous step further along the road from simile (*Tenne d'angel sembianza*) to metaphor (*d'umiltà vestuta*): from assimilation to, to appropriation of, the divine». TEODOLINDA BAROLINI, *Dante and the Lyric Past*, in *The Origins of Italian Literary Culture*, p. 33.

gentle one» (*gentilissima*) with the angels in terms of operations has received little critical attention. The literary project of a 'figural' character (historical and allegorical at the same time), representing divine intervention in the life of the protagonist, and mirroring the universal operations of Intelligences in the individual dimension of the protagonist, started in Dante's juvenile work and culminated in the *Commedia*. Key to the development of Beatrice's literary figure was Dante's elaboration of his angelology in the *Commedia*, which allowed him to transform Beatrice into an 'analogical model' of the angel without contradicting theological doctrine, and without overstepping the threshold of heretical propositions.[84]

The *Vita Nuova* differs from the *Commedia* in the type of analogical relation that Dante established between the *gentilissima* and the angel, which included Christ and the Virgin. While in the *libello* Dante formulated this analogy in terms of Beatrice's exceptionally virtuous nature, in the *Commedia* this analogy relied on her angelic operations. Similar to an angel, Beatrice mediates between natural and supernatural worlds by reflecting on Dante the light of divine wisdom. This mirroring effect enables the poet to progress toward his final vision in canto XXXIII. By fulfilling Dante's desire for truth, Beatrice's illumination is crucial in rendering him capable of participating in the eschatological cosmogony centered on Christ's redemptive work displayed in the *Commedia*. The light of knowledge that Beatrice conveys to Dante allows him to see the truth, and transforms his erotic and intellectual desire into enlightened and morally enlightening writing.[85]

---

[84] Maria Corti showed that the application of analogical models was inherent to the *forma mentis* of medieval poets and that Dante analogically adopted the mystical model in the *Commedia*. She noticed that the application of an analogical model has three characteristics: «a) la funzione strutturante; b) la funzione nobilitante; c) la presenza di vocaboli-segnale dell'ambito da cui si assume il modello», [«a) the structuring function; b) the nobilitating function; c) the presence of word-signals of the context from which the model derives»]. All these characteristics can be found in the assumption of the analogy of attribution among Beatrice, the angels, Mary, and Christ. In particular, Beatrice's exceptionality casts a nobilitating light on Dante himself as the one chosen to be the object of the sign of divine love represented by Beatrice. MARIA CORTI, *Il modello analogico nel pensiero medievale e dantesco*, in *Dante e le forme dell'allegoresi*, M. Picone ed., Ravenna, Longo, 1987, p. 15.

[85] The tropological end of the *Commedia* is clearly stated in Dante's *Epistle XIII* to Cangrande della Scala: «Est ergo subiectum totius operis, litteraliter tantum accepti, status animarum post mortem simpliciter sumptus; nam de illo et circa illum totius operis versatur processus. Si vero accipiatur opus allegorice, subiectum est homo prout merendo et demerendo per arbitrii libertatem iustitie premiandi et puniendi obnoxius est». [«And we must therefore consider the subject of this work as literally understood, and then its subject as allegorically intended. The subject of the whole work, then, taken in the literal sense only, is "the state of souls after death", without qualification, for the whole progress of the work hinges on it and about it. Whereas if the work be taken allegorically the subject is "man, as by good or ill desserts, in the exercise of the freedom of his choice, he becomes liable to rewarding or punishing justice"»], *Letter to Cangrande*, Epistle XIII, 24-25. The question on the

a. The Exceptional Virtue of the Gentilissima Beatrice

In the *Vita Nuova*, written around 1294, presumably after her death, Dante adorned Beatrice with the attributes of angelic iconography.[86] The *prosimetron* narrates the story of Dante's juvenile love for this extraordinarily virtuous woman both before and after her death. Dante insists on the exceptional nature of the «angiola giovanissima» and, quoting Homer in chapter II of *Vita Nuova*, affirms that she does not seem born from a mortal, but from God:[87]

Elli [Amore] mi comandava molte volte che io cercasse per vedere questa angiola giovanissima, onde io ne la mia puerizia molte volte l'andai cercando, e vedeala di sì nobili e laudabili portamenti, che certo di lei si potea dire quella parola del poeta Omero: "Ella non parea figliuola d'uomo mortale ma di deo" (*VN* I, 9; II 8).

[many times he (Love) ordered that I tried to see this very young angel, so that during my adolescence many times I went looking for her, and I she seemed to me of such noble and praiseworthy manners, that certainly one could name her with that word of the poet Homer: "she did not appear as the daughter of man but of God"].

The passage recalls a topos of the courtly tradition, whereby Love drives the lover to seek his beloved and admire her nobility so that he may benefit from her almost divine qualities. Dante developed the nobility motive into one

---

authenticity of the *Epistle* is still unsolved. For the longstanding debate among Dante scholars, see ROBERT HOLLANDER, *Dante's Epistle to Cangrande*, Ann Arbor (Michigan), University of Michigan Press, 1993; JOHN AHERN, *Can the Epistle to Cangrande Be Read as a Forgery?*, and ALBERT R. ASCOLI, *Access to Authority: Dante in the Epistle to Cangrande*, in *Atti of the First International Dante Seminar*, Z. Baranski ed., Firenze, Le Lettere, 1997, pp. 279-307 and pp. 309-352 respectively. See also ALBERT R. ASCOLI, *Epistola a Cangrande*, in *The Dante Encyclopedia*, R.H. Lansing ed., New York, Garland, 2000 (Garland Reference Library of the Humanities, vol. 1836), pp. 348-352, and JOHN SCOTT, *Understanding Dante*, Notre Dame, Notre Dame University Press, 2004, pp. 345-347; 409.

[86] For the problematic date of composition of the *Vita Nuova*, see DANTE ALIGHIERI, *Vita Nuova*, G. Gorni ed., Torino, Einaudi, 1996, pp. XVIII-XXI.

[87] The verb 'apparire' is a technical term in the poetry of the *Dolce Stil Nuovo*. It indicates the act of appearing, showing, or being revealed. Original text is from DANTE ALIGHIERI, *Vita Nuova*, G. Gorni ed., English translation by M. Musa, *Vita Nuova*, Oxford, Oxford University Press, 1999. In the quotations, Arabic numbers indicate G. Gorni's division in paragraphs. Traditional division is in brackets, indicated by Roman numbers. For Gorni's philological criteria see DANTE ALIGHIERI, *Vita Nuova*, pp. XXI-XXIX, and GUGLIELMO GORNI, *Dante prima della Commedia*, Firenze, Cadmo, 2001, pp. 111-132; ID., *'Paragrafi' e titolo della «Vita nova»*, «Studi di filologia italiana», LIII, 1995, pp. 203-222. On the issue of paragraph division, see also DINO CERVIGNI, *Segni paragrafali, maiuscole e grafia nella Vita nuova: dal libello manoscritto al libro a stampa*, «Rivista di Letteratura Italiana», XIII, 1-2, 1995, pp. 283-362 and D. CERVIGNI and EDWARD VASTA, *From Manuscript to Print: The Case of Dante's Vita Nuova*, in *Dante Now: Current Trends in Dante Studies*, T. J. Cachey ed., Notre Dame, University of Notre Dame Press, 1995, pp. 83-114. Herafter *Vita Nuova* is indicated with *VN*.

of material divinization of woman into an angelic figure, as seen when the images of Beatrice and the angel overlap in the episode of chapter 23 [XXXIV]. On the anniversary of Beatrice's death Dante draws an angel on a sheet of paper, projecting the image of Beatrice onto that of the angel. Several men approach his desk attracted by the virtue emanating from his sketch.[88] The drawing becomes the emblem of the fictional Beatrice, where the woman and the angel become one thanks to the inspired art of the poet. As a sort of simulacrum, the image Dante sketches acts as a revealing force of the sacred, capable of producing a vision of truth so deep as to guide Dante's hand to word its equivalent in a sonnet. In the final tercet of the sonnet, Love («Amore») moves Dante's sighs («sospiri») to speak to the woman – who died one year earlier – and address her as «noble intellect»: «Oi nobile intelletto, / oggi fa l'anno che nel ciel salisti». [«This day, O intellect sublime, / completes a year since you rose heavenward»] (*VN* 23, 11, [XXXIV, 11], 13-14). While the adjective «noble» reiterates the traditional Stilnovistic motive of nobility, the qualified term «intellect», reproduces the visual association between Beatrice and the angel, as «Intelletto» is a technical word to define angelic Intelligences.

The exceptional virtue of the living Beatrice prefigures the angelic characteristics she develops in the *Commedia*. Her virtue is so exceptional that, as Mary before her, she exceeds the limits of human nature to become, in the words of Dante, «almost an angel» («*quasi* angelo»).[89] Indeed she is a «miracle», a concrete materialization of divine spirit symbolically represented by the number nine (as nine are the angelic choirs).[90] The *Vita Nuova* repeatedly in-

---

[88] «Era venuta nella mente mia / Quella donna gentil cui piange Amore / Entro 'n quel punto che lo suo valore / Vi trasse a riguardar quel ch'eo facea». [«Into my mind had come the gracious image / of the lady for whom Love still sheds tears, / just when you were attracted by her virtue / to come and see what I was doing there»]. *VN* 23, 8 [XXXIV, 8], 1-4.

[89] Dante used the word «a guisa che» equivalent to «quasi» to define Beatrice's relation to the angel. The «quasi» had a technical scholastic meaning and indicated a relation of analogy: «e cotale dico io che è questa donna, sì che la divina virtude, *a guisa che* discende nell'angelo, discende in lei», [«and such, I say, is this lady, for the divine virtue descends into her *just as* it descends into an angel»]. *Convivio* III, vii, 7. See also *Convivio* III, vii, 6 and III, ii, 14. My italics.

[90] For the angelic attributes of Beatrice in the *Vita Nuova* and her relation with the theological and literary notion of miracle, see MARCO SANTAGATA, *La donna del miracolo*, in *Amate e Amanti. Figure della lirica amorosa tra Dante e Petrarca*, Bologna, Il Mulino, 1999, pp. 13-61. Santagata offers an intriguing interpretation of Beatrice, as operating a 'lay' miracle in eliciting love even in those hearts where love does not exists in potency. Her miracle would represent Dante's attempt to universalize his poetry of love-*caritas*, distancing himself from the elitist tradition of the 'gentle hearts' of Guinizzelli and Cavalcanti. Guglielmo Gorni noticed that the number nine is graphically present even in her name, which can be deconstructed as BEATR and IX. GUGLIELMO GORNI, *Il numero di Beatrice*, in *Lettera, nome, numero. L'ordine delle cose in Dante*, Bologna, Il Mulino, 1990 cap. III. See also CARLO VECCE, *Beatrice e il numero amico*, in *Beatrice nell'opera di Dante e nella memoria*

sists on associating this sacred number with Beatrice to mark her supernatural nature. Her first encounter with the poet occurs when they are both in their ninth year of life, and after nine years she greets Dante with her 'saluto' (greeting) in the ninth hour of the day.[91] It is again in the ninth hour of the day that the poet, desolate after Beatrice denies him her «saluto» (greeting), receives a vision where Love invites him to write in her praise.[92] Still in the ninth hour of the day, Beatrice appears in a vision to the misled poet after her death.[93]

Chapter 19 [XXIX] of the *Vita nuova*, illustrates Dante's effort to justify the analogy between Beatrice and the angels by recurring to the symbology of the number nine. Here Dante specifies that when referring to Beatrice, the sacred number properly defines her nature: she 'is' number nine; a real incarnation of a miracle:

> Lo numero del tre è la radice del nove, però che, sanza numero altro alcuno, per se medesimo fa nove [...] Dunque se lo tre è fattore per se medesimo del nove, e lo fattore per se medesimo de li miracoli è tre, cioè Padre e Figlio e Spirito Santo, li quali sono tre e uno, questa donna fue accompagnata da questo numero del nove a dare ad intendere ch'ella era uno nove, cioè uno miracolo, la cui radice, cioè del miracolo, è solamente la mirabile Trinitade (*VN* 19, 6 [XXIX, 3]).

The number three is the root of nine for, without any other number, multiplied by itself, it gives nine: it is quite clear that three times three is nine. Therefore, if three is the sole factor of nine, and the sole factor of miracles is three, that is, Father, Son, and Holy Spirit, who are Three in One, then this lady was accompanied by the number nine so that it might be understood that she was a nine, or a miracle, whose root, namely that of the miracle, is the miraculous Trinity itself.

The miracle is the perfect product of the Holy Trinity, the fruit of God's mind sent as a vision to manifest to humankind the presence of the divine.[94] The

---

europea 1290-1990, Atti del Convegno Internazionale, 10-14 dicembre 1990, Istituto Universitario Orientale di Napoli, Maria Picchio Simonelli ed., Firenze, Cadmo, 1994, pp. 101-35. See also FRANCESCO MAZZONI, *Il 'trascendentale' dimenticato*, in *Omaggio a Beatrice*, R. Abardo ed., Firenze, Le Lettere, 1997, pp. 93-132, and ROSETTA MIGLIORINI-FISSI, *Da Matelda a Beatrice a Maria*, pp. 23-82.

[91] «quasi dal principio del suo nono anno apparve a me, ed io la vidi quasi da la fine del mio nono», [«she appeared to me at about the beginning of her ninth year, and I first saw her near the end of my ninth year»]. *VN* 1, 3 [II, 2]; «Poi che fuoro passati tanti die, che appunto erano compiuti li nove anni appresso l'apparimento soprascritto di questa gentilissima», [«After so many days had passed that precisely nine years were ending since the appearance, just described, of this most gracious lady»], and «L'ora che lo suo dolcissimo salutare mi giunse, era fermamente nona di quello giorno», [«It was precisely the ninth hour of that day, three o'clock in the afternoon, when her sweet greeting came to me.»] *VN* 1, 12 [III, 1] and 1, 13 [III, 2].

[92] *VN* 5, 16, [XII, 9].

[93] *VN* 28, 1 [XXXIX, 1].

[94] Similarly in *Conv*. III, vii, 17: «[ella] fu ordinata nella mente di Dio in testimonio de la fede a

theological numerical argument adds to the astrological foundation of Beatrice's iconic 'magic' we will see below, and both find support in the philosophical justification of virtuous exceptionality that Dante found in Aristotle. In the third treatise of *Convivio* Dante comments on a passage from Aristotle's *Ethics to Nicomachus*, where the Philosopher argued that creatures that were more than human existed:

nell'ordine intellettuale dell'universo si sale e si discende per gradi quasi continui [...] e tra l'angelica natura, che è cosa intellettuale, e l'animo umano non sia grado alcuno [...] è da porre e da credere fermamente che sia alcuno tanto Nobile e di sí alta condizione, che *quasi* non sia altro che angelo[95] (*Convivio* III vii, 6).

[in the intellectual order of the universe one ascends and descends by degrees almost continuous from the lowest form to the highest [...]; and between the angelic nature, which is intellectual, and the human soul there is no step [...] so it is to be asserted and firmly believed that there may be some one so noble and of a condition so exalted that it is almost no other than angel.]

In this passage, Dante shows dialectical prudence by speaking in terms of analogy, and introduces, as in the *Vita Nuova*, the technical term «quasi» to make admissible an otherwise problematic statement.[96] Theologians commonly accepted this proposition, and interpreted the existence of such creatures as necessary to guarantee a *continuum* in the spectrum of substances. Divine and human at the same time, supernaturally virtuous beings constituted a necessary bridge between human and angelic natures. Such had been Mary and the Apostles, whose earthly Beatitude anticipated that of the Blessed in Paradise.

The sacred number finally returns in Dante's description of Beatrice's death, where he explains its meaning in astrological terms, and attributes Beatrice's exceptionality to cosmological influences:

---

coloro che in questo tempo vivono». [she «was ordained in the mind of God in testimony of the faith to those who live in these times»]. See also *Convivio*, III, xiv, 14. For other passages referring to Beatrice as miracle see *VN* 10, 18 [XIX, 7]: «Angelo clama in divino intelletto / e dice: "Sire, nel mondo si vede / maraviglia ne l'atto che procede / d'un anima che 'nfin qua su risplende"», [«The mind of God receives an angel's prayer: "My Lord, there appears to be upon your earth / a living miracle, proceeding from / a radiant soul whose light reaches us here"»]. The «maraviglia ne l'atto» alludes to the incarnated miracle. See also *VN* 176 [XXVI, 7], 7-8: «e par che sia una cosa venuta / da cielo in terra a miracol mostrare», [«she seems to be a creature come from Heaven / to earth, to manifest a miracle»].

[95] ARISTOTLE, *Ethics to Nicomachus* 1145a, 15-30. My italics.

[96] The tercet 76-78 in *Inferno* II in which Virgil hails Beatrice should be read in the same perspective. «O donna di virtù, sola per cui / L'umana specie eccede ogni contento / Di quel ciel che ha minor li cerchi sui», [«O lady of such virtue that by it alone / the human race surpasses all that lies / within the smallest compass of the heavens»]. (*Inf.* II, 76-78).

Perché questo numero fosse in tanto amico di lei, questa potrebbe essere una ragione: con ciò sia cosa che, secondo Tolomeo e secondo la cristiana veritade, nove siano li cieli che si muovono, e, secondo commune oppinione astrologa, li detti cieli adoperino qua giuso secondo la loro abitudine insieme, questo numero fue amico di lei per dare ad intendere che ne la sua generazione tutti e nove li mobili cieli perfettissimamente s'aveano insieme (*VN* 19, 5 [XXIX, 2]).

One reason why this number was in such harmony with her might be this: since, according to Ptolemy and according to Christian truth, there are nine heavens that move, and since, according to widespread astrological opinion; these heavens affect the earth below according to the relations they have to one another, this number was in harmony with her to make it understood that at her birth all nine of the moving heavens were in perfect relationship to one another.

As later in *Convivio*, Dante conjoins judicial astrology and Christian truth to support his literary invention of Beatrice. Astrological influence grants her exceptionality, for she is the miraculous fruit of a perfect allineation of the nine spheres, similar to that present at Christ's birth. In more general terms, Dante will return to the same concept again in the *Convivio*:

E sono alcuni di tale oppinione che dicono, se tutte le precedenti vertuti s'accordassero sopra la produzione d'un anima ne la loro ottima disposizione, che tanto discenderebbe in quella de la deitade, che quasi sarebbe un altro Iddio incarnato (*Conv.* IV, xxi, 10).

There are some who would even claim that if all of the preceding virtues in their best disposition were brought into agreement in the creation of a soul, so much of the Deity would descend into it that it would almost become another God incarnate.

The «some» («alcuni») are the judicial astrologers, who engaged in the science of horoscopes. Most likely Dante is alluding to those who attempted to reconstruct the special astral combination under which Christ was born. Dante cautiously shields his idea of a 'perfect' Beatrice behind their opinion. He wisely again introduces a 'quasi', to suggest the possibility that someone with the same horoscope as Beatrice could be an 'almost' incarnated divinity. For non-deterministic astrologers the special disposition of the stars was only a necessary, but not sufficient, condition to determine the birth of an exceptional nature or cause a unique event such as Incarnation. Only God could realize what the stars dispose as a mere possibility. Dante's 'quasi' reflects his awareness of the difference between deterministic diviners such as Cecco d'Ascoli and the theologians who subjected astral influences to God's will. It is a rhetorical necessity that holds Dante within the limits of admissible poetic license.[97]

---

[97] Dante's mention of Christ's horoscope suggests his familiarity with an 'heretical' branch of

To give authority to his 'miraculous' Beatrice, however, Dante needed to go beyond the traditional figure of the 'angelicated' woman, and give realistic, 'scientific' ground to his poetic invention. Although the 'quasi' indicates Dante's distancing from astrological determinism, at this stage of his literary itinerary to the blessed Beatrice he seems to rely on the influence of the planets to provide such scientific ground.[98] The astrological argument seems to lose its momentum in the *Commedia*, where not nature but operations become central to shape the angelicity of Beatrice, suggesting a change in Dante's conception of her figure.

The project of a literal 'angelic' Beatrice remains in some ways incomplete in the works preceding the *Commedia*. The young lover saw in his beloved a revealing sign of divine presence. However, although Dante had already taken a different direction than his predecessors, she remained an epiphanic object of contemplation that did not interact with her frustrated lover. In this sense, Beatrice maintained continuity with the lyric tradition that preceded the «new style» Bonagiunta Orbicciani later defined in *Purgatorio* XXIV.[99] Dante overcame the metaphorized woman-angel of the tradition only with his invention of the heavenly guide of the *Commedia*, where he gave Beatrice an active role within the context of his general doctrine of angelic operations.

b. *Beatrice's Angelic Operations*

The rhetoric simile «angiola giovanissima» (very young angel) of the *Vita Nuova* anticipated the literary creation of Beatrice as angelic character, and established a figural relation between the idealized 'historical' Beatrice of the *libello* and the heavenly fictional character of the *Commedia*. Beyond the Primum Mobile («la sfera che più alta gira», *VN* 30, 10 [XLI, 10]), Beatri-

---

judicial astrology that presumed the influence of the stars on Christ rather than vice versa. These astrological doctrines, unlike those predicting the great historical and universal events, reintroduced a form of astrological determinism that the Church condemned. Cecco d'Ascoli, the author of the anti-*Commedia* poem l'*Acerba*, was condemned also because he applied horoscope astrology to Christ and the Virgin. For the doctrines on the great conjunctions see GRAZIELLA FEDERICI VESCOVINI, *Medioevo magico*, pp. 225-311. See also Chapter III of this work, pp. 133-150.

[98] Giuseppe Mazzotta observes the importance of the astrological dimension for the poetics of the *Vita Nuova*, in which «To love is to read the stars' prodigious conjunctions, to will to understand the seemingly chance encounter of a boy and a girl within the architecture of magic correspondences». GIUSEPPE MAZZOTTA, *Dante's Vision and the Circle of Knowledge*, p. 172.

[99] On the relevance of contemporary and past lyric tradition for Dante's mature poetics, see TEODOLINDA BAROLINI, *Dante's Poets: Textuality and Truth in the Comedy*, Princeton, Princeton University Press, 1984.

ce's iconical powers are materialized in her operations. As the *messo*, she is sent from heaven, a messenger assisting Dante's voyage in virtue of the Virgin's preemptive intercession. As the *messo*, she harrows hell in search of Dante's infernal chaperon, and similarly to the porter-angel in *Purgatorio*, she presides over Dante's spiritual examination before he can proceed with the rites of initiation in Eden. In *Paradiso*, she becomes the enlightening teacher who makes Dante's itinerary through the heavens possible by mirroring supernatural light onto him. Her blazing eyes stimulate and fulfill at the same time his desire to know and love truth, to pursue his journey toward the contemplation of «God's face».[100] As seen in the preceding chapters, Dante placed the angels' act of mirroring at the core of his angelological design. In insisting on the analogy between Beatrice and the angels, he coherently placed this operation at the heart of the relation between her and Dante the character.[101]

Beatrice's splendor results from the divine rays her eyes reflect as in a mirror, and intensifies as the two travelers advance through the heavens. At each passage, and in moments of intellectual impasse, she mirrors this light onto Dante. These enlightening events are usually preceded or followed by an oral explanation of some major doctrinal point the pilgrim is not yet able to grasp. Step by step, Beatrice's luminous reflection enhances Dante's capacity to penetrate the mystery of the ordered structure of providential cosmos. Dante's ability to see, and therefore to understand, deepens as his eyes look into Beatrice's eyes. The coupling of human and celestial sights intensifies his desire to know. The cognitive moments that Beatrice's enlightening provokes are often accompanied by her epiphanic smile, which is a sign of the perfect *laetitia* that follows the blessed souls' contemplation.[102]

---

[100] This was the formula the mystics used to indicate the highest stage of contemplation. See for example GUILLAUME OF SAINT THIERRY, *The Golden Epistle: A Letter to the Brethren at Mont Dieu*, Collegeville, MN, Cistercian Publications, 1980.

[101] Beatrice's eyes direct Dante to the right path to follow in view of salvation even in the *Vita Nuova*. In Purgatory XXX, Beatrice insists on the effects of her glance while still alive: «Alcun tempo il sostenni col mio volto: / mostrando li occhi giovanetti a lui, / meco il menava in dritta parte vòlto», [«My countenance sustained him for a while; / showing my youthful eyes to him, I led / him with me toward the way of righteousness»]. (*Purg.* XXX, 121-123). In life, Beatrice's splendor comes from her virtue.

[102] Peter S. Hawkins sees in Beatrice's smile the epiphany of the divine that accompanies and uplifts Dante through the heavens. Uplifting, however, is caused by Beatrice's mirroring of light onto Dante, and always occurs whenever she mirrors light. The smile often accompanies enlightenment, but not always, as for example in *Paradiso* XXVIII, where Dante looks in Beatrice's eyes before enjoying the vision of the angelic choirs. See PETER S. HAWKINS, *All Smiles: Poetry and Theology in Dante*.

At the opening of the third canticle (*Par.* I, 49-54), Dante illustrates Beatrice's mirroring effect by using a simile with the optical phenomenon of light reflection. He describes the empowering effect of Beatrice's gaze in improving the protagonist's capacity to bear the luminosity of the sun:

> E sì come secondo raggio suole
> uscir del primo e risalire in suso,
> pur come pelegrin che tornar vuole,
>   così de l'atto suo, per li occhi infuso
> ne l'imagine mia, il mio si fece,
> e fissi li occhi al sole oltre nostr' uso. (*Par.* I, 49-54)

> And as a second ray will issue from
> the first and reascend, much like a pilgrim
> who seeks his home again, so on her action,
>   fed by my eyes to my imagination,
> my action drew, and on the sun I set
> my sight more than we usually do.

The tercets above offer a technical description of the optic phenomenon of light reflection, similar to what anticipated in *Purgatorio* XV, 14-24, where Dante described the same effects in reference to the angel.[103] As soon as Beatrice directs her gaze to the sun, her splendor intensifies and Dante's sight absorbs («per li occhi infuso ne l'imagine mia») her 'act of vision' («atto suo»). The light pours from Beatrice to Dante, thus increasing his capacity to stare at the sun «beyond our wont» («oltre nostr'uso»). By insisting on images of light and instruments of vision (the ray, the sun), and focusing on the physical playing of light from Beatrice's to Dante's eyes, the tercets define a special type of communication between the guide and her pupil. The cognitive uplifting associated with Beatrice's reflection culminates in the passage and return of the luminous ray – the «pilgrim» («pelegrin») refers to the ray and to Dante – from Beatrice to Dante.

A few lines later, Beatrice fixes her eyes on the eternal circles («rote»), thus anticipating the cosmological vision that will unfold in the third canticle, and setting the individual enterprise of the poet within a universal perspective:

> Beatrice tutta ne l'etterne rote
> fissa con li occhi stava; e io in lei
> le luci fissi, di là sù rimote (*Par.* I, 64-66).

---

[103] See chapter II, pp. 82-83, for a discussion on *Purgatorio* XV, 14-24.

> The eyes of Beatrice were all intent
> on the eternal circles; from the sun,
> I turned aside; I set my eyes on her.

As in the preceding tercets, the act of seeing is rendered with reiterated insistence on the verb 'to fix' («fissare») and its variants. The allusion to the compenetration of Beatrice and Dante's sights in a circular movement is stressed by the rhythm of the lines, and accentuated by the strong sound of the double consonants. While Beatrice intensely stares at the source of the primary light – the «lume» recalled few lines below – Dante's capacity of vision enhances, thus inducing in the protagonist a process of progressive 'trans-humanization'.[104] This phenomenon defines a pattern in the relation between Dante and Beatrice. At each passage from a lower to a superior sphere she blooms in beauty and luminosity, thus increasing Dante's capacity to receive the illumination that provokes his noetic leaps.[105] His enhanced power to perceive divine light parallels the effects of angelic enlightenment in the anagogical process to unification with God.[106] Dante's vision through 'specula' (mirrors) ends at the end of his 'trans-humanization'.[107] Only after Bernard's pray to the interceding Virgin, he will be able to receive a direct vision of God. Until Dante reaches the highest sphere of the Empyrean, after contemplating the river of light, he can see only through Beatrice's specular mediation.

Besides the obvious Pauline reference implicit in the concretization of divine illuminating creatures, the angelic image of the mirror, when associated with Beatrice, echoes the Ovidian myth of the Gorgon and Perseus. Beatrice

---

[104] «S'i' era sol di me quel che creasti / novellamente, amor che 'l ciel governi, / tu 'l sai, che col tuo lume mi levasti», [«Whether I only was the part of me / that You created last, You-governing / the heavens-know: it was Your light that raised me»], *Par.* I, 73-75. Here Dante addresses God's governing love, and refers to the light (and metonymically to Beatrice) as the means by which the Creator raised him to the heavens.

[105] See main occurrences in *Par.* II, 107-110; *Par.* V, 7-9; *Par.* VIII, 13-15; *Par.* XIV, 82-84; *Par.* XXI, 1-12; *Par.* XXIII, 10-15; *Par.* XXVI, 76-78; XXVII, 88-96; *Par.* XXXI, 70-72.

[106] See the function of enlightenment in Dionysius's definition of hierarchy in Chapter II, pp. 54-55.

[107] Beatrice exhorts Dante to use his eyes as mirrors of the mirror in which he sees the image of what the heaven of Sun transmits in illuminating the world below. Universe appears in Beatrice's words as a chain of mirrors linked in an infinite circular reflection that brings together illuminating and illuminated realities. This circular reflection is essential to vital and cognitive processes. Her words evidence the crucial function of image reflection as she leads Dante through his initiation to vision, in which he himself must become a mirror: «Ficca di retro a li occhi tuoi la mente, / e fa di quelli specchi a la figura / che 'n questo specchio ti sarà parvente», [«Let your mind follow where your eyes have led, / and let your eyes be mirrors for the figure / that will appear to you within this mirror»]. *Par.* XXI, 16-18.

appears as a reversed type of the Gorgon – her eyes no longer petrify but vivify.[108] The mirror no longer shields the unbearable stare of the goddess, for the woman herself becomes mirror to convey the salvific ray descending from the luminous abyss at the centre of the universe. This mirror abates the overwhelming power of the primary light, just as Perseus's abated the virtue of the Gorgon's mischievous glance.[109] Yet, as anti-Medusa, Beatrice's glance mirrors truth that vivifies, for it allows the poet-hero to see the image it retains without putting his life at risk. The mediation of the reflected image, and metaphorically the mediation of poetry, is necessary to rise toward higher understanding until understanding itself is no longer necessary, and grace grants the power to soar to the direct vision of truth.

The parallel between Beatrice and the mirror emerges in at least three key episodes. The first one occurs in *Purgatorio* XXXI, where Dante, after his immersion in the Letè, looks at the Griffin-Christ through the eyes of Beatrice:

> Mille disiri più che fiamma caldi
> strinsemi li occhi a li occhi rilucenti,
> che pur sopra 'l grifone stavan saldi.
>   Come in lo specchio il sol, non altrimenti
> la doppia fiera dentro vi raggiava,
> or con altri, or con altri reggimenti (*Purg.* XXXI, 118-123).

> A thousand longings burning more than flames
> compelled my eyes to watch the radiant eyes
> that, motionless, were still fixed on the griffin.
>   Just like the sun within a mirror, so
> the double-natured creature gleamed within,
> now showing one, and now the other guise.

---

[108] John Freccero notices the echo of Dante's juvenile *Rime Petrose* in the lines of *Inferno* IX referring to the Gorgon (Medusa). In his interpretation, the Gorgon represents, like the woman of stone of the *Rime*, the «threat of idolatry. In terms of mythological *exempla*, petrification by the Medusa is the real consequence of Pygmalion's folly». As Medusa's anti-type, Beatrice also represents Dante's overcoming of this poetic threat. See JOHN FRECCERO, *Medusa: the Letter and the Spirit*, pp. 175-193; ID., *Dante and the Poetics of Conversion*, p. 130.

[109] Beatrice explains to Dante that her smile would increase her beauty and therefore her brightness, to a point that would be unbearable to his mortal faculties. Therefore, the necessity to temperate her splendor: «ché la / bellezza mia, che per le scale / de l'etterno palazzo più s'accende, / com' hai veduto, quanto più si sale, / se non si temperasse, tanto splende, / che 'l tuo mortal podere, al suo fulgore, / sarebbe fronda che trono scoscende», [«because, as you have seen, my loveliness – / which, even as we climb the steps of this eternal palace, blazes with more brightness – / were it not tempered here, would be so brilliant / that, as it flashed, your mortal faculty / would seem a branch a lightning bolt has cracked»]. *Par.* XXI, 7-12.

The rhetorical duplication of the term «eye» referring to Beatrice and Dante's sights – as in the previously discussed tercets (*Par.* I, 64-66) – stresses the blurred physical boundaries between the poet and his guide's spectrum of vision («strinsemi li occhi a li occhi rilucenti»). As in the first canto of *Paradiso*, the mediated vision allows Dante to bear the brightness of the sun; in Eden it allows him to see the true nature of Christ as unity of human and divine. In Beatrice's glittering eyes the poet can admire the Griffin-Christ radiant like a sun («come in lo specchio il sol»). The repeated stylemes of the eyes, and the simile between the sun and the Christ, establish a textual link between the two cantos and stress the vital effects of Beatrice's mirroring function.

Parallel to this episode is that of *Paradiso* XXIII, where again Dante can set his eyes directly on Christ only after looking at Beatrice. Her glance grants him access to the superior knowledge associated to the heaven of Saturn. Finally, the parallel between the angels and Beatrice's enlightening operation returns in canto XXVIII, which begins with an open simile of Beatrice and the mirror, where Dante sees «Come in lo specchio fiamma di doppiero / vede colui che se n'alluma retro» («just as one who sees a mirrored flame / its double candle stands behind his back», *Par.* XXVIII, 4). The last vision, through the uplifting power of Beatrice's reflecting operation, is in *Paradiso* XXX and XXXI, where her eyes convey to the pilgrim the images of the river of light that from the centre of the universe flows through the Empyrean. Here Beatrice's task is finally accomplished. Without a 'saluto' (literally 'greetings', but metaphorically referring to 'salvation', a power that Beatrice cannot have), Beatrice rejoins the Blessed in the candid rose. Her old lover is left to admire the general form of Paradise in ecstatic stupor. At this point Dante's initiation is complete. From this moment onward, he will be able to see spiritual reality face to face. At the foot of the rose, Bernard tells Dante to raise his eyes and look into the gleaming animated flower of the Blessed to contemplate directly the «pacifica orifiamma» («peaceful oriflamme» *Par.* XXXI, 127). In Canto XXXIII, he will directly set his eyes in the eternal ray.

Beatrice's enlightening operation is instrumental not only in revealing to Dante the sacred order of the cosmos, but also in allowing him a new understanding of his own voyage. The end of the last canticle implicitly answers the initial questions he had posed to Virgil in *Inferno* II: he is a new Aeneas, a new Paul. He is the poet-hero chosen to manifest the eternal truths displayed to him for the spiritual and moral regeneration of Christians. The poem itself is like a mirror that reflects the images of what has been revealed to him. Dante's visionary journey ends when his writing begins, reproducing in the circularity of vision and writing the binary and specular movements of descent and ascent that mark the anagogical process of *exitus-reditus* undertaken by the angels, the Blessed, and the pilgrim himself.¹¹⁰

The poem gives itself as an 'angelic metaphor': it transports us from one realm to another to reconsign us to the historical present. This view, though less self-evident, is made possible by a poetic text in *Purgatorio* II. We remember how the angel arrives on the shores of *Purgatorio* piloting a light and quick boat – a description that suits that of a galley (*galea*) – crowded with souls waiting for purification.¹¹¹ The angel is indeed a '*galeotto*' («allor che ben conobbe il *galeotto*» [«when he had recognized the helmsman», *Purg.* II, 27); a word Dante had memorably used for the 'book' Francesca and Paolo read («galeotto fu 'l libro e chi lo scrisse» [«a Gallehault indeed, that book and he / who wrote it, too» *Inf.* V, 137-138]); a word literally indicating the helmsman, but metaphorically the mediator. Beatrice herself is indeed a '*galeotta*'. She is the light («lume») that mediates between truth and intellect, as Virgil said when announced her to Dante as his guide in *Paradiso*: «che lume fia tra 'l vero e lo 'ntelletto. / Non so se 'ntendi: io dico di Beatrice», [«she, / the light between your mind and truth, will speak – / lest you misunderstand, the she I mean is Beatrice»; *Purg.* VI, 45-46]. Investing his poem with angelic attributes was for Dante a way of inviting his readers to take a journey under the guidance of angels, as he did under the angel-like Beatrice.

c. *The Analogical Model*

The narrative of the *Commedia* could not unfold unless the Blessed, and specifically Beatrice, could assume angelic functions and operate as angelic Intelligences. The question of whether the Blessed could take angelic offices was an aspect of a wider issue concerning the analogical relations between separate substances. In the *Commedia*, the most important element evidencing the analogy between the blessed Beatrice and the angels is her enlightening power, which mediates to Dante the knowledge necessary to complete his journey. Beatrice's analogy (through the angels) with the luminous princes of Dantean cosmology, Christ and the Virgin, also relies on this mediating

---

¹¹⁰ In *Paradiso* I, 103-126 Dante explains that the order of creation tends in an ascending movement toward the Empyrean. In *Paradiso* II, 112-141 he illustrates the opposite movement from God toward the creatures. Corrado Bologna shows that circularity also characterizes the *Commedia* at a linguistic level. See CORRADO BOLOGNA, *Il ritorno di Beatrice. Simmetrie dantesche tra Vita Nuova, "Petrose" e Commedia*, Roma, Salerno, 1998.

¹¹¹ «Da poppa stava il celestial nocchiero, / tal che faria beato pur descripto; / e più di cento spirti entro sediero». [«The helmsman sent from Heaven, at the stern, / seemed to have blessedness inscribed upon him; / more than a hundred spirits sat within»]. *Purg.* II, 43-45.

function. To best situate Beatrice within the network of analogical relations that hold together both spiritual and material realities, we need to explore the terms of this analogy.

As seen in the preceding paragraphs of this section, in the *Vita Nuova* Dante based the analogy between Beatrice and the angels on the exceptional inclinations of the *gentilissima*, and he extended it to Christ and the Virgin (respectively the 'king' and the 'queen' of angels in Christian tradition). Both the Virgin and Christ are figures of mediation, although in different ways: the first as interceding figure, the second as the real and unique mediator of grace.[112] Mediation is thus the general term by which it is possible to think of the analogy linking these figures in the *Commedia*. The term includes enlightening as instrumental mediation for the communication between spiritual and human worlds.

---

[112] Charles Singleton observed that the analogy with Christ was in terms of «advent of light», and considered Beatrice a metaphorical representation of Christ. See CHARLES SINGLETON, *Journey to Beatrice*, p. 81. For an interpretation of Beatrice as a Christological figure representing Dante's «personal Word», which stresses the influence of sapiential Wisdom on Dante's creation of the literary Beatrice, see ANTONIO ROSSINI, *Il Dante sapienziale*, particularly chaps. V-VI. With a different approach, Giuliana Carugati insists on the historical value of Dante's invention of Beatrice as a purely intellectual and literary figure. For the scholar, she is a figure of desire influenced by both Neo-Platonic gneosiology and Christian theology, but coinciding neither with an allegorization of sapiential wisdom or grace, nor with the *anima mundi*. Although these traditions permeate Dante's contemporary medieval thought, Beatrice remains related to Dante's experience – particularly in the *Vita nuova* – and represents the «verace apparizione del nulla che è l'unico, umano e divino, desiderio» [«the true revelation of nothingness, which is the only, human and divine, desire»], the «irriducibile "altro"» [«irreducible "other"»]. See GIULIANA CARUGATI, *Il ragionare della carne*, p. 112 and 206. The figure of Mary, triumphant in canto XXXIII, becomes more and more visible throughout the canticles suggesting a relation of analogy between her and Beatrice. The figurative link with Mary begins in the *Vita Nuova* and culminates in the parallel prayers of Dante and Bernard in cantos XXXI (79-90) and XXXIII (1-39) of *Paradiso* through the crucial passage of *Purgatorio* XXX. Renato Nicodemo, finds that in her descent Beatrice is an image of Mary, although Beatrice cannot be thought of as a literal symbol of the Virgin: «E il "veni, sponsa, de Libano" è tratto dal *Cantico dei Cantici* che il *sensus fidei* della Tradizione patristica e della Liturgia ce lo fanno leggere in senso mariologico, portandoci a contemplare in trasparenza Maria nella "sposa" tutta bella e senza macchia, la quale con la sua purezza originaria – contrapposta all'infedeltà adultera d'Israele – ricapitola e sublima in sé un "nuovo Israele", cioè la Chiesa», [«and the "come spouse of Lebanon" derives from the *Canticle of Canticles*, which the *sensus fidei* of Patristic and liturgical traditions are read in Mariological sense, leading us to see Mary through the beautiful and spotless "spouse," who in her originary purity – as opposed to the adulterous infidelity of Israel – recapitulates and symbolizes a "new Israel," namely the Church»] RENATO NICODEMO, *La vergine Maria nella Divina Commedia. Aspetti del pensiero teologico di Dante Alighieri*, Firenze, Atheneum, 2001, p. 49. R. Nicodemo also observes that all the blessed women in the Holy Rose were interpreted in the Marian liturgy as figures of the Virgin. Rachel was a type of Mary as 'divine shepherdess'; Sara was the 'mother of the victim'; Rebecca represented Mary's intercession even beyond or without request (like in the case of Dante); Giuditta, was an image of Mary as winner over Lucifer; and Ruth represented Mary's intercession in favor of the sinners. For a detailed account of the Biblical sources of these figural relations see RENATO NICODEMO p. 72, n. 44. For the presence of Mary in the *Divine Commedia* see also FRANCESCO TRUSSO, *La Madonna nella Divina Commedia*, Firenze, Propaganda Missionaria, 1965.

In re-telling the story of Dante's enamourment in *Purgatorio* XXXI, Beatrice rebukes Dante for not having sought through her eyes the image of true love, an act that would have activated his willingness to pursue the Good. The iconic, silent, Beatrice of the *Vita Nuova* did not succeed in directing Dante toward the path of virtue.[113] Dominated by erotic love, Dante's mind was unable to 'see' and therefore to receive 'beatitude' from Beatrice's «saluto». Only at the end of the *libello* did Dante have the intuition that another kind of love would elevate him to a new vision and a new type of poetry, that of the *Commedia*, where the tension between knowledge and desire would bridge human and divine. In this new context, Beatrice ceased to be a sign of transcendent reality and assumed an active role that implied a new kind of relation with the angelic dimension: neither metaphorical nor allegorical, but analogical.[114]

In the twelfth and the thirteenth centuries, analogy was a crucial topic in philosophical and theological debates investigating the relation between God and His creatures and the ways language should express it. From Boethius to Cajetan, medieval thought produced various classifications and definitions of analogy, which can be grouped in two major types: the analogy of proportion, and the analogy of attribution.[115] The first type of analogy indicates that two

---

[113] At the beginning, Dante sees Beatrice's light in an imperfect way, for he is not able to recognize the image of love-*caritas* in his love for Beatrice. Although Dante had sanctifying grace, his moral virtue was not yet cooperating with it. He still needed to participate in the knowledge allowed by the gifts of the Holy Spirit that blossom from *caritas*.

[114] In order to explain the conflict between Dante's treatment of the *Donna Gentile* in the *Vita Nuova* and in the *Convivio*, Luigi Pietrobono conjectured the existence of two different versions of the end of the *Vita Nuova*. Some critics such Bruno Nardi and Maria Corti accepted his thesis, while others, including Michele Barbi and Mario Marti, opposed it. According to the sustainers of the hypothesis of the double redaction, the first version ended with the episode of Dante's consolation with the *Donna Gentile* (Philosophy). This would be the 'error' for which Beatrice rebukes Dante in *Purgatorio*. This version of the *Vita Nuova* would have had an ideal continuity with the *Convivio*, dedicated to the importance of Philosophy as the science allowing to understand the existence of God, i.e. *ancilla theologiae*. Later, however, Dante realized his philosophical and poetic mistake and wrote a different ending, in which he affirmed the impossibility of knowledge and happiness without the action of grace, coherently with the perspective present in the *Commedia*. For a brief excursus on the debate on the double redaction of the *Vita Nuova*, and on the pros and cons of the two positions see MARIA CORTI, *Quel rompicapo del finale della vita nuova*, in *La felicità mentale*, republished in *Scritti su Cavalcanti e Dante*, Torino, Einaudi, 1983, pp. 166-175.

[115] Besides the two analogies of proportion and attribution, a third type used by theologians was that of imitation, which referred to a relation of likeness between God and creatures. Attributes such as 'good' would then refer to the creatures in so far as their goodness reflects God's goodness. This type of analogy seems more problematic when referred to operations such as enlightening or mediation. Of the three types of analogy, the analogy of attribution was the one central to medieval debate. Aristotle treats analogy in various works, among these are *Physics* VII, 4 (249 a 22); *Ethics to Nicomachus*, I, 6 (1096 b 26); *Metaphysics* IV, 2 and V, 6 (1017 a 1). AQUINAS, deals with analogy particularly in I *Sententiae* dist. XIX q. 5, a. 2 ad 1, and *Summa Theologiae* I^a, q. 13, a. 6. The cardinal

terms share the same essential properties with a difference only in proportion. The *Vita Nuova* seems to imply this type of analogy, which in rhetorical terms corresponded to the metaphorical representation of Beatrice as angel, and also as a 'figure' of both Christ and Mary.[116] While this kind of analogy could apply to creatures (taking into account their different degrees of perfection), it was however problematic when referred to Christ, for it did not take into account the unique way in which God is in relation with his own attributes. Goodness or luminosity are not predicable of God in the same way as of creatures. God is neither good nor luminous: He is goodness and He is light. In the *Commedia*, where the analogy centers on operations, it seems more appropriate to speak in terms of an analogy of attribution. This type of analogy implies that the analogated terms only participate in the property they share, each according to its different natures and functions. Participation can be extrinsic, when the analogated terms do not formally possess that property, or intrinsic, when they do. In rhetorical terms, the analogy of attribution would still allow for attributing to the members of the analogy images (and language) pertinent to the attributes of the primary analogate, thus producing the linguistic dilatation that characterizes the plurisemic character of Beatrice.[117]

Tommaso de Vio (Cajetan) authored the first synthesis of the complex medieval doctrine of analogy in his *De nominum analogia*, Roma, Institutum Angelicum, 1934 (1498) (*The analogy of names*, transl. by E.A. Bushinski and H.J. Koren, Pittsburgh, Editions E. Nauwelaerts, 1959). See also FRANCO RIVA, *Analogia e univocità in Tommaso de Vio "Gaetano"*, Milano, Vita e Pensiero, 1995. For the relation between metaphorical and allegorical forms with analogy, see JEAN PÉPIN, *Allegoria ED*, pp. 151-165; ANTONINO PAGLIARO, *Similitudine*, in *ED*, pp. 253-259; A. CHOLLET, *Analogie*, in *Dictionnaire de Théologie Catholique*, pp. 1142-1154; and UMBERTO ECO, *La metafora nel medioevo latino*, in *Metafora e conoscenza*, Milano, A.M. Lorusso, 2005, pp. 149-203, and ID., *Dall'albero al labirinto. Studi sul segno e l'interpretazione*, Milano, Bompiani, 2007, pp. 148-158. For a critical exposition of medieval doctrines of analogy see JENNIFER E. ASHWORTH, *Les théories de l'analogie du XIIe au XVIe siècle*, Paris, Vrin, 2008. Essential to situate analogy in the broader context of medieval exegesis is the classical text by HENRI DE LUBAC, *L'exégèse médiéval*, Paris, Aubier, 1959.

[116] Charles Singleton observed that Dante's use of images and language derived from the New Testament's description of Christ's death and ascension to heaven. The relation of analogy between Christ and Beatrice in the terms analyzed in this section admits to transfer language and images from the primary analogate to the others. In particular, the analogy of attribution corresponds to the rhetorical figures of *synecdoche*. Singleton rightly observed that Beatrice is not an allegory. However, he then postulated a relation of analogy of proportion between Christ and Beatrice, which would correspond to the rhetorical figure of the metaphor, on which an allegory is built. See CHARLES SINGLETON, *The death of Beatrice*, pp. 6-24 and *Beatrice dolce memoria*, pp. 110-116, in *An Essay on the Vita Nuova*, Baltimore, The Johns Hopkins University Press, 1949. See also ID., *The Pattern at the Centre*, in *Commedia. Elements of Structure, Dante Studies I*, Cambridge, Harvard University Press, 1954, pp. 45-60.

[117] For the complexity of symbolism and allegory when referring to a historical character. See JEAN PÉPIN, *Allegoria*, p. 152, particularly his comment to ETIENNE GILSON's *Dante et la philosophie*, Paris, Vrin, 1939.

CHAPTER FOUR

The notion of extrinsic participation in the attribute of mediation describes the figural links that tie Beatrice, the angels, and Mary to Christ as main analogate. The first three are figures of 'mediation' as they are instrumental to the realization of the providential design of salvation, and are linked by this common function even though they differ in nature and gifts of grace. They participate externally in this property that is formally predicable of Christ only. To assume this type of analogy rules out the possibility of a supposed ontological continuity between Beatrice and her analogates, while at the same time it maintains intact the literal value of her figure without recurring to notions of allegory or personification.[118] Thinking of Beatrice in terms of an analogy of extrinsic attribution implies the abandonment of the metaphorical woman-angel, and of the divinized Beatrice, to leave space for her literal, 'realistic', angelization.

Beatrice, the angels, and Mary belong to the same category of analogates because the Mariology that Dante displays in the final cantos of the *Commedia* shows the influence of Bernardian rather than Thomistic theology.[119] In Aqui-

---

[118] Francesco Mazzoni, in developing a previous suggestion by Mario Casella, analyzed the character Beatrice in terms of *analogia entis*, and rejected the traditional allegorical reading of the *gentilissima*. Commenting on Dante's declaration in the *Vita Nuova* that Beatrice is an angel by similitude he observes: «l'inciso specificante "per *similitudine* dico" rinvia immediatamente un lettore medievale alla nozione dell'analogia, e non certo, come ritenevano i critici simbolico-allegorici, a quella (ben diversa proprio nelle pratiche implicazioni) d'allegoria». [«the incidental clause "per *similitudine* dico" immediately refers the medieval reader to the notion of analogy, and certainly not to that of allegory (different in its practical implications), as the symbolic – allegorical critics argued»]. FRANCESCO MAZZONI, *Il 'trascendentale' dimenticato*, p. 108.

[119] Aquinas examined the theological importance of Mary in the *Summa contra gentiles* and in the third part of the *Summa Theologiae*, in particular in question 27, article 5: «Et quia Christus, inquantum est homo, ad hoc fuit praedestinatus et electus ut esset *praedestinatus filius Dei in virtute sanctificationis*, hoc fuit proprium sibi, ut haberet talem plenitudinem gratiae quod redundaret in omnes, secundum quod dicitur Ioan. I, *de plenitudine eius nos omnes accepimus*. Sed beata virgo Maria tantam gratiae obtinuit plenitudinem ut esset propinquissima auctori gratiae, ita quod eum qui est plenus omni gratia, in se reciperet; et, eum pariendo, quodammodo gratiam ad omnes derivaret.» [«And since Christ as man was predestinated and chosen to be "predestinated the Son of God in power [...] of sanctification" (Rm. 1:4), it was proper to Him to have such a fulness of grace that it overflowed from Him into all, according to Jn. 1:16: "Of His fulness we have all received". Whereas the Blessed Virgin Mary received such a fulness of grace that she was nearest of all to the Author of grace; so that she received within her Him Who is full of all grace; and by bringing Him forth, she, in a manner, dispensed grace to all»]. Aquinas' cautious «quodammodo» («in a manner») implicitly refers to Mary's role in participating to Christ's mediation of grace as secondary cause, being *theotokos*. For Bernard's Mariology in the *Divine Commedia*, see STEPHEN BOTTERILL, *Dante and the Mystical Tradition: Bernard of Clairvaux in the Commedia*, Cambridge, Cambridge University Press, 1994, pp. 64-118. See also J. RIVIÈRE, *Redemption, Dictionnaire de Théologie Catholique*, pp. 1912-2004. On the figure of Mary in the *Commedia* see PIER GIORGIO RICCI, *Maria Vergine*, in *ED*, pp. 835-839. On the Marian aesthetic of the *Commedia*, and the transformative power of beauty, see BRIAN REYNOLDS, *A Beauty that Transforms: the Marian Aesthetics of Dante's Commedia*, «Fu Jen Studies», XXXV, 2003, pp. 61-89.

nas, the Virgin subordinately cooperates to the mediation of grace as a secondary cause of Incarnation by being a God-bearer (*theotokos*). Bernard instead likened Mary to an *aquaeductus* (aqueduct) through which grace flows but from which it does not originate. Similarly, Beatrice's role as a guide can be interpreted as 'mediation' only in attributive sense, for she can neither intercede nor mediate grace. Mary only can intercede, and mediation, as well as redemption, can be formally predicated of Christ only. Just as the angels and Mary, Beatrice is only extrinsically related to this term. She cannot save Dante, but can mediate to him the knowledge necessary to rise to the supreme vision of divine light. Like the angels, she cannot even intercede for grace, for this function pertains to Mary. The roles of the Virgin, the angels, and Beatrice as mediators consists in their capacity to 'reveal' the divine through vision; to establish a link between material and spiritual worlds to guide humankind along the path designed by divine Providence. The analogy of attribution maintains its validity even if one considers the term enlightenment instead of mediation. Although this function properly pertains to the Christ – Sun, it is also extrinsically predicable of Mary, the angels, and Beatrice. The relation among the various analogates reflects the various degrees of subordination dictated by their different degrees of perfection. Only in this sense does it seem possible to reconcile the textual evidence of the figurative relations linking Beatrice with the angels, Mary and Christ and remain on safe theological ground.

CONCLUSIONS

All figures of the *Divine Commedia*, from prophetic to exemplary, contribute in different ways to the providential design inscribed in the architecture of the poem. This chapter has shown that as angels constitute the model of providential intervention, angelic figures performing angelic functions contribute to the eschatological meaning of the poem. First, the similitude between angels and Blessed allowes Dante to attribute angelic tasks to non-angelic characters. Second, the presence of angels in *Purgatorio* is connected to its structural design as a ladder to purification attainable only with the intervention of the gifts of the Holy Spirit. By connecting gifts and Beatitudes, angels manifest the presence of the Holy Spirit as universal ordering principle in the realm of atonement. Third, the heavenly messenger of *Inferno* IX is identified with a syncretic figure representing the medieval Christianization of the ancient pagan god Mercury, transformed into angelic Intelligence. The mercurial 'veil' of the angel not only has the function of making this figure homogeneous with the infernal figurative rhetoric, but is also necessary to connect Dante's own descent as Christian hero to the tradition of classical pagan descents and their medieval exegesis, and to define Dante's allegorical and hermeneutical strategies.

Against this background, Beatrice's operation of enlightenment, aimed at enhancing and steer understanding toward its natural and salvific end, emerges as her distinctive angelic trait. Like angels, Beatrice performs her illuminating operation within the general framework of providential order and provides the enlightenment that conveys to Dante the knowledge of what is needed to progress in the universal path along which all creatures return to their creator. Beatrice's main function in the *Commedia* is therefore instrumental to Dante's increasing capacity to see beyond human limits, the limits the poet is able to cross. To this end, I have explored the analogical relations that link Beatrice as mediator of knowledge to the angels, Mary and Christ, and argued that Dante constructed the figure of Beatrice as analogical model of the angel. She is no longer the 'angelicated' woman of the lyric tradition, but a blessed soul performing the function of an Intelligence acting as an instrument of divine Providence; a 'mirror' of light that stands besides the angels. I have showed that from the *Vita Nuova* to the *Commedia* Dante re-elaborated the metaphysical profile of Beatrice by reformulating her angelic image on the basis of the operations she performs as enlightening guide of the poet-pilgrim. In mirroring divine light, she has become an angelic character in a literal sense. While in the *Vita Nuova* Beatrice's analogy with the angels relied on her virtuous attributes, in the *Commedia* this analogy hinged on her enlightening operation, revealing an evolution in the literary figure of Beatrice that parallels the development of Dante's angelology.

CONCLUDING REMARKS

This book has shown the innovative traits of Dante's angelology with respect to the Dionysian tradition and mainstream medieval angelology and has demonstrated its relevance for the structure and the interpretation of the poem. From the analysis performed it emerges that the contemplative and active operations of angelic Intelligences are instrumental to connect physical and spiritual worlds and thus essential to reconcile the Christian doctrine of creation with Aristotelian cosmology. The angelology of the *Commedia*, far from being the result of involuntary syncretism and a simple mapping of encyclopedic knowledge, was the result of a conscious and innovative adaptation of different disciplines and angelological traditions that underpin Dante's poetic design of an essentially eschatological cosmology. Angels are the key to decipher this design and its poetic implications. Not only they are essential elements in the structure and the ideology of the *Commedia* but, as shown, they constitute exemplars under which Dante modeled major poetic figures such as the heavenly messenger, the angels of *Purgatorio*, and Beatrice, whom, I argue, stands out as an analogical model of the angel. The investigation of their operations has also provided new insight on the interpretation of the cantos concerning creation and cosmology, determinism and free will.

Focusing on the solutions Dante found to the philosophical issues involved in structuring his angelology, this book has illustrated three main innovative traits of his angelology. The first is Dante's definition of angelic science as naturally extended to the practical sphere so that contemplation immediately transforms into operative intervention in the order of the cosmos. The second is Dante's assumption that all angelic orders receive direct divine enlightenment, and that it conveys the science necessary to their specific ministries. Third, Dante not only assigned to angels the function of celestial movers, but also established an unprecedented exact correspondence between the nine angelic choirs and the nine material spheres. These departures from mainstream medieval angelology constitute the nucleus of Dante's angelology, and are crucial to the cosmological structure of the *Commedia*. By founding his angelology on philosophical premises, Dante reconciled the Dionysian contemplative system of hierarchies with the commonly accepted principle of Aristotelian celestial causality. In so doing, physical time and space could find their ground and their origin in the timeless and space-less dimension of the Empyrean.

Based on these premises, I have shown that Dante's notion of universal order builds on the threefold hierarchical relation among Love, Providence, and angels as ministers of divine Providence to assure the connection between the spiritual and material sides of the universe. In this system, angelic operations are philosophically, theologically, and poetically connected for they assure the relation between the One and the multiplicity, move the celestial bodies keeping in balance predestination and free will, and assist human souls in their path to salvation.

Reading the *Commedia* from the perspective of angelology has evidenced the impossibility of confining Dante's thought to a specific field of influence. Rather than simply Thomist or Bonaventurian, Dante's vision of creation intertwines Neo-Platonic and Aristotelian scholastic interpretations of Dionysian angelology. The structural complexity of his poem results from the effort to organize all branches of knowledge within the all-encompassing design of a Christian universe where micro and macrocosmos mirror each other in the light of the angels.

My investigation of Dante's angelology suggests that the vision of the cosmos that the *Commedia* displays is not a vision of pure transcendence but one in which angels are new demiurgs, created not to forge but to convey the active principles of life descending from the highest point of universe and that make the great mechanism of creation work in unperturbed harmony. The role of angels is not simply to connect visible and invisible, but also to make the invisible visible; if God enlightens, they illuminate to manifest divine truths. They do not convey knowledge of the divine mind, unaccessible even to spiritual creatures, but knowledge that strictly pertains to the governing of creation in the perspective of salvation. In making angels the connecting link of intellectual and physical dimensions of the universe, Dante used and transgressed the hierarchical organization of angelic orders to assure creation against emanationism, and to provide a bridge between creator and creatures.

The analysis contained in this book evidences the eschatological character of Dante's cosmology and shows that the poem itself is comprised within a providential perspective. It has revealed that the cosmos in which the pilgrim undergoes in body and soul his ultramondane voyage is a universe of grace. The vision that from step to step takes us from hell to *Paradiso* is not a vision of the absolute; of the creator beyond creation. What we experience with Dante as author and protagonist is not the infinite dimension of divinity, but the finite dimension of grace, circumscibed within the boundaries of a beginning and an end, where everything has a function and a meaning in view of the completion of the act of creation by restoring unity with God. The three realms Dante shows to us are not in their absolute and eternal state of perfection. As we are suspended between birth and Resurrection so the cosmos is between the first day of creation and the Judgement Day. What will be of Charon, and Minos, and Flegias after that day? What of the sacred mountain of *Purgatorio*? And even the empty seats in the candid rose cast a shadow of temporality on the adamantine surface of eternity. It is the universe of grace and of glory, of enlightening Wisdom, centered on the two fundamental figures of grace, Mary and Christ, where the light that angels reverberate and pervades the cosmos is the light of the incarnated Word. Dante's vision corresponds to the only knowledge accessible to human beings, that of salvation as one of the infinite dimensions of eternal divinity: the only one we can see with the eyes of the mind, through the visionary word of poetry. The blazing point resplending at the center of Empyrean is the manifestation of the eternal being, penetrable only with the superhuman quality we may gain with the aid of grace. It is not by chance that at the climax of his journey, when he finally contemplates face to face the mystery of revelation, the core of brightness opens to show the image of a face: the seal of god in human flesh, and the image of the likeness we must seek. What is beyond the face and the point, the essence of eternity, remains beyond the light, perhaps in the absolute oscurity of which Dionysius spoke, in what cannot be said, and which Dante indeed does not say. The reiterated proclamations of ineffability that punctuate the *Paradiso* are literal affirmations that the poem does not and cannot speak the unspeakable, but it can and does speak within the limits of what can be said and seen, within the limits of creation, where love and knowledge are the breath and the light of the cosmos that the angels bring to us.

WORKS CITED

ABARDO, R. (ed.), *Omaggio a Beatrice*, Firenze, Le Lettere, 1997 («Quaderni degli Studi Danteschi, 11»).

ALAIN OF LILLE, *Anticlaudianus*, J.J. Sheridan ed., Toronto, Pontifical Institute of Medieval Studies, 1973.

— *Anticlaudianus*, R. Bossuat ed., Paris, Vrin, 1955.

AHERN, JOHN, *Can the Epistle to Cangrande Be Read as a Forgery?*, in *Atti of the First International Dante Seminar*, Z. Baranski ed., pp. 279-307.

ALBERT THE GREAT, *Alberti Magni Opera Omnia*, t. XVIII, 1, Monasterii Westfalorum, 1975.

— *The Book of Minerals (De mineralibus et rebus metallicis)*, D. Wyckoff transl., Oxford, Clarendon Press, 1967.

— *Super Dionysium De coelesti hierarchia*, P. Simon and W. Kübel eds., Aschendorff, Monasterii Westafalorum, 1993.

ALEXANDER OF HALES, *Glossa in quatuor libros sententiarum Petri Lombardi*, Florence, Quaracchi, 1952.

ALIGHIERI, DANTE, *La Commedia secondo l'antica vulgata di Dante Alighieri*, Giorgio Petrocchi ed., Milano, Mondadori, 1966-67, republished by Einaudi, Torino, 1977.

— *Convivio*, in *Opere minori*, tome I, part II, C. Vasoli and D. De Robertis eds., Milano-Napoli, Ricciardi, 1988 («La letteratura Italiana. Storia e testi», 5).

— *Dante's "Il Convivio" (The Banquet)*, R.H. Lansing ed. and transl., New York-London, Garland, 1990 («Garland Library of Medieval Literature Series B», 65).

— *Dante's Monarchia*, R. Kay transl., Toronto, Pontifical Institute of Medieval Studies, 1998 («Studies and Texts», 131).

— *The Divine Comedy of Dante Alighieri*, Allen Mandelbaum transl., Berkeley, University of California Press, 3 vols., 1980-1982.

— *Monarchia*, in *Opere minori*, B. Nardi ed., Milano-Napoli, Ricciardi, 1979, pp. 241-503.

— *Vita Nova*, L.C. Rossi ed., Milano, Mondadori, 1999.

— *Vita Nuova*, G. Gorni ed., Torino, Einaudi, 1996.

AMBROSE, SAINT, *Duties of the Clergy (De officiis ministrorum)*, in *Nicene and Post-Nicene Fathers of the Christian Church*, II, P. Shaff and H. Wace eds., Grand Rapids, Michigan, W.B. Eerdmanns, 1955, pp. 1-89.

APPIANO, AVE, *Le forme dell'immateriale*, Torino, Società Editrice Internazionale, 1996.

APULEIUS, LUCIUS, *Le metamorfosi (Metamorphoseon)*, Milano, Rizzoli, 1989.

AQUINAS, THOMAS, *Contra Gentiles*, Vernon J. Bourke transl., Thomas Institut, http://www.thomasinstituut.org.

— *De ente et essentia*, P. Porro ed. and transl., Milano, Bompiani, 2002.

— *De spiritualibus creaturis*, critical ed., L. Keeler ed., Roma, Universitas Gregoriana, 1937. Translation by M.C. Fitzpatrick and J.J.Wellmuth, Milwaukee, Marquette University Press, 1949.

— *On Being and Essence (De ente et essentia)*, transl. Robert T. Miller, 1997, Medieval Sourcebook, Centre for Medieval Studies, Fordham University, http://www.fordham.edu/halsall/basis/aquinas-esse.html.

— *Opera omnia*, iussu Leonis XIII P.M. edita, cura et studio fratrum ordinis praedicatorum, Romae, 1882-.

— *Responsio ad magistrum Joannem de Vercellis de 43 articulis*, in *Opera omnia*, Studium Fratrum Predicatorum, t. XLII, Rome, 1979, pp. 327-335.

— *The Summa Theologica of Saint Thomas Aquinas*, edited by the Fathers of the English Dominican Province, 5 vols., Christian Classics, 1981.

ARIANI, MARCO, *"E sí come di lei bevve la gronda / de le palpebre mie" (Par I.XXX.8)*: Dante e lo Pseudo-Dionigi Areopagita, in *Leggere Dante*, Lucia Battaglia Ricci ed., pp. 131-152.

— *'Metafore assolute': emanazionismo e sinestesie della luce fluente*, in *La metafora in Dante*, M. Ariani ed., pp. 193-219.

— (ed.), *La metafora in Dante*, Firenze, Olschki, 2009.

ARISTOTLE, *Metaphysics*, W.D. Ross transl., in *The Complete Works of Aristotle*, J. Barnes ed., pp. 1552-1728.

— *Physics*, R.P. Hardie and R.K. Gaye transl., in *The Complete Works of Aristotle*, J. Barnes ed., 1984, pp. 315-446.

ASCOLI, ALBERT R., *Access to Authority: Dante in the Epistle to Cangrande*, in *Atti of the First International Dante Seminar*, Z. Baranski ed., pp. 309-352.

— *Epistola a Cangrande*, in *The Dante Encyclopedia*, R.H. Lansing ed., New York, Garland, 2000 (Garland Reference Library of the Humanities, vol. 1836), pp. 348-352.

ASHWORTH, JENNIFER E., *Les théories de l'analogie du XIIe au XVIe siècle*, Paris, Vrin, 2008.

AUGUSTINUS, AURELIUS, *The City of God (De civitate dei)*, transl. H. Bettenson, London, Penguin, 1984 (1972).

— *De sermone domini in monte*, I, 2, in Aurelii Augustini, *Opera Omnia*, http://www.augustinus.it.

— *Enarrationes in Psalmos*, in *Esposizioni sui Salmi* («Opere di Sant'Agostino», III), Roma, Città Nuova, 1982 (1967).

— *Expositions on the Psalms*, vol. V, Oxford, J.H. Parker ed., 1853.

— *La genesi alla lettera*, Roma, Città Nuova (*Opera omnia di Sant'Agostino*), vol. IX/2, 1989.

— *On Genesis*, E. Hill, O.P. and J.E. Rotelle, introduction, transl., and notes, O.S.A., New York, New City Press («The Works of Saint Augustine. A Translation for the 21st Century», I/13), 2002.

— *The Lord's Sermon on the Mount*, J.J. Jepson, transl., Ancient Christian Writers, New York, The Newman Press, 1948.

BALTRUSAITIS, JURGIS, *Cercles atrologiques et cosmographiques à la fin du Moyen Âge*, «Gazette des Beaux-Arts», XXI, 1939, pp. 65-84.

— *L'image du monde celeste du IXe au XIIe siècle*, «Gazette des Beaux-Arts», XX, 1938, pp. 138-148.

BARANSKI, ZYGMUNT, *Dante tra dei pagani e angeli cristiani*, «Filologia e critica», IX, 1984, pp. 298-299.

— (ed.), *Atti of the First International Dante Seminar*, Firenze, Le Lettere, 1997.

BAREILLE, G., *Angeologie d'apres les Pères* (Ange), *Dictionnaire de Théologie Catholique*, A. Vacant, E. Mangenot, E. Amann eds., Paris, Letouzey et Ané, 1899-1972, pp. 1192-1211.

BAREILLE, G., MISKGIAN, J., PARISSOT, J., PETIT, L. and VACANT, A., *Ange*, in *Dictionnaire de Théologie Catholique*, A. Vacant, E. Mangenot, E. Amann eds., Paris, Letouzey et Ané, 1899-1972, pp. 1189-1248.

BARNES, J. (ed.), *The Complete Works of Aristotle*, 2 vols., Princeton, Princeton University Press, 1984.

BAROLINI, TEODOLINDA, *Dante's Poets: Textuality and Truth in the Comedy*, Princeton, Princeton University Press, 1984.

— *The Origins of Italian Literary Culture*, New York, Fordham University Press, 2006.

BAROLINI, T. and STOREY, W.H. (eds.), *Dante for the New Millennium*, New York, Fordham University Press, 2003.

BARSELLA, SUSANNA, *The Mercurial Integumentum of the Heavenly Messenger (Inferno IX, 79-103)*, «Letteratura Italiana Antica», IV, 2003, pp. 371-395.

— *The Role and Function of Angels in Dante Alighieri's Purgatorio*, paper presented at the Graduate Colloquium on French, Hispanic, German and Italian Literature, Culture and Language, The Catholic University of America, Washington, March 1996.

BARTHOLOMEW OF BOLOGNA, *Tractatus de luce Fr. Bartholomaei di Bononia inquisitions et textus*, I. Squadrani ed., «Antonianum», VII, 1932, pp. 201-238; 337-376; 465-494.

BASIL OF CESAREA, *Sulla Genesi (Omelie sull'esamerone)*, M. Naldini ed., Milano, Mondadori (Fondazione Lorenzo Valla), 1990.

BATTAGLIA RICCI, LUCIA (ed.), *Leggere Dante*, Ravenna, Longo, 2003.

BEMROSE, STEPHEN, *Dante's Angelic Intelligences*, Roma, Edizioni di Storia e Letteratura, 1983.

BERNARD OF CLAIRVAUX, *Five Books on Consideration. Advice to a Pope*, Kalamazoo, Cistercian Publications, 1976.

BETTINI, MAURIZIO, *Le orecchie di Hermes. Studi di antropologia e letterature classiche*, Torino, Einaudi, 2000.

Biblia Sacra Vulgatae Editionis, Milano, San Paolo, 1995.

BOBIK, JOSEPH, *Aquinas On Being and Essence*, Notre Dame, Notre Dame University Press, 1970.

BOCCASSINI, DANIELA, *Il volo della mente. Falconeria e sofia nel mondo mediterraneo: Islam, Federico II, Dante*, Ravenna, Longo, 2003.

BOETHIUS, SEVERINUS, *The Consolation of Philosophy*, Oxford, Oxford University Press, 2000.

BOFFITO, GIUSEPPE, *Quali esperienze e leggi dell'ottica furono note a Dante e Cecco d'Ascoli?*, «Rivista di fisica, matematica e scienze naturali», s. 2, IV, 1929-30, pp. 225-228.

— *La teoria della visione in Dante e in Cecco d'Ascoli*, «Rivista di fisica, matematica e scienze naturali», s. 2, IV, 1929-30, pp. 67-73.

BOLOGNA, CORRADO, *Il ritorno di Beatrice. Simmetrie dantesche tra Vita Nuova, "Petrose" e Commedia*, Roma, Salerno, 1998.

BONAVENTURE, SAINT, *Collationes in Hexaëmeron*, in *Opera Omnia*, IX, 10 vols., Quaracchi ed., Firenze, Collegium S. Bonaventura, 1892-1902.

— *Collations on the Six Days*, New Jersey, St. Anthony Guild Press, 1970.

BOTTERILL, STEPHEN, *Dante and the Mystical Tradition: Bernard of Clairvaux in the Commedia*, Cambridge, Cambridge University Press, 1994.

BOUGEROL, J.P., *Saint Bonaventure et la hiérarchie dionysienne*, «Archives d'Histoire Doctrinale et Littéraire du Moyen Âge», XXXVI, 1969, pp. 131-167.

BOUYER, LOUIS, *La spiritualità dei Padri (III-VI secolo) Monachesimo antico e Padri*, in *Storia della spiritualità*, vol. 3/B, Bologna, Edizioni Dehoniane, 1968, pp. 129-156.

BOYDE, PATRICK, *Dante Philomythes and Philosopher: Man in the Cosmos*, Cambridge, Cambridge University Press, 1981.

BOYDE, P. and RUSSO, V. (eds.), *Dante e la scienza*, Ravenna, Longo, 1995.

BRESCHI G. (ed.), *Dante e l'Europa*, Atti del Convegno internazionale di Studi, Ravenna, 29 novembre 2003, Centro Dantesco dei Frati Minori Conventuali, 2004.

BRUGNOLI, GIORGIO, *Nomen Omen. Due nomi parlanti in Dante*, in *I nomi da Dante ai contemporanei. Atti del IV Convegno Internazionale di Onomastica e Letteratura. Pisa, 27-28 febbraio 1998*, Viareggio, Baroni, 1999, pp. 35-45.

BUSNELLI, GIOVANNI, *Il messo del cielo alle porte di Dite*, Roma, Civiltà Cattolica, 1910.

BUSSAGLI, MARCO, *Storia degli Angeli*, Milano, Rusconi, 1995.

CACCIARI, MASSIMO, *L'angelo necessario*, Milano, Adelphi, 1986.

CACHEY, T.J. (ed.), *Dante Now: Current Trends in Dante Studies*, Notre Dame, University of Notre Dame Press, 1995.

CAMPOREALE, SALVATORE, *Lorenzo Valla. Umanesimo e Teologia*, Firenze, Istituto Nazionale di Studi sul Rinascimento, 1972.

CAPPELLI, LUIGI M., *Le gerarchie angeliche e la struttura del Paradiso dantesco*, «Giornale dantesco», VI, 1898, pp. 241-259.

CARUGATI, GIULIANA, *Il ragionare della carne. Dall'anima mundi a Beatrice*, Lecce, Manni, 2004.

CASCIANI, S. (ed.), *Dante and the Franciscans*, Leiden, Brill, 2006 (The Medieval Franciscans, 3).

CASSELL, ANTHONY K., *Dante's Fearful Art of Justice*, Toronto, University of Toronto Press, 1984.

CERVIGNI, DINO, *Segni paragrafali, maiuscole e grafia nella Vita nuova: dal libello manoscritto al libro a stampa*, «Rivista di Letteratura Italiana», XIII, 1-2, 1995, pp. 283-362.

CERVIGNI, DINO and VASTA, EDWARD, *From Manuscript to Print: The Case of Dante's Vita Nuova*, in *Dante Now: Current Trends in Dante Studies*, T.J. Cachey ed., pp. 83-114.

CHANCE, JANE, *Medieval Mythography*. Volume 1: From Roman North Africa to the School of Chartres. AD 433-1177, Gainesville, University Press of Florida, 1994.

CHENU, MARIE-DOMINIQUE, *Involucrum. Le mythe selon les théologiens médiévaux*, «Archives d'Histoire Doctrinale et Litteraire du Moyen Âge», XXII, 1955, pp. 75-79.

— *Nature, Man, and Society in the Twelfth Century. Essays on New Theological Perspectives in the Latin West*, J. Taylor and L.K. Little eds. and transl., Toronto, University of Toronto Press, 1997 (1957).

— *Les réponses de Saint Thomas et de Kilwardby à la consultation de Jean de Verceil*, in *Mélanges Mandonnet*, vol. I, Paris, Vrin, 1930, pp. 191-222.

CHIARINI, GIOACHINO, *Dante e la simbologia classica dei sette pianeti*, in *Dante e la fabbrica della Commedia*, A. Cottignoli, D. Domini, G. Gruppioni eds., pp. 189-197.

CHIAVACCI LEONARDI, ANNA MARIA, *Le Beatitudini e la struttura poetica del Purgatorio*, «Giornale Storico della Letteratura Italiana», CLXI, fasc. 513, 1984, pp. 1-29.

CHOLLET, A., *Analogie*, in *Dictionnaire de Théologie Catholique*, A. Vacant, E. Mangenot, E. Amann eds., Paris, Letouzey et Ané, 1899-1972, pp. 1142-1154.

CIABATTONI, FRANCESCO, *Dante's Journey to Polyphony*, Toronto, University of Toronto Press, 2010.

CICCUTO, MARCELLO, *Michele Scoto e la naturalis philosophia di Dante*, «Tenzone», I, 2000, pp. 31-41.

CICERO, MARCUS TULLIUS, *On Duties*, M.T. Griffin and E.M. Atkins eds., Cambridge, Cambridge University Press, 1991.

COGAN, MARC, *The Design in the Wax. The structure of the Divine Commedia and its Meaning*, Notre Dame, University of Notre Dame Press, 1999.

COLISH, MARCIA L., *Early Scholastic Angelology*, «Recherches de Théologie ancienne et médiévale», LXII, 1995, pp. 80-109.

— *Medieval Foundations of the Western Intellectual Tradition 400-1400*, New Haven, Yale University Press, 1997.

COLLINS, JAMES D., *The Thomistic Philosophy of the Angels*, Washington DC, The Catholic University of America (Philosophical Series), LXXXIX, 1947.

CONTINI, GIANFRANCO (ed.), *Poeti del Duecento*, Milano-Napoli, Ricciardi, 1960.

CORNISH, ALISON, *Planets and Angels in Paradiso XXIX: The First Moment*, «Dante Studies», CVIII, 1990, pp. 1-28.

— *Reading Dante's Stars*, Yale, Yale University Press, 2000.

CORTI, MARIA, *La Commedia di Dante e l'oltretomba Islamico*, in ID., *Scritti su Cavalcanti e Dante*, pp. 365-379.

— *Metafisica della luce come poesia*, in ID., *Percorsi dell'invenzione. Il linguaggio poetico e Dante*, pp. 147-163.

— *Le metafore della navigazione, del volo e della lingua di fuoco nell'episodio di Ulisse (Inferno, XXVI)*, in ID., *Scritti su Cavalcanti e Dante*, pp. 348-364.

— *Il modello analogico nel pensiero medievale e dantesco*, in *Dante e le forme dell'allegoresi*, M. Picone ed., Ravenna, Longo, 1987, pp. 11-20.

— *Percorsi dell'invenzione. Il linguaggio poetico e Dante*, Torino, Einaudi, 1993.

— *Quel rompicapo del finale della vita nuova*, in *La felicità mentale*, republished in *Scritti su Cavalcanti e Dante*, Torino, Einaudi, 1983, pp. 166-175.

— *Scritti su Cavalcanti e Dante. La Felicità mentale, Percorsi dell'invenzione e altri saggi*, Torino, Einaudi, 2003.

COSTA, CRISTINA D'ANCONA, *Recherches sur le Liber de causis*, Paris, Vrin, 1995.

COTTIGNOLI, A. – DOMINI, D. – GRUPPIONI, G. (eds.), *Dante e la fabbrica della Commedia*, Ravenna, Longo, 2008 («Interventi classensi n. 22»).

CRANZ, EDWARD, *The Reorientation of Western Thought c. 1100 A.D. The Break with the Ancient Tradition and its Consequences for Renaissance and Reformation*, paper delivered at the Duke University Center for Medieval Studies, March 24, 1982, pp. 1-24.

CRISTALDI, SERGIO, *Dalle Beatitudini all'Apocalisse: il Nuovo Testamento nella Commedia*, «Letture Classensi», XVII, pp. 23-67.

CRISTIANI, MARTA, *Dionigi Areopagita (Pseudo)*, in *Enciclopedia Dantesca*, 6 vols., vol. II, U. Bosco and G. Petrocchi eds., Roma, Istituto della Enciclopedia Italiana, 1970-78, pp. 460-462.

CURTIUS, ERNST R., *European Literature and the Latin Middle Ages*, Princeton, Princeton University Press, 1990 (1948).

D'ALVERNY, MARIE-THÉRÈSE, *Astrologues et théologiens au XIIe siecle*, in *Mélanges offerts a M-D Chenu*, Paris, Vrin, 1967, pp. 31-50.

Dante e il suo secolo: 15 maggio 1865, Firenze, M. Cellini e C., 1865.

DE LUBAC, HENRI, *L'exégèse médiéval*, Paris, Aubier, 1959.

DE WAELE, FERDINAND J.M., *The Magic Staff or Rod in Graeco-Italian Antiquity*, Ghent, Drukkery Erasmus, 1927.

DI GIOVANNI, VINCENZO, *Gli angeli nella Divina Commedia*, in *Dante e il suo secolo: 15 maggio 1865*, pp. 317-341.

DOEBLER, GIAMPIERO W., *Non mi può far ombra: Le distinzioni fra luce e lume nelle Rime di Dante*, «Tenzone», VII, 2006, pp. 29-50.

DONDAINE, HYACINTHE FRANÇOIS, *Le Corpus Dionysien de l'Université de Paris au XIIIe siècle*, Roma, Edizioni Storia e Letteratura, 1953.

DONDAINE, HYACINTHE FRANÇOIS, et al., *Influence du Pseudo-Denis en Occidént*, in *Dictionnaire de Spiritualité*, Paris, Beauchesne, 1957, III, pp. 318-378.

DUHEIM, PIERRE, *Le Système du Monde. Histoire des Doctrines cosmologiques de Platon à Copernic*, 10 vols., Paris, Hermann, 1913-1959.

ECO, UMBERTO, *Dall'albero al labirinto. Studi sul segno e l'interpretazione*, Milano, Bompiani, 2007.

— *La metafora nel medioevo latino*, in *Metafora e conoscenza*, Milano, A.M. Lorusso, 2005, pp. 149-203.

EICHBERG, BARBARA BRUDERER, *Les neuf choeurs angéliques: origine et évolution du thème dans l'art du Moyen Age*, Poitiers, Centre d'Etudes Supérieures de Civilisation Médiévale, 1998.

ERIUGENA, JOHANNES SCOTUS, *Expositiones in ierarchiam coelestem*, J. Barbet ed., Turnholti, Brepols, 1975.

EUSEBIUS OF CESAREA, *Storia ecclesiastica (Historia Ecclesiastica)*, 2 vols., F. Migliore ed., S. Borzì and F. Migliore transl., Roma, Città Nuova, 2001.

FAES DE MOTTONI, BARBARA, *Bonaventura e le gerarchie angeliche*, «Freiburger Zeitschrift für Philosophie und Theologie», 3, 1993, pp. 312-358.

— *San Bonaventura e la scala di Giacobbe. Letture di angelologia*, Napoli, Bibliopolis, 1995.

FALLANI, GIOVANNI, *Dante poeta teologo*, Milano, Marzorati, 1965.

FASOLINI, DIEGO, *'Illuminating' and 'Illuminated' Light: a Biblical-Theological Interpretation of God-as-Light in Canto XXXIII of Dante's Paradiso*, «Literature and Theology», XIX, 4, 2005, pp. 297-310.

FINAZZI, SILVIA, *La metafora scientifica e la rappresentazione della corporeitas luminosa*, in *La metafora in Dante*, M. Ariani ed., pp. 167-192.

FORTI, FIORENZO, *Matelda*, in *Enciclopedia Dantesca*, 6 vols., vol. III, U. Bosco and G. Petrocchi eds., Roma, Istituto della Enciclopedia Italiana, 1970-78, pp. 854-860.

FOSCA, NICOLA, *Beatitudini e processo di purgazione*, Electronic Bulletin of the Dante Society of America, 2002, http://www.princeton.edu/~dante/ebdsa.

FRANKE, WILLIAM, *Dante's Hermeneutic Rite of Passage, Inferno IX*, «Religion and Literature», XXVI, 2, 1994, pp. 1-26.

— *Dante's Interpretive Journey*, Chicago, University of Chicago Press, 1996.

— *The Place of the Proper Name in the Topographies of the Paradiso*, in ID., *The Veil of Eternity: Language and Transcendence in Dante's Paradiso* (forthcoming).

WORKS CITED

FRECCERO, JOHN, *Dante's Cosmos*, Binghamton, Center for Medieval & Renaissance Studies, State University of New York at Binghamton, 1998.

— *The Neutral Angels*, in *Dante and the Poetics of Conversion*, pp. 110-118.

— *Medusa: the Letter and the Spirit*, in *Dante and the Poetics of Conversion*, pp. 175-193.

— *Dante and the Poetics of Conversion*, R. Jacoff ed., Cambridge, Harvard University Press, 1986.

GARDEIL, A., *Dons du Saint-Esprit*, in *Dictionnaire de Théologie Catholique*, A. Vacant, E. Mangenot, E. Amann eds., Paris, Letouzey et Ané, 1899-1972, pp. 1728-1781.

GARDNER, EDMUND G., *Dante and the Mystics; a Study of the Mystical Aspect of the Divina Commedia and its Relations with Some of its Medieval Sources*, New York, Octagon, 1968 (1913).

— *Dante's Ten Heavens, A Study of the Paradiso*, New York, Haskell, 1970 (1898).

GARIN, EUGENIO, *Lo zodiaco della vita. La polemica sull'astrologia dal Trecento al Cinquecento*, Bari-Roma, Laterza, 1996.

GHISALBERTI, ALESSANDRO, *La cosmologia del Duecento e Dante*, «Letture Classensi», XIII, Ravenna, Longo, 1984, pp. 33-48.

GILSON, ÉTIENNE, *The Christian Philosophy of St. Thomas Aquinas*, Notre Dame, University of Notre Dame Press, 1994 (1956).

— *La philosophie de Saint Bonaventure*, Paris, Vrin, 1953.

GILSON, SIMON, *Dante and the "Metaphysics of Light": A Reassessment*, in ID., *Medieval Optics and Theories of Light in the Works of Dante*, pp. 151-169.

— *Light Reflection, Mirror Metaphors, and Optical Framing in Dante's Commedia: Precedents and Transformations*, «Neophilologus», LXXXIII, 1999, pp. 241-252.

— *Medieval Optics and Theories of Light in the Works of Dante*, Lewinston, Edwin Mellen Press, 2000.

GORNI, GUGLIELMO, *La Beatrice di Dante, dal tempo all'eterno*, in DANTE ALIGHIERI, *Vita Nova*, L.C. Rossi ed., Milano, Mondadori, 1999, V-XL.

— *Dante prima della Commedia*, Firenze, Cadmo, 2001.

— *Lettera, nome, numero. L'ordine delle cose in Dante*, Bologna, Il Mulino, 1990.

GRAGNOLATI, MANUELE, *Experiencing the Afterlife. Soul and Body in Dante and Medieval Culture*, Notre Dame, University of Notre Dame Press, 2005 («The William and Katherine Devers Series in Dante Studies», VII).

GRANT, EDWARD, *The Foundations of Modern Science in the Middle Age*, Cambridge, Cambridge University Press, 1996.

— *Planets, Stars, and Orbs: The Medieval Cosmos, 1200-1687*, Cambridge, Cambridge University Press, 1996.

GRAVES, ROBERT, *The Greek Myths*, London, Penguin, 1990 (1955).

GREGORY THE GREAT, *Hom. XXIV in Evangelium*, in *Opere*, II, *Omelie sui Vangeli*, Roma, Città Nuova, 1994.

GREGORY, TULLIO, *"Anima mundi." La filosofia di Guglielmo di Conches e la scuola di Chartres*, Firenze, Sansoni, 1955.

— *Astrologia e teologia nella cultura medievale*, in ID., *Mundana Sapientia*, pp. 291-328.

— *I cieli, il tempo, la storia*, in ID., *Speculum naturale*, pp. 69-91.

— *Mundana sapientia. Forme di conoscenza nella cultura medievale*, Roma, Edizioni di Storia e Letteratura, 1992.

— *Natura e Qualitas planetarum*, in ID., *Speculum naturale*, pp. 47-68.
— *Nature au Moyen Âge*, in ID., *Speculum naturale*, pp. 1-14.
— *Riscoperta della natura e nuove scienze nel secolo XII*, in ID., *Speculum naturale*, pp. 15-33.
— *Speculum naturale. Percorsi del pensiero medievale*, Roma, Storia e Letteratura, 2007.
GROSSATESTA, ROBERTO, *De luce seu inchoatione formarum*, in *Metafisica della luce*, P. Rossi ed., pp. 109-123.
GUARDINI, ROMANO, *L'Angelo nella Divina Commedia*, in ID., *Studi su Dante*, pp. 13-130.
— *Studi su Dante*, Brescia, Morcelliana, 1967.
GUIDUBALDI, EGIDIO, *Bartolomeo da Bologna*, in *Enciclopedia Dantesca*, 6 vols., I, U. Bosco and G. Petrocchi eds., Roma, Istituto della Enciclopedia Italiana, 1970-78, pp. 526-527.
— *Dal "De luce" di R. Grossatesta all'isalmico "Libro della scala". Il problema delle fonti arabe una volta accettata la mediazione oxfordiana*, Firenze, Olschki, 1978.
GUILLAUME (WILLIAM) DE SAINT THIERRY, *The Golden Epistle: A Letter to the Brethren at Mont Dieu*, Collegeville, MN, Cistercian Publications, 1980.
HASKINS, CHARLES H., *Studies in the History of Medieval Science*, New York, Frederick Ungar, 1967.
HAVELY, NICK, *Dante and the Franciscans: Poverty and the Papacy in the "Commedia"*, Cambridge, Cambridge University Press, 2004 (Cambridge Studies in Medieval Literature).
HAWKINS, PETER S., *All Smiles: Poetry and Theology in Dante*, «PMLA», CXXI, n. 2, 2006, pp. 371-387.
— *Dante's Testament. Essays in Scriptural Imagination*, Stanford, Stanford University Press, 1999.
HOLLANDER, ROBERT, *Allegory in Dante's Commedia*, Princeton, Princeton University Press, 1969.
— *Dante's Epistle to Cangrande*, Ann Arbor (Michigan), University of Michigan Press, 1993.
— *Le opere di Virgilio nella Commedia*, in *Dante e la «Bella scola» della poesia. Autorità e sfida poetica*, A.A. Iannucci ed., pp. 247-343.
— *The Tragedy of Divination in Inferno XX*, in ID., *Studies in Dante*, pp. 131-218.
— *Studies in Dante*, Ravenna, Longo, 1980.
HUGH OF SAINT VICTOR, *Commentary on the Celestial Hierarchy of Saint Dionysius the Aeropagite*, in *Patrologiae cursus completus. Series Latina*, Jacques-Paul Migne ed., Paris, Garnier fratres, 1844-1903, PL 175, 925-1154.
IANNUCCI, AMILCARE A.A. (ed.), *Dante e la «Bella scola» della poesia. Autorità e sfida poetica*, Ravenna, Longo («Studi Danteschi»), 1993.
— *The Harrowing of Dante from Upper Hell*, in *Inferno. Lectura Dantis*, A. Mandelbaum, A. Oldcorn, C. Ross eds., pp. 123-135.
— *Virgil's Erichthean Descent and the Crisis of Intertextuality*, «Forum Italicum», XXXIII, n. 1, 1999, pp. 13-26.
IRIBARREN, I. and LENZ, M. (eds.), *Angels in Medieval Philosophical Inquiry. Their Function and Significance*, Hampshire, Ashgate, 2008 («Ashgate Studies in Medieval Philosophy»).
ISIDORE OF SEVILLE, *The Etymologies of Isidore of Seville*, by S.A. Barney, W.J. Lewis, J.A. Beach, O. Berghof transls., Cambridge, Cambridge University Press, 2006.
JEWISS, VIRGINIA, *On Men and of Angels: The Poetic of Angelology in the Works of Dante Alighieri*, Ph.D Dissertation, Yale University, 1995.
KECK, DAVID, *Angels and Angelology in the Middle Ages*, Oxford, Oxford University Press, 1998.

KERÉNY, KÁROLY, *Gli Dei e gli Eroi della Grecia*, Milano, Garzanti, 1978 (1958).

KEY, RICHARD, *Dante's Christian Astrology*, Philadelphia, University of Pennsylvania Press, 1994.

LECLERCQ, JEAN, *Influence and Noninfluence of Dionysius in the Western Middle Ages*, in *Pseudo-Dionysius. The Complete Works*, Colm Luibheid, Paul Rorem, Rene Roques eds., Introductions by Jaroslav Pelikan, Jean Leclercq, Karlfried Froehlich, New York, Paulist Press, 1987, pp. 25-32.

LEMAY, RICHARD, *Abu Ma'shar and Latin Aristotelianism in the Twelfth Century. The Recovery of Aristotle's Natural Philosophy through Arabic Astrology*, Beirut, 1962.

L'idea e l'immagine dell'universo nell'opera di Dante, Atti del convegno internazionale di Studi, Ravenna, 12 novembre 2005, Ravenna, Longo, 2008.

LILLA, SALVATORE, *Dionigi l'Aeropagita e il platonismo cristiano*, Brescia, Morcelliana, 2005.

LITT, THOMAS, *Les corps célestes dans l'univers de Thomas d'Aquin*, Louvain, Publications Universitaires, 1963.

LUSCOMBE, DAVID E., *Angels as Exemplar of World Order: the Hierarchies in the Writings of Alain de Lille, William of Auvergne, and St. Bonaventure*, in *Angels in Medieval Philosophical Inquiry*, I. Iribarren and M. Lenz eds., pp. 15-28.

LUSCOMBE, DAVID E. and MARENBON, JOHN, *Two Medieval Ideas: Eternity and Hierarchy*, in *The Cambridge Companion to Medieval Philosophy*, A.S. McGrade ed., pp. 51-72.

MANDELBAUM, A. - OLDCORN, A. - ROSS, C. (eds.), *Inferno. Lectura Dantis*, Berkeley, University of California Press, 1998.

— *Purgatorio. Lectura Dantis*, Berkeley, University of California Press, 2008.

MANESCALCHI, ROMANO, *Una nuova interpretazione del Catone dantesco*, «Critica Letteraria», XXXVI, n. 140, fasc. 3, 2008, pp. 419-446.

MARCHESI, SIMONE, *Dante, Virgilio (e Agostino) di fronte ai sette candelabri: Purgatorio 29.43-57*, Electronic Bulletin of the Dante Society, 2002, http://www.princeton.edu/~dante/ebdsa/index.html.

MAZZEO, JOSEPH, *Cultural Tradition in Dante's "Commedia"*, Ithaca, Cornell University Press, 1960.

— *Light, Love, and Beauty in the Paradiso*, «Romance Philology», XI, 1957-58, pp. 1-18.

MAZZONI, FRANCESCO, *Dante misuratore di mondi*, in *Dante e la scienza*, P. Boyde and V. Russo eds., pp. 25-53.

— *Il trascendentale dimenticato*, in *Omaggio a Beatrice*, R. Abardo ed., pp. 93-132.

MAZZOTTA, GIUSEPPE, *Cosmology and the Kiss of Creation (Paradiso I, 27-29)*, «Dante Studies», CXXIII, 2005, pp. 1-21.

— *Dante, Poet of the Desert: History and Allegory in the Divine Commedia*, Princeton, Princeton University Press, 1979.

— *Dante's Vision and the Circle of Knowledge*, Princeton, Princeton University Press, 1993, pp. 197-218.

— *The Heaven of the Sun: Dante between Aquinas and Bonaventure*, in *Dante for the New Millennium*, T. Barolini and H. Wayne Storey eds., pp. 152-168.

— *La metafisica della creazione*, in *L'idea e l'immagine dell'universo nell'opera di Dante*, pp. 61-81.

— *Spettacolo e geometria della giustizia (Paradiso XVIII-XX): L'Europa e l'universalità di Roma*, in *Dante e l'Europa*, G. Breschi ed., pp. 59-77.

MCEVOY, JAMES, *Medieval Cosmology and Modern Science*, in *Philosophy and Totality*, J. McEvoy ed., Belfast, 1977, pp. 91-110.

MCGRADE, A.S. (ed.), *The Cambridge Companion to Medieval Philosophy*, Cambridge, Cambridge University Press, 2003.

MELLONE, ATTILIO, *Angelo*, in *Enciclopedia Dantesca*, 6 vols., vol. I, U. Bosco and G. Petrocchi eds., Roma, Istituto della Enciclopedia Italiana, 1970-78, pp. 268-271.

— *Il canto XXIX del Paradiso. Una lezione di angelologia*, «Nuove Letture Dantesche», VII, 1974, pp. 198-200.

— *La dottrina di Dante Alighieri sulla prima crezione*, Nocera, Convento di Santa Maria degli Angeli, 1950.

— *Empireo*, in *Enciclopedia Dantesca*, 6 vols., vol. II, U. Bosco and G. Petrocchi eds., Roma, Istituto della Enciclopedia Italiana, 1970-78, pp. 668-671.

— *Gerarchia angelica*, in *Enciclopedia Dantesca*, 6 vols., III, U. Bosco and G. Petrocchi eds., Roma, Istituto della Enciclopedia Italiana, 1970-78, pp. 122-124.

— *Luce*, in *Enciclopedia Dantesca*, 6 vols., vol. III, U. Bosco and G. Petrocchi eds., Roma, Istituto della Enciclopedia Italiana, 1970-78, pp. 706-713.

MELLONE A. (ed.), *I primi undici canti del Paradiso*, Roma, Bulzoni, 1992.

MIGLIO, MASSIMO, *Alfragano*, in *Enciclopedia Dantesca*, 6 vols., vol. I, U. Bosco and G. Petrocchi eds., Roma, Istituto della Enciclopedia Italiana, 1970-78, pp. 122-123.

— *Alpetragio*, in *Enciclopedia Dantesca*, 6 vols., vol. I, U. Bosco and G. Petrocchi eds., Roma, Istituto della Enciclopedia Italiana, 1970-78, pp. 180-181.

MIGLIORINI-FISSI, ROSETTA, *Da Matelda a Beatrice a Maria*, in *Omaggio a Beatrice*, R. Abardo ed., pp. 23-82.

MOEVS, CHRISTIAN, *The Metaphysics of Dante's Comedy*, Oxford, Oxford University Press, 2005.

MOORE, EDWARD, *Studies in Dante. First series. Scripture and classical authors in Dante*, Oxford, 1955.

— *Studies in Dante*. Second series, *Miscellaneous Essays*, Oxford, 1968.

MORGAN, ALISON, *Dante and the Medieval Other World*, Cambridge University Press, 1990.

MORPURGO, PIERO, *Michele Scoto e Dante: una continuità di modelli culturali?*, in *Filosofia scienza e astrologia nel Trecento europeo*, Graziella Federici Vescovini and F. Barocelli eds., pp. 79-94.

MORRISON, MOLLY, *Looking at God: Imagery for the Divinity in Dante's Paradiso*, «Forum Italicum», XXXV, 2, 2001, pp. 307-317.

NARDI, BRUNO, *La caduta di Lucifero e l'autenticità della "quaestio de acqua et terra"*, in *"Lecturae" e altri studi danteschi*, R. Abardo ed., pp. 227-265.

— *Le citazioni dantesche del "Liber de causis"*, in ID., *Saggi di filosofia Dantesca*, pp. 81-109.

— *Dal Convivio alla Commedia. Sei saggi danteschi*, Roma, Istituto Storico Italiano per il Medio Evo, 1992 (1960).

— «*Dal Convivio alla Commedia*», in ID., *Dal Convivio alla Commedia*, pp. 37-150.

— *Dante e Alpetragio*, in ID., *Saggi di filosofia Dantesca*, pp. 139-166.

— *Dante e la cultura medievale*, Roma-Bari, Laterza, 1983 (1942).

— *Dante e Pietro d'Abano*, in ID., *Saggi di filosofia Dantesca*, pp. 40-62.

— *La dottrina delle macchie lunari nel secondo canto del Paradiso*, in ID., *Saggi di filosofia Dantesca*, pp. 3-39.

— *La dottrina dell'Empireo e la sua genesi storica e nel pensiero dantesco*, in ID., *Saggi di filosofia dantesca*, pp.167-215.

— *"Lecturae" e altri studi danteschi*, R. Abardo ed., Firenze, Le Lettere, 1959.

— *Nel mondo di Dante*, Roma, Edizioni di Storia e Letteratura, 1944.

— *Saggi di filosofia Dantesca*, Firenze, La Nuova Italia, 1967 (1930).

— *"Sì come rota c'igualmente è mossa"*, in *Nel mondo di Dante*, pp. 337-350.

— *Sull'origine dell'anima umana*, in ID., *Dante e la cultura medievale*, pp. 207-224.

— *Tutto il frutto ricolto del girar di queste spere*, in ID., *Dante e la cultura medievale*, pp. 245-264.

NASTI, PAOLA, *Favole d'amore e "saver profondo". La tradizione salomonica in Dante*, Ravenna, Longo, 2007.

NICODEMO, RENATO, *La vergine Maria nella Divina Commedia. Aspetti del pensiero teologico di Dante Alighieri*, Firenze, Atheneum, 2001.

NYGREN, ANDERS, *Eros e Agape. La nozione cristiana dell'amore e le sue trasformazioni*, Bologna, Edizioni Dehoniane, 1990.

OLIVIERO, ADRIANA, *La composizione dei cieli in Restoro d'Arezzo e Dante*, in *Dante e la scienza*, P. Boyde and V. Russo eds., pp. 351-362.

OLSCHKI, LEONARDO, *Sacra dottrina e Teologia mystica. Il canto XXX del Paradiso*, «Giornale dantesco», XXXVI, n.s., IV, 1933, pp. 3-25.

ORIGEN, *The commentary of Origen on S. John's Gospel: the text revised with a critical introduction and indices (In Iohannis Evangelium)*, Alan England Brooke ed., BiblioBazaar, 2009.

OSSERMAN, ROBERT, *Poetry of the Universe: a Mathematical Exploration of the Cosmos*, New York, Doubleday, 1996 (1995).

PADOAN, GIORGIO, *Ercole*, in *Enciclopedia Dantesca*, 6 vols., vol. II, U. Bosco and G. Petrocchi eds., Roma, Istituto della Enciclopedia Italiana, 1970-78, pp. 717-718.

— *Teseo*, in *Enciclopedia Dantesca*, 6 vols., vol. V, U. Bosco and G. Petrocchi eds., Roma, Istituto della Enciclopedia Italiana, 1970-78, pp. 595-596.

— *Traduzione e fortuna del commento all'Eneide di Bernardo Silvestre*, «Italia Medievale e Umanistica», III, 1960, pp. 227-240.

PAGLIARO, ANTONINO, *Similitudine*, in *Enciclopedia Dantesca*, 6 vols., vol. V, U. Bosco and G. Petrocchi eds., Roma, Istituto della Enciclopedia Italiana, 1970-78, pp. 253-259.

PALACIOS, MIGUEL ASÍN, *La Escatologia musulmana en la "Divina Comedia"*, Madrid, Instituto Hispano-Árabe de Cultura, 1961.

PANOFSKY, ERWIN, *Renaissance and Renascences in Western Art*, Copenhagen, Russak & Company Ltd., 1960.

PARRONCHI, ALESSANDRO, *Perspettiva*, in *Enciclopedia Dantesca*, 6 vols., vol. IV, U. Bosco and G. Petrocchi eds., Roma, Istituto della Enciclopedia Italiana, 1970-78, pp. 438-439.

— *La perspettiva Dantesca*, «Studi danteschi», XXXVI, 1959, pp. 5-103.

PASQUAZI, SILVIO, *All'eterno dal tempo. Studi danteschi*, Roma, Bulzoni, 1985 (1966).

— *Il messo celeste angelo del Limbo*, in ID., *All'eterno dal tempo. Studi danteschi*, pp. 49-116.

PATRIDES, C.A., *Renaissance Thought on the Celestial Hierarchy: The Decline of a Tradition*, «Journal of the History of Ideas», XX, vol. 2, 1959, pp. 155-166.

PELIKAN, JAROSLAV, *The Odyssey of Dionysian Spirituality*, in *Pseudo-Dionysius. The Complete Works*, C. Luibheid, P. Rorem, R. Roques eds., introductions by J. Pelikan, J. Leclercq, K. Froehlich, pp. 11-24.

PÉPIN, JEAN, *Allegoria*, in *Enciclopedia Dantesca*, 6 vols., vol. 1, U. Bosco and G. Petrocchi eds., Roma, Istituto della Enciclopedia Italiana, 1970-78, pp. 151-165.

— *Dante et la tradition de l'allégorie*, Paris, Librairie J. Vrin, 1970.

PERTILE, LINO, *La punta del disio: semantica del desiderio nella Commedia*, Firenze, Cadmo, 2005.

— *La puttana e il gigante. Dal Cantico dei Cantici al Paradiso Terrestre di Dante*, Ravenna, Longo, 1998.

PETERSON, MARK, *Dante and the 3-sphere*, «American Journal of Physics», XLVII, 1979, pp. 103-135.

PETIT, L., PARISOT, J., MISKGIAN, J. and VACANT, A., *Angélologie*, in *Dictionnaire de Théologie Catholique*, A. Vacant, E. Mangenot, E. Amann eds., Paris, Letouzey et Ané, 1899-1972, pp. 1248-1271.

PIANA, CELESTINO, *Le questioni inedite «De glorificatione Beatae Mariae Virginis» di Bartolomeo da Bologna O. F. M. e le concezioni del Paradiso Dantesco*, «L'Archiginnasio», XXXIII, 1938, pp. 247-262.

PICONE, MICHELANGELO, *Canto II*, in *Lectura Dantis Turicensis. Paradiso*, G. Güntert and M. Picone eds., Firenze, Cesati, 2002, pp. 35-52.

PICONE, M. (ed.), *Dante e le forme dell'allegoresi*, Ravenna, Longo, 1987.

PROTO, ENRICO, *La dottrina Dantesca delle macchie lunari*, in *Scritti vari di erudizione e critica in onore di Rodolfo Reiner*, Torino, Reiner, 1912, pp. 197-213.

PSEUDO-DIONYSIUS, THE AREOPAGITE, *La hiérarchie celeste*, Introd. by R. Roques; critical text edited by G. Heil; translation and notes by M. de Gandillac, Paris, Les Éditions du Cerf, 1958.

— *Pseudo-Dionysius. The Complete Works*, Colm Luibheid, Paul Rorem, Rene Roques eds., Introductions by Jaroslav Pelikan, Jean Leclercq, Karlfried Froehlich, New York, Paulist Press, 1987.

RAIMONDI, EZIO, *Ritual and Story*, in *Purgatorio*, A. Mandelbaum, A. Oldcorn, C. Ross eds., pp. 1-10.

RÉAU, LOUIS, *Iconographie de l'art Chrétien*, Paris, Presses Universitaire de France, 1956.

RENUCCI, PAUL, *Dante disciple et juge du monde gréco-latin*, Paris, Les Belles Lettres, 1958.

REYNOLDS, BRIAN, *A Beauty that Transforms: the Marian Aesthetics of Dante's Commedia*, «Fu Jen Studies», XXXV, 2003, pp. 61-89.

REYNOLDS, L.D, TARRANT, R.J. and WINTERBOTTOM M. (eds.), *Texts and Transmission. A Survey of the Latin Classics*, Oxford, Oxford University Press, 1984.

RICCI, PIER GIORGIO, *Maria Vergine*, in *Enciclopedia Dantesca*, 6 vols., vol. III, U. Bosco and G. Petrocchi eds., Roma, Istituto della Enciclopedia Italiana, 1970-78, pp. 835-839.

RICHARD OF SAINT VICTOR, *Beniamin Major*, in *Patrologiae cursus completus. Series Latina*, Jacques-Paul Migne ed., Paris, Garnier fratres, 1844-1903, PL 196, 63-202.

— *De Arca Noe morali libri IV*, in *Patrologiae cursus completus. Series Latina*, Jacques-Paul Migne ed., Paris, Garnier fratres, 1844-1903, PL 176, 617-680 D.

— *De Arca mystica*, in *Patrologiae cursus completus. Series Latina*, Jacques-Paul Migne ed., Paris, Garnier fratres, 1844-1903, PL, 176, 681-704 A.

RIPA, CESARE, *Iconologia*, New York, Georg Olms Verlag Hildesheim, 1970 (Roma, 1593).

RIVA, FRANCO, *Analogia e univocità in Tommaso de Vio "Gaetano"*, Milano, Vita e Pensiero, 1995.

RIVIÈRE, J., *Redemption*, in *Dictionnaire de Théologie Catholique*, A. Vacant, E. Mangenot, E. Amann eds., Paris, Letouzey et Ané, 1899-1972, pp. 1912-2004.

ROSS, CHARLES, *The Ritual Keys*, in *Purgatorio*, A. Mandelbaum, A. Oldcorn, C. Ross eds., 2008, pp. 85-94.

ROSSI, P. (ed.), *Metafisica della luce*, Milano, Rusconi, 1986.

ROSSINI, ANTONIO, *Il Dante sapienziale. Dionigi e la bellezza di Beatrice*, Pisa, Serra, 2009.

SANTAGATA, MARCO, *La donna del miracolo*, in *Amate e Amanti. Figure della lirica amorosa tra Dante e Petrarca*, Bologna, Il Mulino, 1999.

SBACCHI, DIEGO, *La presenza di Dionigi l'Aeropagita nel «Paradiso» di Dante*, Firenze, Olschki («Biblioteca di "Lettere Italiane". Studi e Testi», LXVI), 2006.

SCAZZOSO, PIERO, *I nomi di Dio nella "Divina Commedia" e il "De divinis nominibus" dello Pseudo-Dionigi*, «La Scuola Cattolica», LXXXVI, 1958, pp. 198-213.

SCOTT, JOHN, *Understanding Dante*, Notre Dame, Notre Dame University Press, 2004.

SEZNEC, JEAN, *The Survival of the Pagan Gods*, Princeton, Princeton University Press, 1972.

SILVESTRIS, BERNARDUS, *The Commentary on the First Six Books of the Aeneid of Vergil Commonly Attributed to Bernardus Silvestris*, J.W. Jones and E.F. Jones eds., Lincoln, University of Nebraska Press, 1977.

— *Commentum Bernardi Silvestris super sex libros Eneidos Virgilii*, W. Riedel ed., Gryphiswaldae (Greifswald), Abel, 1924.

SINGLETON, CHARLES, *Beatrice dolce memoria*, in ID., *An Essay on the Vita Nuova*, pp. 110-116.

— *The Death of Beatrice*, in ID., *An Essay on the Vita Nuova*, pp. 6-24.

— *An Essay on the Vita Nuova*, Baltimore, Johns Hopkins University Press, 1949.

— *Journey to Beatrice*, Baltimore, Johns Hopkins University Press, 1977.

— *The Pattern at the Centre*, in *Commedia. Elements of Structure, Dante Studies I*, Cambridge, Harvard University Press, 1954, pp. 45-60.

STABILE, GIORGIO, *Cosmologia e teologia nella Commedia: la caduta di Lucifero e il rovesciamento del mondo*, «Letture classensi», XII, Ravenna, Longo, 1983, pp. 139-173.

— *Cosmologia, teologia e viaggio dantesco*, in *L'idea e l'immagine dell'universo nell'opera di Dante*, Centro Dantesco dei Frati Minori Conventuali Ravenna, pp. 21-59.

SUÁREZ, FRANCISCO, *Opera Omnia*, Paris, Ludovicum Vives, 1856.

SUAREZ-NANI, TIZIANA, *Les anges et la philosophie. Subjectivité et function cosmologique des substances séparées à la fin du XIIIe siècle*, Paris, Vrin («Études de Philosophie Médiévale», LXXXII), 2002.

— *Connaissance et language des anges selon Thomas d'Aquin et Gilles de Rome*, Paris, Vrin, 2002 («Études de Philosophie Médiévale», LXXXV).

THOMPSON, DAVID, *Dante and Bernard Silvestris*, «Viator», I, 1, pp. 201-206.

THORNDIKE, LYNN, *The History of Magic and Experimental Science*, v. 4, New York, Columbia University Press, 1958.

— (ed.), *The Sphere of Sacrobosco and Its Commentators*, Chicago, University of Chicago Press, 1949, pp. 1-58.

TOLAN, JOHN, *Reading God's Will in the Stars. Petrus Alfonsi and Raymond de Marseille defend the New Arabic Astrology*, «Revista Espanola de Filosofia Medieval», 7, 2000, pp. 13-30.

TOMMASO DE VIO (CAJETAN), *The analogy of names (De nominum analogia)*, transl. E.A. Bushinski and H.J. Koren, Pittsburgh, Editions E. Nauwelaerts, 1959 (1498).

TOOMER, GERALD J., *Ptolemy's Almagest*, Princeton, Princeton University Press, 1998.

TRUSSO, FRANCESCO, *La Madonna nella Divina Commedia*, Firenze, Propaganda Missionaria, 1965.

VALLA, LORENZO, *Opera*, E. Garin ed., Torino, Bottega d'Erasmo, 1962, vol. 1.

VAN DER MEERSCH, J., *Grace*, in *Dictionnaire de Théologie Catholique*, A. Vacant, E. Mangenot and E. Amann eds., Paris, Letouzey et Ané, 1903-1976, VI, pp. 1553-1687.

VASOLI, CESARE, *Dante, Alberto Magno e la scienza dei peripatetici*, in *Dante e la scienza*, P. Boyde and V. Russo eds., pp. 55-70.

— *Dante e l'immagine del mondo nel Convivio*, in *L'idea e l'immagine dell'universo nell'opera di Dante*, Centro Dantesco dei Frati Minori Conventuali Ravenna, pp. 83-102.

— *Il canto II del Paradiso*, in *I primi undici canti del Paradiso*, A. Mellone ed., pp. 27-51.

VASTA, EDWARD and CERVIGNI, DINO, *From Manuscript to Print: The Case of Dante's Vita Nuova*, in *Dante Now: Current Trends in Dante Studies*, T.J. Cachey ed., Notre Dame, University of Notre Dame Press, 1995, pp. 83-114.

VECCE, CARLO, *Beatrice e il numero amico*, in *Beatrice nell'opera di Dante e nella memoria europea 1290-1990*, Atti del Convegno Internazionale, 10-14 dicembre 1990, Istituto Universitario Orientale di Napoli, Maria Picchio Simonelli ed., Firenze, Cadmo, 1994, pp. 101-135.

VESCOVINI, GRAZIELLA FEDERICI, *"Arti" e filosofia nel secolo XIV. Studi sulla tradizione aristotelica e i moderni*, Firenze, Nuovedizioni Enrico Vallecchi, 1983.

— *Astrologia e scienza. La crisi dell'aristotelismo sul cadere del Trecento e Biagio Pelacani da Parma*, Firenze, Nuovedizioni Enrico Vallecchi, 1979.

— *L'astrologia tra magia, religione e scienza*, in ID., *"Arti" e filosofia nel secolo XIV*, pp. 171-193.

— *Medioevo magico*, Torino, Utet, 2008.

— *Studi sulla prospettiva medievale*, Torino, Giappichelli, 1965.

— *Le teorie della luce e della visione ottica dal IX al XV secolo. Studi sulla prospettiva medievale e altri saggi*, Perugia, Morlacchi, 2003.

VESCOVINI, G. FEDERICI and BAROCELLI F. (eds.), *Filosofia scienza e astrologia nel Trecento europeo*, Padova, Il Poligrafo, 1992.

VIRGIL, *The Aeneid*, Robert Fagles, transl., New York, Penguin, 2006.

VON BALTHASAR, HANS UR, «Dante», in ID., *The Glory of the Lord. A Theological Aesthetics*, vol. III («*Studies in Theological Style: Lay Style*»), San Francisco, Ignatius Press, 1986 (1962), pp. 9-104.

WINTHROP, WETHERBEE, *Platonism and Poetry in the Twelfth Century: The Literary Influence of the School of Chartres*, Princeton, Princeton University Press, 1972.

ZAMBELLI, PAOLA, *The "Speculum astronomiae" and its Enigma*, Dordrecht-Boston-London, 1992.

ZANINI, CARLO, *Gli angeli nella Divina Commedia. In relazione ad alcune fonti sacre*, Milano, Cogliati, 1908.

ZINGARELLI, NICOLA, *I tempi, la vita e le opere di Dante. Storia letteraria d'Italia*, Milano, Vallardi, 1939.

INDEX

Abelard, Peter, 85n. 31, 86
activity. *See energheia*
Aeneas, 93, 146, 146nn. 39, 40, 149, 151nn. 51, 53, 152-157, 158, 175. *See also* Virgil: *Aeneid*
Ahern, John, 165n. 85
Albert the Great (Albertus Magnus): on angelic hierarchy, 45-46n. 32, 53, 53n. 45, 55, 107n. 68; on astrology, 75n. 9, 110; on beatific vs. hierarchical illumination, 53, 53n. 45; on the Blessed, 53n. 45; on celestial motion, 81; influence on Dante, 2, 4n. 3, 5n. 5, 28, 28n. 1, 30n. 5, 53, 56, 68n. 69, 73n. 2, 100, 102n. 59, 108n. 69, 109n. 70, 115n. 83; and John of Vercelli's questionnaire, 80-81; on Mercury, 159; and separation of philosophy and theology, 82; on separation between physical and spiritual realms, 84
Albumasar (Abū Ma'shar), 100, 118; *Introductorium in Astronomiam*, 75n. 9, 108n. 70
Alexander of Hales, 7n. 10, 8, 8n. 14
Alfraganus (Ahmad ibn Muhammad ibn al Farghanî): *Liber de aggregatione scientiae stellarum et principiis coelestium motuum*, 73, 74, 108n. 70
Alhazen (Abū 'Alī al-Hasan ibn al-Hasan ibn al-Haytham), 118; *Book of Optics*, 36n. 16
Alighieri, Dante: angelology (angelological system of *Commedia*), IX-XVI, 1, 2-3, 5, 6, 12-13, 15, 19, 26, 30, 34, 35, 36, 61, 66, 69, 77, 83, 100, 122-123, 146, 163, 164, 171, 182, 183-184; attempt to reconcile Aristotelianism and Neo-Platonism, 72, 84, 184; attempt to reconcile Aristotelianism and Christian doctrine (harmonization of philosophy and theology), 18, 19, 20, 26, 27, 31, 35, 69, 76, 82, 100, 123, 183; attribution of angels as celestial movers to Dionysian hierarchies, 71, 77-84; on the Blessed, XV, 53, 54n. 47, 87, 88, 105, 126-133; definition of angels as pure acts, 15-17; on inclinations, 111-115, 122, 123-124; influence of Albert the Great on, 2, 4n. 3, 5n. 5, 28, 28n. 1, 30n. 5, 53, 56, 68n. 69, 73n. 2, 100, 102n. 59, 108n. 69, 115n. 83; influence of angelological debates on, IX, X-XI, XII, XIV, 1-26, 183-184 (*see also* angelology); influence of Aquinas on, XIV, 2, 20, 21n. 36, 28, 30-31, 32, 106n. 66, 140-143; influence of Aristotle on, XII, XIII, XIV, 2, 6, 19, 20, 20n. 35, 21n. 36, 22, 26, 27, 31, 77, 100, 123, 139, 168, 184; influence of Averroist philosophy on, 14, 15-16, 18n. 32, 19, 21, 21n. 36, 23, 79n. 18, 99; influence of Avicenna on, 38n. 20, 97, 102n. 58; influence of Biblical sources on, 2, 2, 19-20n. 34, 20, 24-25, 28-35, 42-43, 69, 125, 136n. 19, 139, 149n. 47, 179n. 116; influence of Bonaventure on, XIV, 2, 28, 29-32, 47, 47n. 37, 47-48n. 38, 49, 125n. 1, 132-133; influence of Dionysius on, XII, XIV, 2, 5, 5n. 5, 6, 26, 27, 28, 30, 31, 34, 34n. 14, 51-52, 53-54n. 46, 61, 69, 77, 183-184; influence of Gregorian angelology on, 35, 46, 69, 130-131n. 9; influence of Pauline theology on, XIV, 66, 68, 69, 88n. 35, 90n. 38, 149n. 47, 173; influence of science on, X, XII, XV, 36, 40, 71, 72, 73, 73n. 2, 101, 108-109n. 70, 124; as 'maker' of *Commedia*, XI; reinterpretation of Dionysian angelology, XIV, 8-9, 20, 26, 30-31, 32, 35, 44-45, 46-47, 54-56, 69, 71, 102n. 58, 123, 183-184; syncretism and synthesis of different traditions, IX, XII, XIV, XV, 6, 21n. 36, 26, 35, 76-77, 90, 125, 146, 146n. 41, 183 (*see also* mythology: Christianization of). Works: *Commedia* (*see under Commedia*); *Convivio* (*see under Convivio*); *De vulgari eloquentia*, 1, 93; *Epistles*, 11, 28, 66n. 67, 75n. 8, 164-165n. 85; *Monarchia*, 1, 16-17, 22-24, 66n. 67, 113n. 80; *Ri-*

— 201 —

me, 40, 40n. 23, 174n. 108; *Vita Nuova*, 49, 136, 144n. 36, 163-170, 170n. 98, 171n. 101, 177, 177n. 112, 178, 178n. 114, 179, 180n. 118, 182. *See also* Dante (poet-pilgrim in *Commedia*)

Almagest. *See under* Ptolemy

Alpetragius (Nur el-Din al-Betrugi): *Liber de motibus coelorum*, 73, 108-109n. 70

analogia entis, 51, 66n. 67, 180n. 118

analogy: analogical interpretation of Nature and Scriptures, 73-74; analogical language of 'divine names' in *Paradiso*, 68; angels' analogous mediation of grace and divine wisdom, 3-4; of attribution, 164n. 84, 178, 178n. 115, 179-181; Beatrice and the analogical model, XV, 126, 131n. 10, 163-164, 166n. 89, 167-168, 171, 176-182, 183; between angelic hierarchies and the Blessed, XV, 127, 127-133, 176, 181; between angelic imitation of God and human imitation of Christ, 60n. 58; between angels and humans, 25, 34n. 13, 123; between divine and material lights, 39; between earthly and celestial orders, 116-117; Bonaventure's use of, 48n. 38; characteristics of the analogical model, 164n. 84; medieval doctrine of analogy, 178, 178-179n. 115; of proportion, 178-179, 179n. 116

angels: angelic choirs, Xn. 2, 2, 24, 50, 51, 90, 92n. 41, 106, 129, 131, 166, 171n. 102, 183; angelic vision, 13, 49-50, 51, 52, 53, 54-55, 59, 61, 62, 63, 97, 105, 127-128, 175; 'Angels' (in hierarchy), 34n. 13, 46nn. 33, 34, 47n. 36, 48, 83; as archetypes of human behaviour, 34n. 13; and astrology, XII-XIII, XIII-XIV, XV, 14, 72, 77, 110-122, 123; as bees, 125, 126-127; as chronocrators, XIII, 35, 83, 88, 125; creation of, XIII, 1, 6-7, 9, 10, 12n. 21, 13, 15, 16, 64; enlightening function of, 3, 32n. 9, 34, 35, 45, 46n. 33, 54n. 47, 63, 67, 77, 125, 136, 137, 138, 163, 173, 173n. 106, 175, 176, 178n. 115, 182; etymological meaning of 'angel,' 33; functional interpretation of, 31, 32-35; as guardians and guards, IX, 7, 34, 126, 133-134, 135, 160, 160n. 78; as guides, IX, 34, 48, 52, 69, 138, 155, 170, 181, 182; and illumination, XI, XII, 26, 34, 34n. 14, 42, 45, 45-46n. 32, 48, 48n. 40, 50, 52-57, 58, 60-64, 69, 122-123, 184 (*see also* angels: enlightening function of); as images of God, 34n. 14, 65-66; in *Inferno*, IX, XV, 2, 34, 125, 126, 144-148; as instruments of Providence, XI, XIII, XIV, XV, 3, 23n. 38, 30, 77, 86, 88, 89-90, 99, 111, 126, 137-138, 181, 182, 184; as intermediaries between divine and human worlds, XII, 1-4, 9, 26, 55, 69, 76, 183, 184; ministerial functions of, XIV, XV, XVI, 4, 7, 15, 16, 20, 30-31, 32-34, 35, 46n. 33, 65, 77, 135, 141, 143, 183; as mirrors, XII, XIII, 3, 26, 34n. 14, 45, 53-54, 65-69, 91, 92, 105, 123, 125, 171, 173; as movers of celestial spheres, IX, XIII, XIV, XV, 2, 12-13, 18, 18n. 32, 19n. 33, 21-22, 24, 31, 32, 35, 53-54, 57n. 51, 69, 71-124, 183; neutral, 125-126, 126n. 2; in *Paradiso*, IX, XIII, 1, 19, 24-25, 126-133; in *Purgatorio*, IX, X, 34, 88, 126, 133-143, 181, 183; relation to the Blessed, 127, 127-133, 176, 181; relation to the Holy Spirit, 24, 47, 47n. 37, 48, 48n. 38, 137-138, 139, 140, 141, 143, 181; role in creation, XI, 4, 77, 92, 129, 131n. 9, 184; role in salvation, 3, 7, 13, 32, 34, 46n. 33, 48, 88, 125, 138, 180, 184; speculative (contemplative) and active functions of, XIV, 7, 7n. 10, 18-25, 26, 35, 52, 123, 125n. 1, 183. *See also* Alighieri, Dante: angelology; angelic operations; angelology; *Commedia*; hierarchy (angelic); Intelligences

angelic operations: Beatrice and, XII, XV, 163-164, 170-176, 182; in *Commedia*, IX-XVI, 3, 17, 24, 25, 30, 31, 32, 67, 183, 184; Dante on, 1, 14, 18, 19, 26, 69; debates on, XIV, 4, 18-26; functional interpretation of angels, 32-35; and grand design of salvation, 13, 34; and hierarchical ordering, 46, 48, 51, 62; and light transmission (in *Commedia*), 40, 42, 44, 61, 62, 71, 122, 123, 124, 125, 126; relation between understanding and operation, 17; speculative (contemplative) and active operations, XIV, 7, 7n. 10, 18-25, 26, 35, 52, 71, 123, 125n. 1, 183. *See also* angels *for individual operations*

angelology (medieval): angelic nature (debates on), XIV, 1, 9-18, 26, 61; angelic operations (debates on), XIV, 9, 18-26; angels as celestial movers (debates on), 71, 77-84; influence of angelological debates on Dante, IX, X-XI, XII, XIV, 1-26, 183-184; number of angels (debates on), 18, 19-20, 79, 79n. 18; Patristic traditions, Xn. 1, XII, 5, 21, 31, 32, 58n. 53, 125n. 1; relation to cosmology, 71, 72-108; studies on, Xn. 2. *See also* Alighieri,

INDEX

Dante; Aquinas; Aristotle; Augustine; Bonaventure; Dionysius; Gregory the Great
anima mundi (soul of the world), 84-86, 177n. 12
Anselm of Laon: *Glossa Interlinearis*, 6
Apollodorus, 154n. 59, 155, 158n. 71
Appiano, Ave, 160n. 77
Apuleius, Lucius, 157
Aquinas, Thomas: on angelic hierarchy, 28n. 2, 82, 82n. 25, 83, 116-117, 129; on angelic nature, 10-11, 14n. 25, 16, 17, 61; angelological system, 8; on angels as celestial movers, 31, 81-82, 82n. 25, 83; on angels and Intelligences, 81; on angels as mediators of grace, 3; on astrology, 120-121n. 93; on the Blessed, 54, 129-130, 132; on celestial motion, 106n. 66; on direct vs. indirect illumination, 54-55; distinction between act and potency in angels, XIV, 9, 10, 11, 13, 15-16, 21. 36; as *Doctor angelicus*, IX; doctrine of 'ens' and 'essentia,' 10, 16; influence of Aristotle on, 31, 32; influence on Dante's angelology, XIV, 2, 20, 21n. 36, 28, 30-31, 32, 106n. 66, 140-143; influence of Dionysian angelology on, 27, 31; influence of Platonism and Neo-Platonism on, 73; and John of Vercelli's questionnaire, 80-81; on light, 39-40, 42; and *Liber de causis*, 75n. 8; on Mary, 180n. 119; in *Paradiso*, 28, 30; and reconciliation of philosophy and theology, 82, 84, 84n. 27; on simultaneous creation, 13-14; on three distinctions of Grace, 63-64n. 62. Works: *De ente et essentia*, 10n. 17, 11n. 19, 15n. 27; *On spiritual substances*, 28n. 2; *Summa contra Gentile*, 10n. 17, 15n. 27, 21n. 36, 28n. 2, 83n. 26, 180n. 119; *Summa Theologiae*, 2, 10n. 17, 13n. 23, 28, 28n. 2, 47, 54-55, 64nn. 62, 64, 106n. 66, 129-130, 141, 141nn. 28, 29, 30, 178n. 115, 180n. 119
Archangels, 34n. 13, 46nn. 33, 34, 47n. 36, 48, 83, 116. *See also* Gabriel; hierarchy (angelic)
Areopagite. *See* Dionysius
Ariani, Marco, 4n. 3, 5n. 5, 37n. 17, 65n. 66
Aristotle: on angels as celestial movers, 12-13, 18, 18n. 32, 31, 72, 79; Aristotelian causality, XIII, 75, 100, 109, 184; doctrine of the Intelligences, Xn. 2, XII, XIV, 2, 9n. 15, 14, 16, 16n. 29, 18, 20n. 35, 77-84; influence on Aquinas, 31, 32; influence on Dante's angelology, XII, XIII, XIV, 2, 6, 19, 20, 20n. 35, 21n. 36, 22, 26, 27, 31, 77, 100, 123, 139, 168, 184; *Liber de causis* attribution, 72-73, 75, 75n. 8; scientific thought, 72-73, 74-75, 76, 78, 80, 97, 101-102. Works: *Ethics to Nicomachus*, 7n. 11, 20, 22, 168, 178n. 115; *Metaphysics*, 7n. 11, 14, 16, 16n. 28, 72, 75, 89, 99n. 54, 178n. 115; *Meteors*, 100; *On Heavens (De coelo)*, 72, 73n. 2, 75, 80, 102n. 59, 109n. 70; *Physics*, 72

Ascoli, Albert R., 165n. 85
Ashworth, Jennifer, 179n. 115
astrology: and angels, XII-XIII, XIII-XIV, XV, 14, 72, 77, 110-122, 123; astrological determinism (debate on), 72, 75-76, 80, 108-122, 169-170n. 97; astrological influence and creation, 14, 77, 123-124; astrological texts, XV, 73, 79n. 20, 108, 108-109n. 70, 118-119; Augustine on, 118; in *Commedia*, XIII-XIV, 14, 77, 100, 103, 110-122, 123-124, 168-169; in *Convivio*, 109-110n. 73, 168-169; defense of, 85-86, 86n. 32; distinguished from astronomy, 109, 109n. 71; doctrine of inclinations, XV, 72, 111-115, 121-122, 123-124; hammer metaphor, 103; and medieval iconography of the cosmos, 85n. 30; *Picatrix*, 119n. 89; and theology, 76nn. 10, 11, 85-86, 109-110, 118-120, 168-169, 169-170n. 97. *See also* astronomy; celestial spheres; science
astronomy, 101; astronomical texts, XV, 72n. 1, 108, 108-109n. 70; and debates about the number of angels, 19; distinguished from astrology, 109, 109n. 71. *See also* astrology; celestial spheres; science
Augustine: on astrology, 118; on creation, 13n. 9; doctrine of the gifts, 140-143; doctrine of illumination, 37, 42-44; and the functional interpretation of angels, XIV, 33-35; on the identification of the *anima mundi* with the Holy Spirit, 85; influence on interpreters of Dionysius, 6; on Mercury, 158, 158-159n. 73, 162; *regio dissimilitudinis*, 154, 154n. 60. Works: *Confessiones*, 154n. 60; *De civitate dei*, 159n. 73; *De Genesi ad litteram*, 42-43; *De Trinitate*, 29n. 3; *Enarrationes in Psalmos*, 33
Averroes (Abū l-Walīd Muhammad ibn Ahmad Muhammad ibn Rushd), 82, 84, 97; Averroist 'divinization' of the Intelligences, XIV; influence of Averroist philosophy on Dante, 14, 15-16, 18n. 32, 19, 21, 21n. 36, 23, 79n. 18, 99

— 203 —

INDEX

Avicenna (Abū Alī al-Husayn ibn 'Abd Allāh ibn Sīnā): 84; on angels, 22, 82, 82n. 25; influence on Dante, 38n. 20, 97, 102n. 58

Bacon, Roger, 38, 76n. 10, 110; *Perspettiva*, 36n. 16
Baeumker, Clemes, 36n. 15
Baranski, Zygmunt, Xn. 2, 165n. 85
Barbi, Michele, 178n. 114
Bareille, G., Xn. 2, 7n. 10, 34n. 13
Barolini, Teodolinda, 29n. 3, 99n. 53, 134n. 15, 163n. 83, 170n. 99
Barsella, Susanna, 140n. 26, 146n. 41, 154n. 59
Bartholomew of Bologna, 44; *De luce*, 38, 38n. 20
Basil of Cesarea: on light, 42-43n. 27; *Hexaëmeron*, 42
Beatrice: allegorical interpretations of, 131n. 10, 180, 180n. 118; and the analogical model, XV, 126, 131n. 10, 163-164, 166n. 89, 167-168, 171, 176-182, 183; angelization of, XV, 166, 170, 180; Beatrice's smile, 171, 171n. 102; as Blessed, 131, 131n. 11, 175, 176, 182; on celestial spheres, 100-103, 107; in *Convivio*, 166n. 89, 169, 178n. 114; and Dionysius, 30, 30n. 6, 46; dissertation on ontological principles, 86; expositions on angels, XIII; Grace bestowed on, 63n. 62; as heavenly messenger, 171; on the motion of the Primum Mobile, 89; and myth of Gorgon and Perseus, 173-174; and number nine, 49, 163n. 83, 166-169; relation to angels, X, XII, 131-132, 136-137, 163-182, 183 (*see also* Beatrice: and the analogical model; angelization of); on the 'schiere' of blessed and angels, 88; as theology, 131n. 10; and vision of angelic choirs, 24, 171n. 102; in the *Vita Nuova*, 49, 163-170, 177, 178-179, 180n. 118, 182; warning to Dante, 95-96
Bemrose, Stephen, Xn. 2, XII, 77n. 12, 146n. 39
Bernard of Clairvaux, 86; on angelic functions, 7n. 10; *De considerazione V (On Consideration V)*, 2, 7n. 10, 8n. 14, 34n. 13; theory of Marian intercession, 3
Bettini, Maurizio, 157n. 69
Bible: angels in, 31, 77; Beatitudes, 133-143, 181; Book of Daniel, 19n. 34; Dante's scriptural sources, 2, 2, 19-20n. 34, 20, 24-25, 28-35, 42-43, 69, 125, 136n. 19, 139, 149n. 47, 179n. 116; *Ecclesiasticus*, 43-44, 44n. 28; *Genesis*, 6, 12n. 21, 38n. 19, 42-43, 129; light in, 42-43; New Testament (angels in), 33n. 11, 127n. 3, 136n. 19; Old Testament (angels in), IX-X, 3, 5, 33n. 11, 136, 136n. 19; scriptural vision of the cosmos, 73-74, 76
Blessed, 54, 105n. 64; Albert the Great on, 53n. 45; analogy between angelic hierarchies and the Blessed, XV, 127, 127-133, 176, 181; Aquinas on, 54, 129-130, 132; Beatrice as, 131, 131n. 11, 175, 176, 182; Cato as, 134; Bonaventure on, 56n. 48, 129, 132-133; Dante on, XV, 53, 54n. 47, 87, 88, 105, 126-133; Gregory the Great on, 129. *See also* Bible: Beatitudes
Bobik, Joseph, 10n. 17
Boccaccio, Giovanni, 145, 163n. 82
Boccassini, Daniela, 97n. 51
Boethius, Severinus: *De consolatione philosophiae (The Consolation of Philosophy)*, 60n. 58, 85, 154n. 61; *De Trinitate*, 29n. 3
Boffito, Giuseppe, 36n. 16
Bologna, Corrado, 176n. 110
Bonatti, Guido, 118, 118n. 88, 119-120
Bonaventure of Bagnoregio: angelic hierarchy, 31-32, 48, 48n. 38, 55-56, 56n. 48, 129, 132; on angelic nature, 10, 10n. 16; angelological system, 8; on angels' subordinate role, 3; on the Blessed, 56n. 48, 129, 132-133; as *Doctor seraficus*, IX; influence on Dante's angelology, XIV, 2, 28, 29-32, 47, 47n. 37, 47-48n. 38, 49, 125n. 1, 132-133; influence of Dionysian angelology on, 27, 28n. 2, 32, 48; influence of Gregorian angelology on, 28n. 2, 32, 129; influence of Platonism and Neo-Platonism on, 73; on light, 39, 39n. 21, 55-56; in *Paradiso*, 28; on the Trinity, 29n. 3, 30, 31, 48, 48n. 38, 55-56. Works: *Collationes in Hexaëmeron*, 2, 10n. 16, 28, 28n. 2, 29n. 3, 31, 44, 47n. 37, 55, 132n. 12; *Praenotata* to *Distinctio IX*, 28n. 2
Book of Optics. See under Alhazen
Book of Splendor, 39n. 21
'book of stars,' 74
Botterill, Stephen, 60n. 58, 180n. 119
Bouyer, Louis, 6n. 7, 45, 58n. 53
Boyd, Patrick, 72n. 1
Busnelli, Giovanni, 159n. 77
Bussagli, Marco, Xn. 2, 33n. 11, 34n. 14

Cacciari, Massimo, 160n. 80
Cajetan (Tommaso de Vio), 5n. 4, 178; *De nominum analogia*, 178-179n. 115

— 204 —

Callippus, 79, 79n. 18
Camporeale, Salvatore, 5n. 4,
Capella, Martianus: *De nuptiis mercurii et philologia*, 73, 149n. 48, 156, 156n. 65, 158, 163n. 82
Cappelli, Luigi M., 48n. 38
caritas, 86, 87, 106, 107n. 68, 118, 138n. 22, 166n. 90, 178n. 113
Carugati, Giuliana, 84n. 28, 177n. 112
Cassell, Anthony, 156n. 66, 160n. 80
Casella, Mario, 180n. 118
Cato, 133-135
Cavalcanti, Guido, 134n. 15, 166n. 90
celestial spheres: angels as movers of, IX, XIII, XIV, XV, 2, 12-13, 18, 18n. 32, 19n. 33, 21-22, 24, 31, 32, 35, 53-54, 57n. 51, 69, 71-124, 183; Beatrice on, 100-103, 107; correspondence between angelic choirs and, 51, 90, 183; and John of Vercelli's questionnaire, 80-81; ninth sphere, 77, 79, 79-80n. 20, 103, 105. *See also* astrology; astronomy; Empyrean; Primum Mobile
Cervigni, Dino, 165n. 87
Chance, Jane, 152n. 56, 155n. 62, 156n. 65, 163n. 82
Chenu, Marie-Dominique, 8, 149n. 48
Cherubim, 34n. 13, 46nn. 33, 34, 47n. 36, 48, 49n. 42, 62, 83, 106-107, 107n. 68. *See also* hierarchy (angelic)
Chiarini, Gioachino, 101n. 56
Chiavacci Leonardi, Anna Maria, 139, 139nn. 24, 25
Christ, XV, 3, 30, 34, 35, 38, 44, 47, 60-61, 88, 91, 92, 125n. 1, 133n. 14, 144-145, 146, 150, 154, 156n. 66, 162, 163n. 83, 164, 164n. 84, 169, 169-170n. 97, 174, 175, 176, 177, 177n. 112, 179-181, 182, 185. *See also* Christological doctrines
Christological doctrines: approach to light, 38, 44; Christ's work of mediation, 3, 13, 35, 90n. 38, 177, 180-181; influence on *Commedia*, 3, 13, 24, 30, 35, 38, 44, 88, 90-91, 164 (*see also* Beatrice: and the analogical model). *See also* mythology, Christianization of; Trinitarian theology
Chollet, A., 179n. 115
chronocrators. *See under* angels
Church Fathers. *See* angelology: Patristic traditions
Ciabattoni, Francesco, 92n. 41

Ciccuto, Marcello, 118n. 88
Cicero, Marcus Tullius: *De officiis*, 113, 113nn. 79, 80
coelum Trinitatis, 98n. 52
Cogan, Marc, 5ln. 44
Colish, Marcia L., 4n. 3, 6n. 8, 7n. 9, 12n. 21
Collationes in Hexaëmeron. See under Bonaventure
Collins, James D., 10n. 17, 28n. 2, 110n. 74
Commedia: as 'angelic metaphor,' 176; angels and astrology in, 72, 77, 110-122, 123; astrology in XIII-XIV, 14, 77, 100, 103, 110-122, 123-124, 168-169; angelic operations in, IX-XVI, 3, 17, 24, 25, 30, 31, 32, 67, 183, 184; angels as celestial movers in, XIII, XIV, XV, 2, 12-13, 24, 31, 32, 35, 53-54, 57n. 51, 69, 71-124, 183; compared to a mirror, 175; concept of free will in, 72, 108-124; concept of imitation in, 27, 58-61, 91; concept of merit in, 27, 61-65; correspondence between angelic choirs and celestial spheres in, 2, 51, 90, 183; cosmological design of, XII, 9, 26, 27, 28, 29-32, 42, 57, 69, 71, 74n. 7, 76-77, 84, 88, 89-90, 93, 93n. 43, 94, 97, 98, 113, 120, 122-124, 137, 140, 183, 184; critical studies of, IX, X, Xn. XII-XIII; as epic poem, 93-97; eschatological perspective of, XIII, 13, 35, 69, 71, 88, 88n. 35, 90, 90n. 38, 93, 164, 181, 183, 184; influence of Christological doctrines on, 3, 13, 24, 30, 35, 38, 44, 88, 90-91, 164 (*see also* Beatrice: and the analogical model); influence of Trinitarian theology on, 3, 27, 29, 29n. 3, 30, 35, 40, 41, 4ln. 25, 50, 58, 86-87, 90, 137, 140, 163; geographical names in, 57-58n. 52; hammer metaphor in, 102-103; heavenly messenger (*messo celeste*) in, X, XV, 33, 125, 126, 144-163, 171, 181, 183; hierarchy (angelic) in, XII, XIII, XIV, XV, 2, 14-15, 17, 2ln. 36, 26, 27, 31-32, 35, 44-45, 46, 49-56, 59, 62-63, 65, 71, 77, 83, 102-104, 115-116, 129, 184; light in, 27, 36, 38, 39-44, 50-51, 54, 56, 57-69, 91-92, 104-105, 123, 128, 134, 172, 175, 182, 185; love and knowledge in, 71, 91-97; mirrors (angelic) in, XII, XIII, 3, 26, 45, 65, 66-69, 91, 92, 105, 123, 125, 171, 173; as microcosm, XII; as poetry of light, 57; as 'poiesis,' XI; reconciliation of Aristotelianism and Neo-Platonism in, 84; reconciliation of science and Providence in, 71; references to writing in, 35, 50-51; relation

— 205 —

between speculative (contemplative) and active functions in, XIV, 18-25, 26, 35, 52, 71, 123, 125n. 1, 183; relation between the unity and multiplicity in, XIII, 50, 50n. 43, 68, 88, 90, 99, 123, 184; role of angels in, IX-XI, 3, 34, 35, 69, 88, 126, 137-138; salvific (soteriological) perspective of, IX, XII, 3, 13, 34, 34n. 14, 38, 69, 88, 122, 136, 144, 182; sources, XII, XIII, XIV, 2, 28, 74n. 7; temporal dimensions in, XII, 2; Trinitarian theology in, 3, 27, 29, 29n. 3, 30, 35, 47, 86-87, 90, 137, 163, 167; virtuous descent in, 154-163; wings metaphor in, 96, 96n. 51. *See also* Alighieri, Dante; *Inferno*; *Purgatory*; *Paradiso*

Contini, Gianfranco, 97n. 51

Convivio, 26, 37, 56, 75n. 8, 93, 93n. 44, 99, 148-149n. 46, 149n. 48, 159, 167-168n. 94, 168-169; angels and angelic hierarchies in, 1, 2n. 1, 18-22, 23, 34, 46, 47-48n. 38, 49, 49n. 42, 53-54, 56n. 48, 115, 115n. 83, 130; Beatrice in, 166n. 89, 169, 178n. 114; celestial movers in, 72n. 1, 92n. 40; correspondence between heavens and disciplines in, 109-110n. 73, 115; doctrine of 'secondary causes' in, 79n. 18; heavens as 'book of stars' in, 73-74; Hercules's stories in, 154n. 59; light in, 38, 38n. 20, 40; *perspettiva* in, 37n. 18; reconciliation of Aristotelian and theological doctrines in, 18, 20, 20n. 35; theory of Restoration in, 130-131n. 9; Trinitarian theology in, 47, 47n. 37, 47-48n. 38, 49, 49n. 42

Cornish, Alison, XII, 111n. 76

Corti, Maria, 36n. 15, 37n. 16, 38n. 20, 94n. 46, 164n. 84, 178n. 114

cosmology: Aristotelian-Ptolemaic, 79-80; cosmic order and hierarchy, 29-32, 51; cosmological design of *Commedia*, XII, 9, 26, 27, 28, 29-32, 42, 57, 69, 71, 74n. 7, 76-77, 84, 88, 89-90, 93, 93n. 43, 94, 97, 98, 113, 120, 122-124, 137, 140, 183, 184; harmonization of science and Christian doctrine concerning (medieval debate on), 71, 72-77; medieval iconography of the cosmos, 85n. 30; relation to angelology, 71, 72-108; three principles of cosmological order, XV

Costa, Cristina D'Ancona, 75n. 8

Costanza of Altavilla, 95-96

Cranz, Edward, 7n. 11, 72n. 1

creation: of angels, XIII, 1, 6-7, 9, 10, 12n. 21, 13, 15, 16, 64; and astrology, 14, 77, 123-124; Augustine on, 131n. 9; Biblical exegesis of, 43-44, 73, 76, 129; doctrine of creation in *Commedia*, 11-15, 26, 72n. 1, 80, 87, 90, 92-93, 93n. 43, 95, 98-99, 104, 112, 123-124, 176n. 110, 184-185; *ex-nihilo*, 12, 29n. 3, 39; incompatibility of scientific and scriptural explanations, 76; light and, 42-44; role of angels in, XI, 4, 77, 92, 129, 131n. 9, 184; simultaneous (*simul*), 12, 13-14; Trinitarian doctrine of, 29-31.

Cristaldi, Sergio, 139n. 25

Cristiani, Marta, 4n. 3, 57, 57n. 51

Curtius, Ernst R., 149n. 48, 163n. 82

da Buti, Francesco, 17, 100, 103-104, 105n. 64

da Imola, Benvenuto, 17, 19-20n. 34, 146

d'Alverny, Marie Thérèse, 76n. 11, 109n. 70

Damian, Peter, 87, 128

Daniello, Bernardino, 17, 100, 145, 145-146n. 39

Dante (poet-pilgrim in *Commedia*): as Aeneas and Paul, 175; Dante's epic voyage, 91-97; Dante's journey, XIII, 8, 58, 184-185; in *Inferno*, 144-148; in *Paradiso*, 24-25, 90-91, 171-173, 175-176; in *Purgatorio*, 67-68, 133-138, 171n. 101, 172, 174; transhumanization of, XV, 58, 60, 60n. 58, 173; and vision of angelic choirs, 24, 50, 106, 171n. 102; as writer, 24, 47, 175-176. *See also* Alighieri, Dante

da Parma, Biagio Pelacani, 110n. 74

d'Arezzo, Ristoro: *La composizione del mondo*, 74, 74n. 7

da Romano, Cunizza, 117

d'Ascoli, Cecco (Francesco degli Stabili), 118n. 88, 119-120, 120n. 91, 169, 169-170n. 97

De coelesti hierarchia. See under Dionysius

De coelo (*On Heavens*). *See under* Aristotle

De considerazione V (*On Consideration V*). *See under* Bernard of Clairvaux

De consolatione philosophiae (*The Consolation of Philosophy*). *See under* Boethius, Severinus

De divinis nominibus. See under Dionysius

De ecclesiastica hierarchia. See under Dionysius

De intelligentiis, 37, 38

de Lille, Alain, 31n. 8; *Anticlaudianus*, 66n. 68, 74; *Hierarchia*, 31

De Lubac, Henri, 179n. 115

della Lana, Jacopo, 108n. 69

De luce. See under Bartholomew of Bologna; Grosseteste

De mystica theologia. See under Dionysius
De nominum analogia. See under Cajetan
de Serravalle, Johannis, 19n. 34, 146, 146n. 40
De Waele, Ferdinand J.M. 157n. 69
di Dante, Pietro, 17, 100, 145, 158n. 70
di Marseille, Folchetto, 88n. 36
Dionysius (Pseudo-Dionysius the Areopagite), 2n. 2, 27; on angelic mirroring, 27, 34n. 14, 65-66, 69; on cataphatic vs. apophatic theology, 57; on cherubim, 107, 107n. 68; Dante's attribution of angels as celestial movers to Dionysian hierarchies, 71, 77-84; Dante's reinterpretation of, XIV, 8-9, 20, 26, 30-31, 32, 35, 44-45, 46-47, 54-56, 69, 71, 102n. 58, 123, 183-184; Dionysian *Corpus*, 2n. 2, 4-5n. 4, 5-6; Dionysian model of cosmic harmonization, XIV, 69, 77; doctrine of angelic hierarchies, 2n. 1, 3, 4, 6, 7-8, 26, 27, 28n. 2, 31-32, 34, 34n. 13, 36, 44-69, 71, 76, 77, 116-117; on *dynamis*, 59n. 54; on illuminative function of angels, 32n. 9, 34n. 14, 54-57, 69; on imitation, 27, 56-57, 58, 59, 59n. 54, 59-60, 59-60nn. 54, 55, 56, 65; on light, 27, 36, 37, 55; influence on Dante's angelology, XII, XIV, 2, 5, 5n. 5, 6, 26, 27, 28, 30, 31, 34, 34n. 14, 51-52, 53-54n. 46, 61, 69, 77, 183-184; on merit, 27, 61, 62, 63; on the Principalities, 116, 116n. 84; references to in *Paradiso*, 28, 30; symbolic theology of, 6, 8, 29; on the Thrones, 115. Works: *De coelesti hierarchia*, XIV, 2, 4-5, 28, 28n. 1, 30, 32n. 9, 46, 48, 53, 59, 59n. 54, 59-60n. 55, 60n. 56, 65, 107n. 68; *De divinis nominibus*, 5, 28, 65n. 66, 66; *De ecclesiastica hierarchia*, 5, 31; *De mystica theologia*, 5, 57; *Letters*, 5
Divine Comedy. See Commedia
Doebler, Giampiero W., 40n. 23
Dominations, 28n.1, 34n. 13, 46nn. 33, 34, 47n. 36, 48, 83. *See also* hierarchy (angelic)
Donati, Piccarda, 86-87, 95-96, 127, 132
Donatism, 5, 5n. 6
Dondaine, Hyacinthe F., 4n. 3, 5n. 5
Duheim, Pierre, 81n. 22

Eco, Umberto, 179n. 115
Eichberg, Barbara Bruderer, Xn. 2, 85n. 30
emanatism, 9, 37, 38n. 19, 39, 42, 50, 58, 80, 84, 102n. 58, 104, 184
Empyrean (Dante's conception of), XII, 38n. 20, 40, 42, 67, 69, 80n. 20, 89, 91, 98, 98n. 52, 100-101, 101n. 57, 103, 104-105, 105n. 64, 106n. 66, 125, 129, 132-133, 173, 175, 176n. 110, 184
energheia (angelic), 45, 52, 56-61
'ens' and 'essentia,' 10, 16
episteme (angelic), 45, 51-56
Eriugena, John Scotus, 5n. 5, 6, 45n. 30, 88n. 35, 90n. 38
Ethics to Nicomachus. See under Aristotle
Eudoxus, 79, 79n. 18
Eusebius of Cesarea, 2n. 2
exitus and *reditus*, 49, 58, 76, 153n. 58, 176

Faes de Mottoni, Barbara, 28n. 2
Fasolini, Diego, 4n. 25
Finazzi, Silvia, 39n. 22, 135n. 17
Forti, Fiorenzo, 134n. 15
Fortuna, 2, 161, 161n. 81
Fosca, Nicola, 140n. 26
Francesca, 95-96; and Paolo, 96
Franke, William, 57-58n. 52, 148n. 45
Freccero, John, 98n. 52, 126n. 2, 144n. 35, 148n. 45, 149n. 47, 174n. 108
free will: 96n. 51; and predestination (determinism), XIV, XV, 71, 72, 108-124, 183, 184

Gabriel, 129
Gardeil, A., 138n. 22
Gardner, Edmund G., XII, 4n. 3, 48n. 38
Garin, Eugenio, 5n. 4, 119n. 89
Genesis. See under Bible
Ghisalberti, Alessandro, 100n. 56
Giles of Rome, 11, 22, 22n. 37
Gilson, Etienne, 18n. 31, 28n. 2, 82n. 24, 179n. 117
Gilson, Simon, 36nn. 15, 16, 68n. 69
Gorni, Guglielmo, 154n. 61, 165n. 87, 166n. 90
Gragnolati, Manuele, 113n. 77
Grant, Edward, 72n. 1, 80n. 20
Gregory the Great: angelic doctrine of, 2, 27, 34-35, 69; on angelic ministerial functions, XIV; on astrology, 118; on the Blessed, 129; and the functional interpretation of angels, 34-35; Gregorian hierarchical ordering, 34, 34n. 13, 46, 51, 129; Gregorian theory of Restoration, 129, 131; influence on Dante's angelology, 35, 46, 69, 130-131n. 9; in *Paradiso*, 52; on the Thrones, 115
Gregory, Tullio, 72n. 1, 73n. 3, 74n. 6, 75n. 9,

76nn. 10, 11, 84n. 28, 85n. 31, 86n. 32, 88n. 35, 90n. 38, 109n. 72, 110n. 75, 119n. 90
Grosseteste, Robert, 5n. 5, 38; *De luce*, 38n. 19
Guardini, Romano, xn. 2, XII, 136n. 19
Guidubaldi, Egidio, 36n. 15, 38n. 20
Guillaime (William) de Saint Thierry, 85n. 31, 86, 171n. 100

hammer metaphor, 102-103
Haskins, Charles H., 86n. 32, 118n. 88
Havely, Nick, 29n. 3
Hawkins, Peter, 47n. 35, 134n. 15, 171n. 102
Hercules, 94, 146, 148n. 44, 152, 152nn. 54, 55, 56, 153, 153-154n. 58, 154n. 59, 155, 156, 156n. 66, 161-162, 163n. 82
Hexaëmeron. *See under* Basil of Cesarea
hierarchy (angelic), 29-32, 44-69; Albert the Great on, 45-46n. 32, 53, 53n. 45, 55, 107n. 68; Aquinas on, 28n. 2, 82, 82n. 25, 83, 116-117, 129; association between hierarchical order and angelic functions, 3, 7n. 10, 15, 28n. 1, 31-32, 34, 34n. 13, 45-46n. 32, 47, 48, 62, 65, 80, 83, 90, 107n. 68, 115-116, 126; Bonaventure on, 31-32, 48, 48n. 38, 55-56, 56n. 48, 129, 132; in *Commedia*, XII, XIII, XIV, XV, 2, 14-15, 17, 21n. 36, 26, 27, 31-32, 35, 44-45, 46, 49-56, 59, 62-63, 65, 71, 77, 83, 102-104, 115-116, 129, 184; in *Convivio*, 2n. 1, 46, 47, 47-48n. 38, 49, 49n. 42, 73-74; Dionysian doctrine of, 2n. 1, 3, 4, 6, 7-8, 26, 27, 28n. 2, 31-32, 34, 34n. 13, 36, 44-69, 71, 76, 77, 116-117 (*see also* Dionysius: *De coelesti hierarchia*); Gregorian, 34, 34n. 13, 46, 51, 129; hierarchical illumination, 53-56; as immutable, 46; term 'hierarchy,' 45; three constitutive elements of, 45; triadic hierarchies, 7, 27, 31-32, 46, 47-48, 49, 55, 143; William of Auxerre on, 8n. 14. *See also* angels; Archangels; Cherubim; Dominations; Powers; Principalities; Seraphim; Thrones; Virtues
Hilduin of Saint Denis, 2n. 2, 5n. 5, 6
Hollander, Robert, 149n. 48, 151n. 53, 158n. 71, 165n. 85
Holy Spirit: associated with the *anima mundi*, 84-86; and creation, 86; as inspiring divine love, XV, 86, 142; reference to in *Paradiso*, 86-88; relation to angels and angelic operations, 24, 47, 47n. 37, 48, 48n. 38, 137-138, 139, 140, 141, 143, 181; relation to Providence, XV, 71, 86, 87-88, 90, 137-138; seven gifts of, XV, 87-88, 133, 137, 138, 138n. 22, 139, 140-143, 178n. 113, 181. *See also* Trinitarian theology
Holy Trinity. *See* Trinitarian theology
Holywood, John (Sacrobosco), 118n. 88, 120n. 92; *De sphaera*, 119n. 90, 120, 120nn. 91, 92
Hugh of Saint Victor, 2, 5n. 5, 6, 8, 48, 48n. 39, 139, 139n. 25. *See also* Victorine theology
hylomorphism, 9-10, 10nn. 16, 17, 16, 25

Iannucci, Amilcare, 148n. 45, 151nn. 52, 53
Incarnation, 3, 4, 5n. 6, 35, 42, 43, 55, 69, 88n. 35, 90-91, 92, 100, 130, 132, 139n. 25, 169, 181
inclinations (doctrine of), 110-115, 121-122, 123-124
Inferno: angels in, 2, 34; attack on astrological determinism in, 119-120; fallen angels in, IX, 125-126; heavenly messenger (*messo celeste*) in, XV, 33, 125, 126, 144-163, 181; poetic impasse in, XV, 34, 144, 150-154
Innocent II, 85n. 31, 86
integumentum, 146n. 39, 149-150, 157
Intelligences: Albert the Great on, 81, 82; Aquinas on, 81-82; Aristotelian theory of, xn. 2, XII, XIV, 2, 9n. 15, 14, 16, 16n. 28, 18, 18n. 32, 20n. 35, 32, 35, 76, 77-79, 79n. 18, 123; Averroes on, XIV, 14, 16, 16n. 28, 82; Avicenna on, 82; Beatrice as 'Intelletto,' 166; in *Commedia*, IX, 2-3, 23, 24, 26, 29, 63, 67, 77 126, 183; Dante on, 1, 13, 16-17, 19, 19n. 33, 20, 20-21n. 35, 22-24, 86; Fortuna as angelic Intelligence, 2, 161, 161n. 81; identification with angels, 77-84; Kilwardby on, 81, 82; Mercury as angelic Intelligence, 157, 161, 161n. 81, 181; and planets, 103, 103n. 60, 108, 118; Plato on, 19, 19n. 33. *See also* angels
Introductorium in Astronomiam. See under Albumasar
Isidore of Seville, 2n. 1, 109. Works: *De natura rerum*, 74; *Etymologie*, 34n. 13, 74, 109n. 71, 159n. 73

James, Saint, 24-25
Jewiss, Virginia, xn. 2
John of Scythopolis, 4-5n. 4, 5
John of Vercelli: questionnaire, 80-81, 81n. 22

Kay, Richard, XII
Keck, David, xn. 2, 9n. 15, 10n. 16, 28n. 2, 125n. 1
Kerény, Károly, 158n. 71, 163n. 82
Kilwardby, Robert, 8n. 21; on celestial mo-

tion, 81; and John of Vercelli's questionnaire, 80-81; and separation of philosophy and theology, 82

Latini, Brunetto, 2, 2n. 1, 37n. 16
Leclercq, Jean, 5n. 5, 6n. 7, 99n. 53
Lemay, Richard, 75n. 9, 86n. 32
Liber de aggregatione scientiae stellarum et principiis coelestium motuum. See under Alfraganus
Liber de causis, 37, 38, 38n. 19, 72, 73n. 2, 75, 75n. 8, 91, 93n. 44, 102n. 58
Liber de motibus coelorum. See under Alpetragius
light: angels and illumination, XI, XII, 26, 34, 34n. 14, 42, 45, 45-46n. 32, 48, 48n. 40, 50, 52-57, 58, 60-64, 69, 122-123, 184; Aquinas on, 39-40, 42; beatific illumination, 53; Bonaventure on, 39, 39n. 21, 55-56; Christological approach to, 38, 44; in *Commedia*, 27, 36, 38, 39-44, 50-51, 54, 56, 57-69, 91-92, 104-105, 123, 128, 134, 172, 175, 182, 185; and creation, 42-44; Dionysius on, 27, 36, 37, 55; direct vs. indirect illumination, 54-55; divine (*lux*) vs. material (*lumen*), 39, 39n. 21, 40n. 23; hierarchical illumination, 53-56; 'illuminating grace,' 63-64, 63-64n. 62; mechanics of light transmission, 67-68; medieval theories of, 27, 36-44; 'metaphysics of light,' 36, 36n. 15, 38
Lilla, Salvatore, 59n. 54
Litt, Thomas, 82n. 24 and 25
Lombardo, Marco, 120-122
Lucifer, 62, 64, 64n. 64, 129, 144, 160, 162, 177n. 112
Luscombe, David E., 7n. 10, 31nn. 7, 8, 48nn. 39, 40

Macrobius: 155n. 62, 163n. 82; *Somnium scipionis*, 66n. 68, 73, 85
Manescalchi, Romano, 133n. 14
Mandelbaum, Allen, 12n. 20
Marchesi, Simone, 14n. 27
Marenbon, John, 3n. 7
Martello, Carlo (Charles Martel), XIII-XIV, 111n. 76, 114, 115, 117, 118, 122
Marti, Mario, 178n. 114
Mary, XV, 3, 35, 38n. 20, 41, 51, 88, 127, 131, 136, 139, 139n. 24, 141n. 30, 164, 164n. 84, 166, 168, 176-181, 182, 185

Matelda, 134, 134n. 15
Maurus, Hrabanus, 73n. 4; *De universo*, 73n. 4
Maximus the Confessor, 4-5n. 4, 5
Mazzoni, Francesco, 66n. 67, 101n. 56, 167n. 90, 180n. 118
Mazzeo, Joseph, 36n. 15
Mazzotta, Giuseppe, XIII, 29n. 3, 37n. 18, 51n. 44, 92n. 42, 93n. 43, 95, 95n. 49, 149n. 48, 170n. 98
McEvoy, James, 38n. 19
Mellone, Attilio, Xnn. 1, 2, 34nn. 13, 14, 36n. 15, 38n. 19, 48n. 38, 98n. 52, 125n. 1
Mercury, 145-146, 146n. 39, 147, 153, 154n. 59, 155-163, 181-182
merit, 7, 7n. 10, 27, 35, 50, 61-65, 69, 127
Metamorphoses. See under Ovid
Metaphysics. See under Aristotle
Michael (archangel), 157n. 68, 159-160, 161, 162
Miglio, Massimo, 73n. 2
Migliorini-Fissi, Rosetta, 167n. 90
mirrors (angelic): in *Commedia*; XIII, 3, 26, 27, 42, 45, 53-54, 65-69, 92, 102n. 58, 105, 123, 125, 135n. 17, 138, 144, 171-176; *Commedia* as mirror, XII, 93, 93n. 43; Dante's voyage as, XIII; Dionysius on mirroring and angelic mirrors, 27, 31, 34n. 14, 56, 65-66, 69; and light reflection, 34n. 14, 41, 42, 45, 66, 68, 91, 105, 138, 171-172, 182; 'speculum,' 66; universe as mirror image of the divine mind, 74
Moevs, Christian, XII, 15n. 27, 80n. 20, 91n. 39, 98n. 52, 101n. 57, 102n. 59
Monarchia. See under Alighieri, Dante
Monophysitism, 5, 5n. 6
moon spots, 86, 96, 99-100, 108
Moore, Edward, 113n. 80
Morgan, Alison, 139n. 25
Morpurgo, Piero, 118n. 88
Morrison, Molly, 4n. 3
multiplicity and unity (relation between), 76, 84, 85; in *Commedia*, XIII, 50, 50n. 43, 68, 88, 90, 99, 123, 184
mythology: Christianization of, Xn. 2, 2, 97, 144-163, 181-182. *See also* Aeneas; Hercules; Mercury; Orpheus; Ovid; pagan tradition: influence on *Commedia*; Ulysses

Nardi, Bruno, XII, 14, 15-16, 16nn. 28, 29, 17n. 30, 18-19, 21n. 36, 36n. 15, 72n. 1, 73n. 2, 75nn. 8, 9, 78n. 17, 79nn. 18, 19, 80n. 20, 82n. 24, 99n. 55, 101nn. 56, 57, 102n. 58, 109n. 70, 113n. 77, 129n. 6, 131n. 9, 178n. 114

Nasti, Paola, 24n. 39
Neo-Platonism: doctrine of angelic Intelligences, XII; doctrines of the One and of the Good, 7-8, 58; and the image of the mirror, 66, 66n. 68; influence on cosmology, 73, 76; influence on Dante, XII, 2, 26, 35, 36, 36n. 15, 72, 80, 86, 102n. 58, 125, 177n. 112, 184; and Dionysius, 2-3n. 2, 6, 7-8, 27, 30, 37, 184; metaphysics of light, 36, 36n. 15, 37, 38, 39, 39n. 21, 40, 66n. 68; reconciliation of Aristotelian and Biblical cosmologies, 76; reconciliation of Plato and Aristotle's cosmological views, 84, 86, 184; reinterpretation of pagan mythology, 153-154, 154n. 61, 156n. 66; School of Chartres, 84n. 28, 140, 148, 149n. 48, 156n. 65; and tradition of *anima mundi*, 84-85. See also emanatism; Plato
Nestorianism, 5n. 6
Nicodemo, Renato, 177n. 112
Nygren, Anders, 57n. 50

Olschki, Leonardo, 38n. 20
Oliviero, Adriana, 74n. 7
On Heavens (*De coelo*). See under Aristotle
optics. See perspettiva
order. See taxis
Origen, 66n. 67, 129n. 6, 131n. 9
Orpheus, 152n. 56, 153-154nn. 58, 61, 154
Osserman, Robert, 98n. 52
Ottimo (Andrea Lancia), 108n. 69, 150n. 49, 151n. 50
Ovid: *Metamorphoses*, 145, 145n. 37, 157, 157n. 69, 158nn. 70, 71; myth of Gorgon and Perseus, 173-174

Padoan, Giorgio, 152n. 56, 155n. 62, 156n. 66
pagan tradition: angels in, 16, 19, 33n. 11; influence on *Commedia*, XV, 144-163, 181-182. See also mythology
Pagliaro, Antonino, 179n. 115
Panofsky, Erwin, 147, 147n. 43, 147-148n. 44, 157n. 68
Parronchi, Alessandro, 36n. 16
Paradiso: angelic hierarchies in, 29-32, 49-52, 102-103; angels in, IX, XII, I, 19, 24-25, 126-133; angels and astronomy in, 110-122; analogical language of 'divine names' in, 68; Aquinas in, 28, 30; Beatrice's dissertation on ontological principles in, 86; Beatrice's expositions on angels in, XIII; Beatrice's revelation in, 30; Bonaventure in, 28; and the desire of motion, 97-98; dialectics of merit and illumination in, 61-65; Dionysius in, 28, 30; discussion of the motion of the Primum Mobile in, 89; distinction between love and desire in, 98-99; and doctrine of animation, 82-83; hammer metaphor in, 102-103; idea of imitation in, 59-61, 91; light in, 40-42, 59, 91-92, 128, 172; love and knowledge in, 91-97; mirrors in, 68, 91, 92, 105, 171-176; order and ontology in, 71, 99-106; reference to the Holy Spirit in, 86-88; Saint James in, 24-25; simile of the optical phenomenon of light reflection in, 172; term 'ministero' in, 30-31; vision of Incarnation in, 90-91
Pascoli, Giovanni, 146n. 40
Pasquazi, Silvio, 154n. 59, 159n. 77, 160, 160nn. 77, 78
Patrides, C.A., 2n. 1, 4n. 3, 5n. 5, 34n. 13
Paul, Saint, 2n. 2, 4, 30n. 6, 47, 51, 115, 127n. 3; and the functional interpretation of angels, XIV, 32-34; influence on Dante's angelology, XIV, 66, 68, 69, 88n. 35, 90n. 38, 149n. 47, 173; and the notion of 'speculum,' 66, 66n. 67; Pauline concept of collaboration, 60-61, 60n. 57
Pelikan, Jaroslav, 5n. 5 and 6, 45n. 30
Pépin, Jean, 149n. 48, 179nn. 115, 117
perspettiva (optics), 36n. 16, 37, 37n. 18, 38, 40, 42, 44, 68, 73, 92, 105, 123, 138, 172
Perspettiva. See under Bacon, Roger
Pertile, Lino, 24n. 39, 99n. 53
Peter, Saint, 87
Peterson, Mark, 98n. 52
Petrocchi, Giorgio, 12n. 20
Philoponus, John, 79, 79-80n. 20
Physiologus, 74
Piana, Celestino, 38n. 20
Picatrix, 119n. 89
Picone, Michelangelo, 94n. 45
Pietrobono, Luigi, 178n. 114
Plato: doctrine of the *anima mundi*, 84-85; doctrine of ideas, 78; on inclinations, 113, 113n. 78; on Intelligences, 19, 19n. 33; and *integumentum*, 149n. 48. Works: *Republic*, 113, 113n. 78; *Timaeus*, 73, 77, 85, 85n. 31. See also Neo-Platonism
Pliny: *Natural History*, 74
poetry: compared with theological and rational discourse, 57
Powers, 34n. 13, 46nn. 33, 34, 47n. 36, 48, 49n. 42, 83. See also hierarchy (angelic)

INDEX

predestination (determinism), 14, 75, 80, 91, 169-170, 170n. 97; and free will, XIV, XV, 71, 72, 108-124, 183, 184. *See also* astrology

Primum Mobile, 24, 89, 92, 100-101, 104, 105, 105nn. 64, 65, 106, 170

Principalities, 34n. 13, 46nn. 33, 34, 47n. 36, 48, 83, 116, 116n. 84, 118, 160. *See also* hierarchy (angelic)

Proclus, 2n. 2, 4, 79n. 20; *Liber de causis*, 37, 38, 38n. 19, 72, 73n. 2, 75, 75n. 8, 91, 93n. 44, 102n. 58

Proto, Enrico, 99n. 55

Providence: angels as instruments of, XI, XIII, XIV, XV, 3, 23n. 38, 30, 77, 86, 88, 89-90, 99, 111, 126, 137-138, 181, 182, 184; and astrology, 111-112, 118; as the governing principle of divine love, XV; as the ordering principle of the universe, 87; reconciliation of science and Providence in *Commedia*, 71; relation to celestial motion, 86; relation to the Holy Spirit, XV, 71, 86, 87-88, 90, 137-138

Pseudo-Dionysius the Areopagite. *See* Dionysius

Pseudo-Grossatesta, 39n. 21

Ptolemy, Claudius (Tolomeus), 37n. 18, 79-80, 110n. 73, 169. Works: *Almagest*, 73, 79n. 20, 108-109n. 70; *Tetrabiblos*, 79n. 20

Purgatorio: angels in, X, XV, 34, 67-68, 126, 133-143, 181, 183; angels as guardians in, IX, 133; angels as guides in, IX, 138; angels' relation to the Holy Spirit in, 88, 139, 181; attack on astrological determinism in, 119, 120-122; explanation of the mechanics of light transmission in, 67-68; explanation of the soul in, 113, 113n. 77; luminosity and mirrors in, 135n. 17, 138; symbols in, 135-136n. 18

Raimondi, Ezio, 133n. 14

Raymond de Marseilles, 85-86, 110; *Tables of Marseille*, 85, 85n. 31

Réau, Louis, 157n. 68, 159n. 77

reditus. *See exitus* and *reditus*

Reynolds, Brian, 180n. 119

Renucci, Paul, 146n. 39

Restoration (theory of), 129, 129n. 6, 130, 130-131n. 9, 131

Ricci, Pier Giorgio, 180n. 119

Ricci, Lucia Battaglia, 4n. 3,

Richard of Saint Victor, 6, 8, 8n. 12, 139, 139n. 25. *See also* Victorine theology

Ripa, Cesare, 158n. 72

Riva, Franco, 179n. 115

Rivière, J., 180n. 119

Ross, Charles, 136n. 18

Rossini, Antonio, 4n. 3, 24n. 39, 44n. 28, 177n. 112

Salonius of Geneve, 129n. 6

salvation: role of angels in, 3, 7, 13, 32, 34, 46n. 33, 48, 88, 125, 138, 180, 184. *See also Commedia*: salvific (soteriological) perspective of

Santagata, Marco, 166n. 90

Sarrazin, (John), 6

Sbacchi, Diego, XII, 4n. 3, 28n. 1, 45-46n. 32, 65n. 65, 66n. 68, 115n. 83

Scazzoso, Piero, 28n. 1

Scholasticism: and the theory of Restoration, 129-130; and emanatism, 102n. 58; and 'form' and 'act,' 9, 15; influence on Dante, 3, 6, 15, 15n. 27, 17, 35, 125, 125n. 1; influence of Dionysian *Corpus* on, 6, 27, 30n. 5; positions on celestial motion, 81; reconciliation of science and Christian doctrine, 40; Scholastic angelology, 1, 3, 4n. 3, 6, 27, 125n. 1, 132; theories on light, 39, 76. *See also* Abelard; Aquinas; Bonaventure; Magnus; Scotus; William of Ockham

science: angels' contemplation and communication of divine science, 2, 8, 15, 26, 45, 48, 50, 51, 53, 55, 56, 57, 61, 63, 65; Aristotelian scientific thought, 72-73, 74-75, 76, 78, 80, 97, 101-102; cherubim and *multitudo scientiae*, 107, 107n. 68; and debates about the number of angels, 19; and definition of hierarchy, 51-56; harmonization of science and Christian doctrine, 40, 71, 72-77, 85n. 30, 111n. 76, 123; influence on Dante, X, XII, XV, 36, 40, 71, 72, 73, 73n. 2, 101, 108-109n. 70, 124; scientific debates about angels, IX; 'scientific' theology, 7, 73; simile of the mechanics of light refraction in *Purgatorio*, 67-68; translation and transmission of ancient scientific texts, 37, 72-77, 108, 108-109n. 70, 118-119. *See also* astrology; astronomy; *perspettiva*

Scot, Michael, 118, 118n. 88, 119-120

Scotus, Duns, 11

Seraphim, 34n. 13, 46nn. 33, 34, 47n. 36, 48, 49n. 42, 62, 83, 105, 105n. 65, 106, 107n. 68, 135n. 17, 160. *See also* hierarchy (angelic)

Serravalle, Johannis da, 19n. 34, 146
Servius, 154n. 59, 155, 163n. 82
Silvestris, Bernardus, 149-150, 152n. 56, 154-155, 155n. 62, 156, 156n. 65, 163n. 82; *De mundi universitate*, 74
Singleton, Charles, 134n. 15, 177n. 112, 179n. 116
Stabile, Giorgio, XII, 100n. 56, 104n. 63
Statius, 113, 113n. 77, 151nn. 51, 53, 157, 158n. 70
Suarez-Nani, Tiziana, Xn. 2, 10n. 16, 22n. 37, 23n. 38, 80n. 21, 81, 81n. 22, 82n. 25, 110n. 74
Summa Theologiae. See under Aquinas
Suarez, Francisco, 160n. 79

taxis (divine), 45, 46-51
Tempier, Etienne, 79n. 18
tenth order, 74, 129, 129n. 6, 131n. 9, 132, 132n. 12, 133
Theodoric of Chartres: *De sex dierum operibus*, 85n. 31
Theodoric of Strasburg, 85
Thorndike, Lynn, 75n. 9, 120n. 92
Thrones, 28n. 1, 34n. 13, 46nn. 33, 34, 47n. 36, 48, 83, 115, 115-116n. 83, 117-118. *See also* hierarchy (angelic)
Tolan, John, 76n. 11
Tolomeus, Claudio. *See* Ptolemy
Trinitarian theology: Bonaventure on the Trinity, 29n. 3, 30, 31, 48, 48n. 38, 55-56; in *Commedia*, 3, 27, 29, 29n. 3, 30, 35, 40, 41, 41n. 25, 50, 58, 86-87, 90, 137, 140, 163; in *Convivio*, 47, 47n. 37, 47-48n. 38, 49, 49n. 42; Hugh of Saint Victor on the Trinity, 48n. 39; in *Vita Nuova*, 167. *See also* Holy Spirit
Trusso, Francesco, 177n. 112

Ulysses, 94-95, 95nn. 48, 49, 97, 119-120
understanding. *See episteme*
unity. *See* multiplicity and unity

Valla, Lorenzo, 4, 4-5n. 4
Van der Meersch, J., 64n. 62
Vasoli, Cesare, XII-XIII, 18n. 32, 20n. 35, 38n. 20, 47n. 37, 74n. 7, 75n. 9, 79n. 18, 102nn. 58, 59
Vasta, Edward, 165n. 87
Vecce, Carlo, 166n. 90
Vellutello, Alessandro, 105, 105n. 65
Vescovini, Graziella Federici, 36n. 16, 72n. 1, 110n. 74, 118n. 88, 119n. 89, 120n. 91, 158n. 71, 170n. 97
Victorine theology, 1, 8, 27, 140, 148. *See also* Hugh of Saint Victor; Richard of Saint Victor
Vincent of Beauvais, 5n. 5, 47n. 37 and 38, 151n. 52
Virgil: *Aeneid*, 85, 146n. 39, 149, 149n. 48, 151nn. 51, 53, 152-157 (*see also* Aeneas); in *Commedia*, 96, 119, 133, 134, 135, 142n. 31, 144, 147, 147n. 42, 148n. 45, 150-151 158n. 71, 162, 168n. 96, 175, 176
Virtues, 34n. 13, 46nn. 33, 34, 47n. 36, 48, 82, 82n. 25, 83, 83n. 26. *See also* hierarchy (angelic)
Vita Nuova. See under Alighieri, Dante
Von Balthasar, Hans Urs, 13, 13n. 22

Wetherbee, Wintrop, 149n. 48
William of Auvergne, 31, 31n. 8
William of Auxerre, 8, 8n. 14
William of Conches, 84n. 28, 85, 85n. 31, 149n. 48, 155n. 62, 156n. 65; *Philosophia*, 85n. 31
William of Ockham, 11

Zambelli, Paola, 75n. 9
Zanini, Carlo, Xn. 2, XII
Zingarelli, Nicola, 146n. 39

CITTÀ DI CASTELLO • PG
FINITO DI STAMPARE NEL MESE DI AGOSTO 2010

BIBLIOTECA DELL'«ARCHIVUM ROMANICUM»

Serie I: Storia - Letteratura - Paleografia

1. Bertoni, G. *Guarino da Verona fra letterati e cortigiani a Ferrara (1429-1460)*. 1921. (esaurito)
2. —— *Programma di filologia romanza come scienza idealistica*. 1922. (esaurito)
3. Verrua, P. *Umanisti ed altri «studiosi viri» italiani e stranieri di qua e di là dalle Alpi e dal mare*. 1924, 234 pp., 2 tavv.
4. Cino da Pistoia, *Le rime*. 1925. (esaurito)
5. Zaccagnini, G. *La vita dei maestri e degli scolari nello Studio di Bologna nei secoli XIII e XIV*. 1926. (esaurito)
6. Jordan, L. *Les idées, leurs rapports et le jugement de l'homme*. 1926, X-234 pp.
7. Pellegrini, C. *Il Sismondi e la storia della letteratura dell'Europa meridionale*. 1926, 168 pp.
8. Restori, A. *Saggi di bibliografia teatrale spagnola*. 1927, 122 pp., 3 cc.
9. Santangelo, S. *Le tenzoni poetiche nella letteratura italiana dalle origini*. 1928. (esaurito)
10. Bertoni, G. *Spunti, scorci e commenti*. 1928, VIII-198 pp.
11. Ermini, F. *Il «dies irae»*. 1928, VIII-158 pp.
12. Filippini, F. *Dante scolaro e maestro. (Bologna - Parigi - Ravenna)*. 1929, VIII-224 pp.
13. Lazzarini, L. *Paolo de Bernardo e i primordi dell'Umanesimo in Venezia*. 1930. (esaurito)
14. Zaccagnini, G. *Storia dello Studio di Bologna durante il Rinascimento*. 1930, X-348 pp., 42 ill.
15. Catalano, M. *Vita di Ludovico Ariosto*. 2 voll. 1931. (esaurito)
16. Ruggieri, J. *Il canzoniere di Resende*. 1931, 238 pp.
17. Döhner, K. *Zeit und Ewigkeit bei Chateaubriand*. 1931. (esaurito)
18. Troilo, S. *Andrea Giuliano politico e letterato veneziano del Quattrocento*. 1932. (esaurito)
19. Ugolini, F. A. *I Cantari d'argomento classico*. 1933. (esaurito)
20. Berni, F. *Poesie e prose*. 1934. (esaurito)
21. Blasi, F. *Le poesie di Guilhem de la Tor*. 1934, XIV-78 pp.
22. Cavaliere, A. *Le poesie di Peire Raimond de Tolosa*. 1935. (esaurito)
23. Toschi, P. *La poesia popolare religiosa in Italia*. 1935. (esaurito)
24. Blasi, F. *Le poesie del trovatore Arnaut Catalan*. 1937. (esaurito)
25. Gugenheim, S. *Madame d'Agoult et la pensée européenne de son époque*. 1937. (esaurito)
26. Lewent, K. *Zum Text der Lieder des Giraut de Bornelh*. 1938. (esaurito)
27. Kolsen, A. *Beiträge zur Altprovenzalischen Lyrik*. 1938. (esaurito)
28. Niedermann, J. *Kultur. Werden und Wandlungen des Begriffs und seiner Ersatzbegriffe von Cicero bis Herder*. 1941.
29. Altamura, A. *L'Umanesimo nel mezzogiorno d'Italia*. 1941. (esaurito)
30. Nordmann, P. *Gabriel Seigneux de Correvon, ein schweizerischer Kosmopolit. 1695-1775*. 1947. (esaurito)
31. Rosa, S. *Poesie e lettere inedite*. 1959. (esaurito)
32. Panvini, B. *La leggenda di Tristano e Isotta*. 1952. (esaurito)
33. Messina, M. *Domenico di Giovanni detto il Burchiello. Sonetti inediti*. 1952. (esaurito)
34. Panvini, B. *Le biografie provenzali. Valore e attendibilità*. 1952. (esaurito)
35. Moncallero, G. L. *Il Cardinale Bernardo Dovizi da Bibbiena umanista e diplomatico*. 1953. (esaurito)
36. D'Aronco, G. *Indice delle fiabe toscane*. 1953, 236 pp.
37. Branciforti, F. *Il canzoniere di Lanfranco Cigala*. 1954. (esaurito)
38. Moncallero, G. L. *L'Arcadia*. Vol. I: *Teorica d'Arcadia*. 1953.
39. Galanti, B. M. *Le villanelle alla napolitana*. 1954. (esaurito)
40. Crocioni, G. *Folklore e letteratura*. 1954. (esaurito)
41. Vecchi, G. *Uffici drammatici padovani*. 1954, XII-258 pp., 73 tavv. esempi mus.
42. Vallone, A. *Studi sulla Divina Commedia*. 1955. (esaurito)
43. Panvini, B. *La scuola poetica siciliana*. 1955. (esaurito)
44. Dovizi, B. *Epistolario di Bernardo Dovizi da Bibbiena*. Vol. I (1490-1513). 1955. (esaurito)
45. Collina, M. D. *Il carteggio letterario di uno scienziato del Settecento (Janus Plancus)*. 1957, VIII-174 pp., 5 tavv. f.t.
46. Spaziani, M. *Il canzoniere francese di Siena (Biblioteca Comunale HX 36)*. 1957. (esaurito)
47. Vallone, A. *Linea della poesia foscoliana*. 1957. (esaurito)
48. Crinò, A. M. *Fatti e figure del Seicento anglo-toscano. (Documenti inediti sui rapporti letterari, diplomatici e culturali fra Toscana e Inghilterra)*. 1957. (esaurito)
49. Panvini, B. *La scuola poetica siciliana. Le canzoni dei rimatori non siciliani*. Vol. I. 1957. (esaurito)
50. Crinò, A. M. *John Dryden*. 1957, 406 pp., 1 tav. f.t.
51. Lo Nigro, S. *Racconti popolari siciliani. (Classificazione e Bibliografia)*. 1958. (esaurito)
52. Musumarra, C. *La sacra rappresentazione della Natività nella tradizione italiana*. 1957. (esaurito)
53. Panvini, B. *La scuola poetica siciliana. Le canzoni dei rimatori non siciliani*. Vol. II. 1958. (esaurito)
54. Vallone, A. *La critica dantesca nell'Ottocento*. 1958, 240 pp. Ristampa 1975.
55. Crinò, A. M. *Dryden, poeta satirico*. 1958. (esaurito)
56. Coppola, D. *Sacre rappresentazioni aversane del sec. XVI, la prima volta edite*. 1959, XII-270 pp., ill.
57. Piramus et Tisbè. *Introduzione - Testo critico - Traduzione e note a cura di F. Branciforti*. 1959. (esaurito)
58. Gallina, A. M. *Contributi alla storia della lessicografia italo-spagnola dei secoli XVI e XVII*. 1959, 336 pp.
59. Piromalli, A. *Aurelio Bertola nella letteratura del Settecento. Con testi e documenti inediti*. 1959. Ristampa 1998.
60. Gamberini, S. *Poeti metafisici e cavalieri in Inghilterra*. 1959, 270 pp.

61. BERSELLI AMBRI, P. *L'opera di Montesquieu nel Settecento italiano*. 1960. (esaurito)

62. *Studi secenteschi*, vol. I (1960). 1961, 220 pp.

63. VALLONE, A. *La critica dantesca del '700*. 1961. (esaurito)

64. *Studi secenteschi*, vol. II (1961). 1962, 334 pp., 7 tavv. f.t.

65. PANVINI, B. *Le rime della scuola siciliana*. Vol. I: Introduzione - Edizione critica - Note. 1962, LII-676 pp. Rilegato.

66. BALMAS, E. *Un poeta francese del Rinascimento: Etienne Jodelle, la sua vita - il suo tempo*. 1962, XII-876 pp., 12 tavv. f.t.

67. *Studi secenteschi*, vol. III (1962). 1963, IV-238 pp. 4 tavv. f.t.

68. COPPOLA, D. *La poesia religiosa del sec. XV*. 1963, VIII-150 pp.

69. TETEL, M. *Étude sur la comique de Rabelais*. 1963. (esaurito)

70. *Studi secenteschi*, vol. IV (1963). 1964, VI-238 pp., 5 tavv.

71. BIGONGIARI, D. *Essays on Dante and Medieval Culture*. 1964. (esaurito)

72. PANVINI, B. *Le rime della scuola siciliana* - Vol. II: Glossario. 1964, XVI-180 pp. Rilegato.

73. BAX, G. *«Nniccu Furcedda», farsa pastorale del XVIII sec. in vernacolo salentino*, a cura di Rosario Jurlaro. 1964, VIII-108 pp., 12 tavv.

74. *Studi di letteratura, storia e filosofia in onore di Bruno Revel*. 1965, XXII-666 pp., 3 tavv.

75. BERSELLI AMBRI, P. *Poemi inediti di Arthur de Gobineau*. 1965, 232 pp., 3 tavv. f.t.

76. PIROMALLI, A. *Dal Quattrocento al Novecento. Saggi critici*. 1965, VI-190 pp.

77. BASCAPÈ, A. *Arte e religione nei poeti lombardi del Duecento*. 1964, 96 pp.

78. GUIDUBALDI, E. *Dante Europeo, I. Premesse metodologiche e cornice culturale*. 1965. (esaurito)

79. *Studi secenteschi*, vol. V (1964). 1965, 192 pp., 2 tavv. f.t.

80. VALLONE, A. *Studi su Dante medioevale*. 1965, 276 pp.

81. DOVIZI, B. *Epistolario di Bernardo Dovizi da Bibbiena*. Vol. II (1513-1520). 1965. (esaurito)

82. *La Mandragola* di Niccolò Machiavelli per la prima volta restituita alla sua integrità. 1965. (esaurito)
Edizione di lusso numerata da 1 a 370, su carta grave, con 2 tavv. f.t.

83. GUIDUBALDI, E. *Dante Europeo, II. Il paradiso come universo di luce (la lezione platonico-bonaventuriana)*. 1966, VIII-462 pp., 2 tavv. f.t.

84. LORENZO DE' MEDICI IL MAGNIFICO, *Simposio*, a cura di Mario Martelli. 1966, 176 pp., 2 riproduzioni.

85. *Studi secenteschi*, vol. VI (1965). 1966, IV-310 pp., 1 tav. f.t.

86. *Studi in onore di Italo Siciliano*. 1966, 2 voll. di XII-1240 pp. compless. e 6 tavv. f.t.

87. ROSSETTI, G. *Commento analitico al "Purgatorio" di Dante Alighieri*. Opera inedita a cura di Pompeo Giannantonio. 1966, CIV-524 pp.

88. PIROMALLI, A. *Saggi critici di storia letteraria*. 1967.(esaurito)

89. *Studi di letteratura francese*, vol. I. 1967, XVI-176 pp.

90. *Studi secenteschi*, vol. VII (1966). 1967, VI-166 pp., 6 tavv. f.t.

91. PERSONÈ, L. M. *Scrittori italiani moderni e contemporanei. Saggi critici*. 1968, IV-340 pp.

92. *Studi secenteschi*, vol. VIII (1967). 1968, VI-230 pp., 1 tav. f.t.

93. TOSO RODINIS, G. *Galeazzo Gualdo Priorato, un moralista veneto alla corte di Luigi XIV*. 1968, VI-226 pp., 9 tavv. f.t.

94. GUIDUBALDI, E. *Dante Europeo, III. Poema sacro come esperienza mistica*. 1968, VIII-736 pp., 24 tavv. f.t. di cui 1 a colori.

95. DISTANTE, C. *Giovanni Pascoli poeta inquieto tra '800 e '900*. 1968, 212 pp.

96. RENZI, L. *Canti narrativi tradizionali romeni. Studi e testi*. 1969, IV-170 pp.

97. VALLONE, A. *L'interpretazione di Dante nel Cinquecento. Studi e ricerche*. 1969, 306 pp.

98. PIROMALLI, A. *Studi sul Novecento*. 1969. (esaurito)

99. CACCIA, E. *Tecniche e valori dal Manzoni al Verga*, 1969, X-286 pp.

100. GIANNANTONIO, P. *Dante e l'allegorismo*. 1969. (esaurito)

101. *Studi secenteschi*, vol. IX (1968). 1969, IV-384 pp., 9 tavv. f.t.

102. TETEL, M. *Rabelais et l'Italie*. 1969, IV-314 pp.

103. REGGIO, G. *Le egloghe di Dante*. 1969, X-88 pp.

104. MOLONEY, B. *Florence and England. Essays on cultural relations in the second half of the eighteenth century*. 1969, VI-202 pp., 4 tavv. f.t.

105. *Studi di letteratura francese*, vol. II (1969). 1970, VI-360 pp., 11 tavv. f.t.

106. *Studi secenteschi*, vol. X (1969). 1970, VI-312 pp.

107. *Il Boiardo e la critica contemporanea* a cura di G. Anceschi. 1970, VIII-544 pp.

108. PERSONÈ, L. M. *Pensatori liberi nell'Italia contemporanea. Testimonianze critiche*. 1970, IV-290 pp.

109. GAZZOLA STACCHINI, V. *La narrativa di Vitaliano Brancati*. 1970, VIII-160 pp.

110. *Studi secenteschi*, vol. XI (1970). 1971, IV-292 pp. con 9 tavv. f.t.

111. BARGAGLI, G. (1537-1587), *La Pellegrina*. Edizione critica con introduzione e note di F. Cerreta. 1971, 228 pp. con 2 ill. f.t.

112. SAROLLI, G. R. *Prolegomena alla Divina Commedia*, 1971, LXXII-454 pp. con 9 tavv. f.t. Ristampa 2002.

113. MUSUMARRA, C. *La poesia tragica italiana nel Rinascimento*. 1972, IV-172 pp. Ristampa 1977.

114. PERSONÈ, L. M. *Il teatro italiano della «Belle Époque». Saggi e studi*. 1972, 410 pp.

115. *Studi secenteschi*, vol. XII (1971). 1972, IV-516 pp. con 2 tavv. f.t.

116. LOMAZZI, A. *Rainaldo e Lesengrino*. 1972, XIV-222 pp. con 2 tavv. f.t.

117. PERELLA, R. *The critical fortune of Battista Guarini's «Il Pastor Fido»*. 1973, 248 pp.

118. *Studi secenteschi*, vol. XIII (1972). 1973, IV-372 pp. con 11 tavv. f.t.

119. DE GAETANO, A. *Giambattista Gelli and the Florentine Academy: the rebellion against Latin*. 1976, VIII-436 pp. e 1 ill.

120. *Studi secenteschi*, vol. XIV (1973). 1974, IV-300 pp. con 4 tavv. f.t.

121. DA POZZO, G. *La prosa di Luigi Russo*. 1975, 208 pp.

122. PAPARELLI, G. *Ideologia e poesia di Dante*. 1975, XII 332 pp.

123. *Studi di letteratura francese*, vol. III (1974). 1975, 220 pp.

124. COMES, S. *Scrittori in cattedra*. 1976, XXXII-212 pp. con un ritratto e 1 tav. f.t.

125. TAVANI, G. *Dante nel Seicento. Saggi su A. Guarini, N. Villani, L. Magalotti*. 1976, 176 pp.

126. *Studi secenteschi*, vol. XV (1974). *Indice generale dei voll. I-X (1960-1969)*. 1976, 188 pp.

127. PERSONÈ, L. M. *Grandi scrittori nuovamente interpretati: Petrarca, Boccaccio, Parini, Leopardi, Manzoni*. 1976, 256 pp.

128. *Innovazioni tematiche, espressive e linguistiche della letteratura italiana del novecento* - Atti dell'VIII Congresso dell'Associazione internazionale per gli studi di lingua e letteratura italiana. 1976, XII-300 pp.

129. *Studi di letteratura francese*, vol. IV (1975). 1976, 180 pp. con 2 ill.

130. *Studi secenteschi*, vol. XVI (1975). 1976, IV-244 pp.

131. CASERTA, E. G. *Manzoni's Christian Realism*. 1977, 260 pp.

132. TOSO RODINIS, S. *Dominique Vivant Denon. I fiordalisi, Il berretto frigio, La sfinge*. 1977, 232 pp. con 10 ill. f.t.

133. VALLONE, A. *La critica dantesca nel '900*. 1976, 480 pp.

134. FRATANGELO, A. e M. *Guy De Maupassant scrittore moderno*. 1976, 180 pp.

135. COCCO, M. *La tradizione cortese e il petrarchismo nella poesia di Clément Marot*. 1978, 320 pp.

136. MASTROBUONO, A. C. *Essays on Dante's Philosophy of History*. 1979, 196 pp.

137. *Primo centenario della morte di Niccolò Tommaseo (1874-1974)*. 1977, 224 pp.

138. SICILIANO, I. *Saggi di letteratura francese*. 1977, 316 pp.

139. SCHIZZEROTTO, G. *Cultura e vita civile a Mantova fra '300 e '500*. 1977, 148 pp. con 9 ill. f.t.

140. *Studi secenteschi*, vol. XVII (1976). 1977, 184 pp., con 5 tavv. f.t.

141. GAZZOLA STACCHINI, V. - BIANCHINI, G. *Le Accademie dell'Aretino nel XVII e XVIII secolo*. 1978, XVIII-598 pp. con 18 ill. n.t. e 24 f.t.

142. FRIGGIERI, O. *La cultura italiana a Malta. Storia e influenza letteraria e stilistica attraverso l'opera di Dun Karm*. 1978, 172 pp. con 5 ill. f.t.

143. *Studi secenteschi*, vol. XVIII (1977). 1978, 276 pp.

144. VANOSSI, L. *Dante e il «Roman de la Rose» Saggio sul «Fiore»*. 1979, 380 pp.

145. RIDOLFI, R. *Studi Guicciardiniani*. 1978, 344 pp.

146. ALLEGRETTO, M. *Il luogo dell'Amore. Studio su Jaufre Rudel*. 1979, 104 pp.

147. MISAN, J. *L'Italie des doctrinaires (1817-1830). Une image en élaboration*. 1978, 204 pp.

148. TOAFF, A. *The Jews in medieval Assisi 1305-1487. A social and economic history of a small Jewish community in Italy*. 1979, 240 pp. con 14 ill. f.t.

149. TROVATO, P. *Dante in Petrarca. Per un inventario dei dantismi nei «Rerum vulgarium Fragmenta»*. 1979, X-174 pp.

150. FIORATO, A. C. *Bandello entre l'histoire et l'écriture. La vie, l'expérience sociale, l'évolution culturelle d'un conteur de la Renaissance*. 1979, XXII-686 pp.

151. *Studi secenteschi*, vol. XIX (1978). 1979, 260 pp.

152. BOSISIO, P. *Carlo Gozzi e Goldoni. Una polemica letteraria con versi inediti e rari*. 1979, 444 pp.

153. ZANATO, T. *Saggio sul «Comento» di Lorenzo de' Medici*. 1979, 340 pp.

154. *Studi di letteratura francese*, vol. V. 1979, 204 pp.

155. PIROMALLI, A. *Società, cultura e letteratura in Emilia Romagna*. 1980, 180 pp.

156. ACCADEMICI INTRONATI DI SIENA, *La Commedia degli Ingannati*. 1980, 248 pp.

157. *Studi di letteratura francese*, vol. VI. 1980, 176 pp.

158. HARRAN, D. *«Maniera» e il Madrigale - Una raccolta di poesie musicali del Cinquecento*. 1980, 124 pp.

159. *Studi secenteschi*, vol. XX (1979). 1980, VI-214 pp.

160. USSIA, S. *Carteggio Magliabechi. Lettere di Borde, Arnaud e associati lionesi ad A. Magliabechi*. 1980, 244 pp.

161. DA COL, I. *Un romanzo del Seicento. La Stratonica di Luca Assarino*. 1981, 244 pp. con 24 tavv. f.t.

162. *Studi secenteschi*, vol. XXI (1980). 1981, 294 pp.

163. *Studi di letteratura francese*, vol. VII. 1981, 224 pp.

164. CASTELLETTI, C. *Stravaganze d'amore. «Comedia»*. 1981, 172 pp.

165. *Carteggio inedito fra N. Tommaseo e G. P. Vieusseux*. I: (1835-1839). A cura di V. Missori. 1981, 688 pp.

166. *Studi secenteschi*, vol. XXII (1981). *Indice generale dei voll. XI-XX (1970-1979)*. 1981, 184 pp.

167. *Il Rinascimento. Aspetti e problemi attuali*. Atti del X Congresso dell'Associazione internazionale per gli studi della lingua e letteratura italiana. 1982, VI-700 pp.

168. *Stendhal e Milano*. Atti del XIV Congresso internazionale Stendhaliano. 1982, 2 tomi di complessive XXVI-972 pp. e 2 tavv. a colori.

169. *Studi secenteschi*, vol. XXIII (1982). 1982, 328 pp. con 1 tav. f.t.

170. *Studi di letteratura francese*, vol. VIII. 1982, 208 pp.

171. *Studi di letteratura francese*, vol. IX. 1983, 274 pp.

172. AONIO PALEARIO, *Dell'economia o vero del governo della casa*. 1983, 120 pp. con 4 tavv. f.t.

173. DALLA PALMA, G. *Le strutture narrative dell'«Orlando Furioso»*. 1984, 228 pp.

174. *Studi secenteschi*, vol. XXIV (1983). 1983, 324 pp.

175. RAUGEI, A. M. *Bestiario valdese*. 1984, 362 pp. con ill. n.t.

176. DA POZZO, G. *L'ambigua armonia. Studio sull'«Aminta» del Tasso*. 1983, 336 pp.

177. *Studi di letteratura francese*, vol. X. 1983, 208 pp.

178. *Miscellanea di studi in onore di V. Branca*. Vol. I: *Dal Medioevo al Petrarca*. 1983, XII-492 pp. con 1 tav. f.t.

179. —— Vol. II: *Boccaccio e dintorni*. 1983, VI-450 pp.

180. —— Vol. III: *Umanesimo e Rinascimento a Firenze e Venezia*. 1983, 2 tomi di complessive XII-848 pp.

181. —— Vol. IV: *Tra Illuminismo e Romanticismo*. 1983, 2 tomi di complessive XII-900 pp.

182. —— Vol. V: *Indagini Otto-Novecentesche*. 1983, VI-390 pp.

183. RIZZO, G. *Tommaso Briganti. Inedito poeta romantico*. 1984, 274 pp.

184. POLIAGHI, N. F. *Stendhal e Trieste*. 1984, VI-202 pp. con 22 ill.

185. MICHELANGELO BUONARROTI IL GIOVANE, *La Fiera. Redazione originaria (1619)*. 1984, 162 pp. con 4 tavv. f.t.

186. *I cantari. Struttura e tradizione*. 1984, 200 pp.

187. BIANCHINI, G. *Federico Nomi. Un letterato del '600. Profilo e fonti manoscritte*. 1984, XVI-338 pp. con 11 tavv. f.t.

188. *Studi secenteschi*, vol. XXV (1984). 1984, 304 pp.

189. ZAMBON, F. *Robert De Boron e i segreti del Graal*. 1984, 132 pp.

190. *Fenoglio a Lecce*. 1984, 248 pp.

191. SCHETTINI PIAZZA, E. *Giuseppe Chiarini. Saggio biobibliografico di un letterato dell'Ottocento*. 1984, X-158 pp. con 1 tav. f.t.

192. *Studi di letteratura francese*, vol. XI. 1985, 362 pp. con 9 tavv. f.t.

193. MISAN, J. *Les lettres italiennes dans la presse française (1815-1824)*. 1985, 210 pp.

194. CAIRNS, C. *Pietro Aretino and the Republic of Venice. Researches on Aretino and his circle in Venice, 1527-1556*. 1985, 272 pp.

195. BERTELÀ, M. *Stendhal et l'Autre. L'homme et l'oeuvre à travers l'idée de féminité*. 1985, 352 pp.

196. PIGLIONICA, A. M. *Dalla realtà all'illusione*: The Tempest *o la parola preclusa*. 1985, 146 pp.

197. *Studi secenteschi*, vol. XXVI (1985), 1985, 352 pp.

198. CERVIGNI, D. S. *Dante's poetry of dreams*. 1986, 230 pp.

199. *Studi di letteratura francese*, vol. XII. 1986, II-282 pp. con 4 tavv. f.t.

200. MARCO POLO, *Il milione*. Edizione del testo toscano («ottimo»). 1986, XII-418 pp.

201. DELMAY, B. *I personaggi della «Divina Commedia». Classificazione e regesto*. 1986, LVI-414 pp.

202. *Patronage and Public in the Trecento*. 1986, 180 pp. con 36 ill. f.t.

203. MITCHELL, B. *The Majesty of the State. Triumphal Progresses of Foreign Sovereigns in Renaissance Italy, 1494-1600*. 1986, VIII-240 pp. con 8 ill. f.t.

204. *Ugo Angelo Canello e gli inizi della filologia romanza in Italia*. 1987, 276 pp. con 4 tavv. f.t.

205. *Studi secenteschi*, vol. XXVII (1986). 1986, IV-348 pp.

206. DÉDÉYAN, C. *Diderot et la pensée anglaise*. 1986, IV-366 pp.

207. *La letteratura e i giardini*. 1987, 436 pp. con 9 tavv. f.t.

208. *Letteratura italiana e arti figurative*. 1988, 3 voll. di complessive VIII-1438 pp. con 60 ill. f.t.

209. *Studi secenteschi*, vol. XXVIII (1987). 1987, IV-332 pp. con 2 ill. f.t.

210. *Dante e la Bibbia*. Atti del convegno internazionale. 1988, 372 pp.

211. *Veronica Gàmbara e la poesia del suo tempo nell'Italia Settentrionale*. Atti del convegno. 1989, 442 pp.

212. *Studi di letteratura francese*, vol. XIII. 1987, 194 pp.

213. COLOMBO, A. *I «Riposi di Pindo». Studi su Claudio Achillini (1574-1640)*, 1988, 228 pp.

214. *Letteratura e storia meridionale. Studi offerti a Aldo Vallone*. 1989, 2 tomi di complessive XVI-960 pp. con 7 tavv. f.t.

215. SABBATINO, P. *La «Scienza» della scrittura. Dal progetto del Bembo al manuale*. 1988, 256 pp.

216. *Studi di letteratura francese*, vol. XIV. 1988, 144 pp.

217. PIRRO SCHETTINO, *Opere edite e inedite*. Edizione critica. 1989, 410 pp. con 4 tavv. f.t.

218. *Giorgio Pasquali e la filologia classica del '900*. Atti del convegno. 1988, VI-278 pp.

219. *Studi secenteschi*, vol. XXIX (1988). 1988, IV-328 pp.

220. LANDONI, E. *La teoria letteraria dei provenzali*. 1989, XXXIV-168 pp.

221. *Il meraviglioso, il verosimile tra antichità e medioevo*. 1989, 360 pp. con 5 tavv. f.t.

222. PROCACCIOLI, P. *Filologia ed esegesi dantesca nel Quattrocento. L'«Inferno» nel «Comento sopra la Comedia» di Cristoforo Landino*. 1989, 266 pp.

223. SANTARCANGELI, P. *Homo Ridens. Estetica, filologia, psicologia, storia del comico*. 1989, VI-452 pp.

224. *Filologia e critica dantesca. Studi offerti a Aldo Vallone*. 1989, XVI-660 pp. con 2 tavv. f.t.

225. *Dantismo russo e cornice europea*. 1989, 2 voll. indivisibili di XXXVI-880 pp. complessive.

226. *Studi di letteratura francese*, vol. XV. 1989, 284 pp. con 1 tav. f.t.

227. *Studi secenteschi*, vol. XXX (1989). 1989, IV-316 pp.

228. *Il tema della fortuna nella letteratura francese e italiana del Rinascimento. Studi in memoria di Enzo Giudici*. 1990, XX-550 pp. con 1 tav. f.t.

229. SEBASTIO, L. *Strutture narrative e dinamiche culturali in Dante e nel «Fiore»*. 1990, 320 pp.

230. *Studi di letteratura francese*, vol. XVI. 1990, 248 pp. con 1 tav. f.t.

231. *Studi di letteratura francese*, vol. XVII. 1990, 156 pp.

232. *Studi di letteratura francese*, vol. XVIII. 1990, 332 pp. con 1 tav. f.t.

233. DOZON, M. *Mythe et symbol dans la «Divine Comédie»*. 1991, XVI-634 pp.

234. VALLONE, A. *Strutture e modulazioni nei canti della «Divina Commedia»*. 1990, 226 pp.

235. COMOLLO, A. *Il dissenso religioso in Dante*. 1990, 154 pp.

236. BENDINELLI PREDELLI, M. *Alle origini del «Bel Gherardino»*. 1990, 362 pp.

237. GUERIN DALLE MESE, J. *Egypte: La mémoire et le rêve. Itineraires d'un voyage, 1320-1601*. 1990, 656 pp. con 7 tavv. f.t.

238. SORELLA, A. *Magia, lingua e commedia nel Machiavelli*. 1990, 264 pp.

239. *Studi secenteschi*, vol. XXXI (1990). 1990, XXVIII-296 pp. con 6 tavv. f.t.

240. *Miscellanea di studi in onore di Marco Pecoraro*. 1991. Vol. I: *Da Dante al Manzoni*, X-398 pp. con 7 tavv. f.t.; Vol. II: *Dal Tommaseo ai contemporanei*, IV-414 pp.

241. *Lingua e letteratura italiana nel mondo oggi*. 1991, 2 tomi di XVI-732 pp. complessive.

242. SABBATINO, P. *L'Eden della nuova poesia. Saggi sulla «Divina Commedia»*. 1991, 232 pp.

243. *Alfonso M. De Liguori e la società civile del suo tempo*. 1990, 2 tomi di VIII-682 pp. complessive.

244. *Famiglia e società nell'opera di Giovanni Verga*. 1991, VI-494 pp.

245. *Studi secenteschi*, vol. XXXII (1991). 1991, IV-332 pp. con 4 tavv. f.t.

246. HEIN, J. *Enigmaticité et messianisme dans la «Divine Comédie»*. 1992, II-654 pp.

247. SANGUINETI WHITE, L. *Dal detto alla figura. Le tragedie di Federico Della Valle*. 1992, 162 pp.

248. GROSSVOGEL, S. *Ambiguity and allusion in Boccaccio's Filocolo*. 1992, 254 pp.

249. *Studi di letteratura francese*, vol. XIX. 1992, 526 pp. con 4 ill. f.t. e figg. n.t.

250. PADOAN, G. *Il lungo cammino del «Poema sacro». Studi danteschi*. 1992, IV-310 pp.

251. *Studi secenteschi*, vol. XXXIII (1992). 1992, IV-210 pp. con 4 tavv. f.t.

252. ANKLI, R. *Morgante iperbolico. L'iperbole nel* Morgante *di Luigi Pulci*. 1993, 422 pp.

253. *Studi secenteschi*, vol. XXXIV (1993). 1993, IV-476 pp. con 1 tav. ripiegata f.t.

254. SABBATINO, P. *Giordano Bruno e la "mutazione" del Rinascimento*. 1993, 230 pp. con 6 figg. f.t. Ristampa 1998.

255. *Studi secenteschi*, vol. XXXV (1994). 1994, IV-286 pp. con 4 tavv. f.t.

256. *Studi di letteratura francese*, vol. XX. 1994, 294 pp. con 1 tav. f.t.

257. SABBATINO, P. - SCORRANO, L. - SEBASTIO, L. - STEFANELLI, R. *Dante e il Rinascimento. Rassegna bibliografica e studi in onore di Aldo Vallone*. 1994, 212 pp.

258. *Italo Svevo scrittore europeo*. A cura di N. Cacciaglia e L. Fava Guzzetta. 1994, VIII-574 pp.

259. SEBASTIO, L. *Il poeta e la storia. Una dinamica dantesca*. 1994, 264 pp.

260. *Le feste dei pastori del Rubicone per Napoleone I Re d'Italia*. Opera inedita a cura di A. Piromalli e T. Iermano. 1994, 152 pp.

261. *Studi secenteschi*. Vol. XXXVI (1995). 1995, IV-302 pp. con 6 tavv. f.t.

262. *Geografia, storia e poetiche del fantastico*. A cura di M. Farnetti. 1995, 244 pp. con 4 ill. f.t.

263. *Studi secenteschi*, vol. XXXVII (1996). 1996, IV-406 pp.

264. IERMANO, T. *Il melanconico in dormiveglia. Salvatore Di Giacomo*. 1995, 270 pp.

265. ARDISSINO, E. *L'«aspra tragedia». Poesia e sacro in Torquato Tasso*. 1996, 236 pp.

266. ZANGHERI, L. *Feste e apparati nella Toscana dei Lorena (1737-1859)*. 1996, 332 pp. con 115 ill. f.t.

267. *Letteratura e industria*. Atti del XV Congresso dell'Associazione internazionale per gli studi di lingua e letteratura italiana. 1997, 2 tomi di XVIII-1288 pp. complessive con 76 ill. f.t.

268. ANGIOLILLO, G. *La nuova frontiera della tanatologia. Le biografie della Commedia*. Vol. I: *Inferno*. 1996, 182 pp.

269. ANGIOLILLO, G. *La nuova frontiera della tanatologia. Le biografie della Commedia*. Vol. II: *Purgatorio*. 1996, 308 pp.

270. ANGIOLILLO, G. *La nuova frontiera della tanatologia. Le biografie della Commedia*. Vol. III: *Paradiso*. 1996, 270 pp.

271. *Studi secenteschi*. Vol. XXXVIII (1997). 1997, IV-444 pp.

272. BENPORAT, C. *Cucina italiana del Quattrocento*. 1996, 306 pp. con 4 figg. f.t. in b. e n. e 8 tavv. f.t. a colori. Ristampa 2001.

273. *Studi di letteratura francese. Rivista europea*, vol. XXI (1996). 1996, 238 pp. con 2 figg. n.t.

274. FRATNIK, M. *Enrico Pea et l'écriture du moi*. 1997, 402 pp.

275. MONTEVECCHI, F. *Il potere marittimo e le civiltà del Mediterraneo antico*. 1997, 596 pp. con 85 figg. n.t.

276. ROSSETTO, S. *Per la storia del giornalismo. Treviso dal XVII secolo all'unità*. 1996, 222 pp. con 10 tavv. f.t.

277. GIRARDI, R. *Incipitario della lirica meridionale e repertorio generale degli autori di lirica nati nel Mezzogiorno d'Italia (secolo XVI)*. 1996, 458 pp.

278. SABBATINO, P. *La bellezza di Elena. L'imitazione nella letteratura e nelle arti figurative del Rinascimento*. 1997, 270 pp. con 1 grafico n.t. e 12 tavv. f.t. Ristampa 2001.

279. PANICARA, V. *La nuova poesia di Giacomo Leopardi. Una lettura critica della* Ginestra. 1997, 148 pp.

280. *Torquato Tasso e la cultura estense*. A cura di G. Venturi, indice dei nomi e bibliografia generale a cura di A. Ghinato e R. Ziosi. 1999, 3 tomi di VIII-1462 pp. complessive con 101 ill. f.t.

281. GAVIOLI, E. *Filologia e nazione: l'«Archivum romanicum» nel carteggio inedito di Giulio Bertoni*. 1997, 202 pp. con 4 ill. f.t.

282. *Studi di letteratura francese. Rivista europea*, vol. XXII (1997). 1997, 330 pp.

283. *Studi secenteschi*. Vol. XXXIX (1998). 1998, IV-368 pp. con 4 tavv. f.t.

284. *Studi secenteschi*. Vol. XL (1999). 1999, IV-390 pp.

285. *Studi di letteratura francese. Rivista europea*, vol. XXIII (1998). «Lire le roman». 1998, 270 pp.

286. *Alfonso M. de Liguori e la civiltà letteraria del Settecento*. Atti del Convegno internazionale per il tricentenario della nascita del Santo (1696-1996). Napoli 20-23 ottobre 1997. A cura di P. Giannantonio. 1999, XX-476 pp.

287. *Leopardi e Bologna*. Atti del Convegno di studi per il Secondo Centenario Leopardiano (Bologna 18-19 maggio 1998). A cura di M. A. Bazzocchi. 1999, XVI-316 pp. con 4 tavv. f.t.

288. *Studi secenteschi*. Vol. XLI (2000). 2000, IV-502 pp.

289. *Studi di letteratura francese. Rivista europea*, vol. XXIV (1999). «L'estranéité». 1999, 246 pp.

290. SMITH, G. *The Stone of Dante and later florentine celebrations of the Poet*. 2000, X-72 pp. con 16 ill. f.t.

291. *L'immaginario contemporaneo*. Atti del Convegno letterario internazionale, Ferrara, 21-23 maggio 1999. A cura di R. Pazzi. 2000, XII-198 pp.

292. *The Poetics of Place. Florence Imagined*. Edited by I. Marchegiani Jones and T. Haeussler. 2001, XIV-220 pp.

293. LAWSON LUCAS, A. *La ricerca dell'ignoto. I romanzi d'avventura di Emilio Salgari*. Traduzione di S. Rizzardi e F. Rusciadelli. 2000, XVI-208 pp. con 1 tav. f.t.

294. *Il castello, il convento, il palazzo e altri scenari dell'ambientazione letteraria*. A cura di M. Cantelmo. 2000, VI-326 pp.

295. *Studi secenteschi*. Vol. XLII (2001). 2001, IV-472 pp. con 20 ill. f.t.

296. *Studi di letteratura francese. Rivista europea*, vol. XXV (2000). 2001, 192 pp.

297. *La lingua e le lingue di Machiavelli*. Atti del Convegno internazionale di studi, Torino 2-4 dicembre 1999. 2001, 352 pp.

298. *Studi secenteschi*. Vol. XLIII (2002). 2002, IV-372 pp. con 9 ill. f.t.

299. *Umanisti bellunesi fra Quattro e Cinquecento*. Atti del Convegno di Belluno, 5 novembre 1999. A cura di P. Pellegrini. 2001, XIV-296 pp. con 24 tavv. f.t.

300. SODINI, C. *L'Ercole tirreno. Guerra e dinastia medicea nella prima metà del '600*. 2001, VI-326 pp. con 16 tavv. f.t. in b. e n. e 9 a colori.

301. *Il tragico e il sacro dal Cinquecento a Racine*. Atti del Convegno internazionale, Torino e Vercelli, 14-16 ottobre 1999. A cura di D. Cecchetti e D. Dalla Valle. 2001, X-330 pp.

302. BENPORAT, C. *Feste e banchetti. Convivialità italiana fra Tre e Quattrocento*. 2001, 290 pp. con 12 tavv. f.t. a colori.

303. *Studi di letteratura francese. Rivista europea*, vol. XXVI (2001). «Théâtre et société au XVIIe siècle». 2002, 254 pp.

304. *La «liquida vertigine»*. Atti delle giornate di studio su Tommaso Landolfi. Prato, Convitto Nazionale Cicognini, 5-6 febbraio 1999. A cura di I. Landolfi. 2002, XXVI-266 pp.

305. *Studi secenteschi*. Vol. XLIV (2003). 2002, IV-340 pp. con 3 tavv. f.t.

306. LEUSHUIS, R. *Le Mariage et l''amitié courtoise' dans le dialogue et le récit bref de la Renaissance*. 2003, XIV-286 pp.

307. FRATNIK, M. *Paysages. Essai sur la description de Federico Tozzi*. 2002, XVI-182 pp.

308. *Alfieri e il suo tempo*. Atti del Convegno internazionale, Torino - Asti, 29 novembre - 1 dicembre 2001. A cura di M. Cerruti, M. Corsi, B. Danna. 2003, XII-488 pp. con 3 figg. n.t. e 5 tavv. f.t. di cui 4 a colori.

309. *Robert Davidsohn (1853-1937). Uno spirito libero tra cronaca e storia*. Tomo I: *Atti della giornata di studio*. Tomo II: *Gli scritti inediti*. Tomo III: *Catalogo della biblioteca*. A cura di W. Fastenrath Vinattieri e M. Ingendaay Rodio. 2003, XXX-812 pp. complessive con 1 fig. n.t. e 30 tavv. f.t.

310. *Studi di letteratura francese. Rivista europea*, vol. XXVII (2002). 2003, 286 pp.

311. *Il volto e gli affetti. Fisiognomica ed espressione nelle arti del Rinascimento*. Atti del Convegno di studi, Torino, 28-29 novembre 2001. A cura di A. Pontremoli. 2003, 314 pp. con 14 tavv. f.t.

312. SICA, P. *Modernist Forms of Rejuvenation. Eugenio Montale and T.S. Eliot*. 2003, X-156 pp.

313. *Studi secenteschi*. Vol. XLV (2004). 2004, IV-484 pp. con 6 tavv. f.t.

314. *Sabba da Castiglione (1480-1554). Dalle corti rinascimentali alla Commenda di Faenza*. Atti del Convegno, Faenza, 19-20 maggio 2000. A cura di A.R. Gentilini. 2004, X-496 pp. con 16 figg. n.t. e 54 tavv. f.t. di cui 6 a colori.

315. SABBATINO, P. *A l'infinito m'ergo. Giordano Bruno e il volo del moderno Ulisse*. 2003, XVI-212 pp. con 15 tavv. f.t.

316. MASTROIANNI, M. *Le Antigoni sofoclee del Cinquecento francese*. 2004, 264 pp.

317. *Francesco di Giorgio alla corte di Federico da Montefeltro*. Atti del Convegno internazionale di studi, Urbino, monastero di Santa Chiara, 11-13 ottobre 2001. A cura di F.P. Fiore. 2004, 2 tomi di complessive XXIV-710 pp. con 296 figg. n.t.

318. *Relazioni letterarie tra Italia e Penisola Iberica nell'epoca rinascimentale e barocca*. Atti del primo Colloquio Internazionale, Pisa, 4-5 ottobre 2002. A cura di S. Vuelta García. 2004, X-178 pp. con 2 figg. n.t.

319. BOZZOLA, S. *Tra Cinque e Seicento. Tradizione e anticlassicismo nella sintassi della prosa letteraria*. 2004, VIII-168 pp.

320. BALMAS, E. *Studi sul Cinquecento*. 2004, XXX-666 pp. con 11 figg. n.t. e 11 tavv. f.t.

321. *Studi di Letteratura francese. Rivista europea*, vol. XXVIII (2003). 2004, 138 pp.

322. FURLAN, F. *La donna, la famiglia, l'amore tra Medioevo e Rinascimento*. 2004, 122 pp.

323. ALFIERI, V. *Esquisse du Jugement Universel*. A cura di G. Santato. 2004, 128 pp. con 2 figg. n.t.

324. *Studi secenteschi*. Vol. XLVI (2005). 2005, IV-386 pp. con 13 tavv. f.t.

325. *Il Capitolo di San Lorenzo nel Quattrocento*. Convegno di studi, Firenze, 28-29 marzo 2003. A cura di P. Viti. 2006, XII-360 pp. con 8 tavv. f.t.

326. MARTELLOTTI, A. *I ricettari di Federico II. Dal «Meridionale» al «Liber de coquina»*. 2005, 284 pp.

327. FOSCOLO, U. *Dell'origine e dell'ufficio della letteratura. Orazione*. 2005, 172 pp.

328. RUGGIERO, R. *«Il ricco edificio». Arte allusiva nella Gerusalemme Liberata*. 2005, XXII-194 pp.

329. *Studi secenteschi*. Vol. XLVII (2006). 2006, IV-368 pp.

330. POZZI, M. - MATTIODA, E. *Giorgio Vasari storico e critico*. 2006, XXII-438 pp.

331. *Leonis Baptistae Alberti Descriptio Vrbis Romae*. Edizione critica di Jean-Yves Boriaud e Francesco Furlan. 2005, 164 pp. con 10 tavv. f.t.

332. *Resultanze in merito alla vita e all'opera di Piero Jahier. Saggi e materiali inediti*. A cura di F. Giacone. 2007, XII-368 pp. con 4 tavv. f.t.

333. CEVOLINI, A. *De arte excerpendi. Imparare a dimenticare nella modernità*. 2006, 460 pp. con 9 figg. n.t.

334. *Studi secenteschi*. Vol. XLVIII (2007). 2007, IV-432 pp.

335. MONTINARO, G. *L'epistolario di Ludovico Agostini. Riforma e utopia*. 2006, 294 pp.

336. *Il mito d'Arcadia. Pastori e amori nelle arti del Rinascimento*. Atti del Convegno internazionale di studi, Torino, 14-15 marzo 2005. A cura di D. Boillet e A. Pontremoli. 2007, XXII-266 pp. con 8 figg. n.t. e 14 tavv. f.t.

337. SEBASTIO, L. *Il Poeta tra Chiesa ed Impero. Una storia del pensiero dantesco*. 2007, 214 pp.

338. *Studi di letteratura francese. Rivista europea*, voll. XXIX-XXX (2004-2005). «Il viaggio francese in Italia». 2007, 226 pp. con 1 fig. n.t.

339. *I linguaggi dell'Altro. Forme dell'alterità nel testo letterario*. Atti del Convegno *I Linguaggi dell'Altro/altro*, Università di Lecce, 21-22 aprile 2005. A cura di A.M. Piglionica, C. Bacile di Castiglione, M.S. Marchesi. 2007, XXIV-228 pp. con 2 figg. n.t.

340. BENPORAT, C. *Cucina e convivialità italiana del Cinquecento*. 2007, 344 pp. con 16 tavv. f.t.

341. *Il cantare italiano fra folklore e letteratura*. Atti del Convegno internazionale di Zurigo, Landesmuseum, 23-25 giugno 2005. A cura di M. Picone e L. Rubini. 2007, XIV-528 pp. con 6 figg. n.t.

342. COVINO, S. *Giacomo e Monaldo Leopardi falsari trecenteschi. Contraffazione dell'antico, cultura e storia linguistica nell'Ottocento italiano*. 2009, I tomo XVI-328 pp. II tomo VI-392 pp. con 2 tavv. f.t.

343. *Studi secenteschi*. Vol. XLIX (2008). 2008, IV-434 pp. con 8 tavv. f.t.

344. *Traduzioni, imitazioni, scambi tra Italia e Portogallo nei secoli*. Atti del primo Colloquio internazionale, Pisa, 15-16 ottobre 2004. A cura di M. Lupetti. 2008, X-172 pp. con 2 figg. n.t. e 15 tavv. f.t. di cui 12 a colori.

345. *L'identità italiana ed europea tra Sette e Ottocento*. A cura di A. Ascenzi e L. Melosi. 2008, XIV-184 pp. con 5 figg. n.t.

346. WILSON, R. *Prophecies and prophecy in Dante's* Commedia. 2007, X-228 pp.

347. *Writing Relations: American Scholars in Italian Archives. Essays for Franca Petrucci Nardelli and Armando Petrucci*. Edited by D. Shemek and M. Wyatt. 2008, XII-242 pp. con 13 figg. n.t. e 2 tavv. f.t.

348. IOLY ZORATTINI, P. *I nomi degli altri. Conversioni a Venezia e nel Friuli Veneto in età moderna.* Con prefazione di M. Massenzio. 2008, XX-388 pp. con 4 tavv. f.t.

349. URRARO, R. *Giacomo Leopardi. le donne, gli amori.* 2008, VIII-378 pp.

350. RABBONI, R. *Speculare sodo, ragionar sostanzioso. Studi sull'abate Conti.* 2008, X-336 pp.

351. TIOZZO, E. *La letteratura italiana e il premio Nobel. Storia critica e documenti.* 2008, VIII-358 pp. con 29 tavv. f.t.

352. CAPECCHI, G. - MARZI, M. G. - SALADINO, V. *I granduchi di Toscana e l'antico. Acquisti, restauri, allestimenti.* 2008, VIII-342 pp. con 78 tavv. f.t. di cui 16 a colori.

353. *Studi secenteschi.* Vol. L (2009). 2008, IV-346 pp. con 2 figg. n.t. e 13 tavv. f.t.

354. *In assenza del re. Le reggenti dal secolo XIV al secolo XVII (Piemonte ed Europa).* A cura di F. Varallo. 2008, XXXII-610 pp. con es. mus. n.t. e 7 tavv. f.t.

355. CELLI, C. *Il carnevale di Machiavelli.* 2009, IV-218 pp.

356. *Iacopo Sannazaro. La cultura napoletana nell'Europa del Rinascimento.* Convegno internazionale di studi, Napoli, 27-28 marzo 2006. A cura di P. Sabbatino. 2009, VIII-430 pp. con 5 figg. n.t. e 14 tavv. f.t.

357. *«La bourse des idées du monde». Malaparte e la Francia.* Atti del Convegno internazionale di studi su Curzio Malaparte, Prato-Firenze, 8-9 novembre 2007. A cura di M. Grassi. 2008, XII-234 pp.

358. *La metafora in Dante.* A cura di M. Ariani. 2009, VI-286 pp.

359. COEN, P. *Il mercato dei quadri a Roma nel diciottesimo secolo. La domanda, l'offerta e la circolazione delle opere in un grande centro artistico europeo.* I. Con una prefazione di E. Castelnuovo. II. Appendice documentaria. 2010, LX-816 pp. con 32 tavv. f.t. a colori.

360. *Saggi di letteratura architettonica, da Vitruvio a Winckelmann.* I. A cura di F.P. Di Teodoro. 2009, VI-372 pp. con 67 figg. n.t. e 21 tavv. f.t.

361. *Don Giovanni nelle riscritture francesi e francofone del Novecento.* Atti del Convegno internazionale di Vercelli, 16-17 ottobre 2008. A cura di M. Mastroianni. 2009, XIII-330 pp.

362. MARCHESI, M.S. *Eliot's Perpetual Struggle. The Language of Evil in* Murder in the Cathedral. 2009, XXXVIII-144 pp.

363. *Studi di letteratura francese. Rivista europea,* voll. XXXI-XXXII (2006-2007). «Dictionnaires et écrivains». In preparazione.

364. *Studi secenteschi.* Vol. LI (2010). 2010, IV-394 pp.

365. *Saggi di letteratura architettonica, da Vitruvio a Winckelmann.* II. A cura di L. Bertolini. 2009, VI-254 pp. con 66 figg. n.t. e 5 tavv. f.t. a colori.

366. FRENQUELLUCCI, C. *Dalla Mancha a Siena al Nuovo Mondo. Don Chisciotte nel teatro di Girolamo Gigli.* 2010, XVI-334 pp.

367. *Giuseppe Ungaretti - Jean Lescure. Carteggio (1951-1966).* A cura di R. Gennaro. 2010, XXVI-252 pp.

368. TESTA, F. *Winckelmann e l'architettura antica.* In preparazione.

369. *Saggi di letteratura architettonica, da Vitruvio a Winckelmann.* III. A cura di H. Burns, F.P. Di Teodoro e G. Bacci. 2010, VI-392 pp. con 126 figg. n.t.

370. BARSELLA, S. *In the Light of the Angels: Angelology and Cosmology in Dante's* Divina Commedia. 2010, XVI-214 pp.

371. DURANTE, E. - MARTELLOTTI, A. *«Giovinetta peregrina». La vera storia di Laura Peperara e Torquato Tasso.* 2010, VI-352 pp. con 2 tavv. f.t. a colori, con CD contenente "Madrigali per Laura Peperara".

372. SQUILLACE, G. *Il profumo nel mondo antico. Con la prima traduzione italiana del «Sugli odori» di Teofrasto.* Prefazione di L. Villoresi. 2010, XX-282 pp. con 8 tavv. f.t. a colori.

373. CEROCCHI, M. *Funzioni semantiche e metatestuali della musica in Dante, Petrarca e Boccaccio.* In preparazione.

374. *La Ronde. Giostre, esercizi cavallereschi e loisir in Francia e Piemonte fra Medioevo e Ottocento.* Atti del Convegno internazionale. Pinerolo, 15-17 giugno 2006. A cura di F. Varallo. In preparazione.

375. *La parola e l'immagine. Studi in onore di Gianni Venturi.* A cura di M. Ariani, A. Bruni. A. Dolfi, A. Gareffi. In preparazione.

376. BERTELLI, S. *La tradizione della «Commedia»: dai manoscritti al testo. I. I codici trecenteschi (entro l'antica vulgata) conservati a Firenze.* In preparazone.

377. *Nascita della storiografia e organizzazione dei saperi.* Atti del Convegno internazionale di studi, Torino, 20-22 maggio 2009. A cura di E. Mattioda. In preparazione.

378. *Studi secenteschi.* Vol. LII (2011). In preparazione.

Serie II: Linguistica

1. Spitzer, L. *Lexikalisches aus dem Katalanischen und den übrigen ibero-romanischen Sprachen*. 1921. VIII-162 pp.

2. Gamillscheg, E. und Spitzer, L. *Beiträge zur romanischen Wortbildungslehre*. 1921, 230 pp., 3 cc.

3. [Schuchardt, U.]. *Miscellanea linguistica dedic. a Ugo Schuchardt per il suo 80° anniv*. 1922, 121 pp., 2 cc.

4. Bertoldi, V. *Un ribelle nel regno dei fiori (I nomi romanzi del «colchicum autunnale L.» attraverso il tempo e lo spazio)*. 1923, VIII-224 pp. con ill.

5. Bottiglioni, G. *Leggende e tradizioni di Sardegna*. (Testi dialettali in grafia fonetica). 1922. (esaurito)

6. Onomastica - I. Paul Aebischer, *Sur la formation des noms de famille dans le canton de Fribourg (Suisse)*. - II. Dante Olivieri, *I cognomi della Venezia Euganea. Saggio di uno studio storico-etimologico*. 1924, 272 pp.

7. Rohlfs, G. *Grichen und Romanen in Unteritalien Ein Beitrag zur Geschichte der unteritalienischen Grazität*. 1923. (esaurito)

8. *Studi di dialettologia alto italiana*. - I. Gualzata, M. *Di alcuni nomi locali del Bellinzonese e Locarnese*. - II. Bläuer-Rini, A. *Giunte al «vocabolario di Bormio»*. 1924, 166 pp.

9. Pascu, G. *Romänische elemente in den Balkansprachen*. 1924, IV-112 pp.

10. Farinelli, A. *Marrano* (Storia di un vituperio). 1925, X-80 pp.

11. Bertoni, G. *Profilo storico del dialetto di Modena. (Con appendice di «Giunte al Vocabolario Modenese»)*. 1925, 88 pp.

12. Bartoli, M. *Introduzione alla neolinguistica* (Principi - Scopi - Metodi), 1926. (esaurito)

13. Migliorini, B. *Dal nome proprio al nome comune*. 1927, VI-358 pp. con LXXVIII pp. di supplemento. Seconda ristampa 1999.

14. Keller, O. *La flexion du verbe dans le patois genevois*. 1928, XXVIII-216 pp., 1 c. ripiegata.

15. Spotti, L. *Vocabolarietto anconitano-italiano*. 1929. (esaurito)

16. Wagner, M. L. *Studien über den sardischen Wortschatz. (I. Die Familie - II. Der menschliche Körper)*. 1930, XVI-156 pp., 15 cc.

17. Soukup, R. *Les causes et l'évolution de l'abreviation des pronoms personnels régimes en ancien français*. 1932, 130 pp.

18. Rheinfelder, H. *Kultsprache und Profansprache in den romanischen Ländern*. 1933. (esaurito)

19. Flagge, L. *Provenzalisches Alpenleben in den Hochtälern des Verdon und der Bléone. Ein Beitrag zur Volkskunde des Basses-Alpes*. 1935. (esaurito)

20. Sainéan, L. *Autour des sources indigènes*. Etudes d'étymologie française et romaine. 1935. (esaurito)

21. Seifert, E. *Tenere «Haben» im Romanischen*. 1935, 122 pp., 4 tavv.

22. Tagliavini, C. *L'Albanese di Dalmazia*. 1937. (esaurito)

23. Bosshard, H. *Saggio di un glossario dell'antico Lombardo*. 1938. (esaurito)

24. Vidos, B. E. *Storia delle parole marinaresche italiane passate in francese*. 1939. (esaurito)

25. Alessio, G. *Saggio di Toponomastica calabrese*. 1939. (esaurito)

26. Folena, G. *La crisi linguistica del 400 e l'«Arcadia» di I. Sannazaro*. 1952. (esaurito)

27. *Miscellanea di studi linguistici in ricordo di Ettore Tolomei*. 1953. (esaurito)

28. Vidos, B. E. *Manuale di linguistica romanza*. Prima edizione italiana completamente aggiornata dall'Autore. 1959, XXIV-440 pp. Terza ristampa 1975.

29. Ruggieri, R. *Saggi di linguistica italiana e italo-romanza*. 1962, 242 pp.

30. Mengaldo, P. V. *La lingua del Boiardo lirico*. 1963, VIII-380 pp.

31. Vidos, B. E. *Prestito espansione e migrazione dei termini tecnici nelle lingue romanze e non romanze*. 1965, VIII-424 pp., 3 ill.

32. Altieri Biagi, M. L. *Galileo e la terminologia tecnico-scientifica*. 1965. (esaurito)

33. Polloni, A. *Toponomastica romagnola*, Prefazione di Carlo Tagliavini. 1966. Ristampa 2002.

34. Ghiglieri, P. *La grafia del Machiavelli studiata negli autografi*. 1969, IV-364 pp.

35. *Linguistica matematica e calcolatori*. A cura di A. Zampolli. 1973, XX-670 pp.

36. *Computational and mathematical linguistics*. Vol. I. A cura di A. Zampolli e N. Calzolari. 1977, 2 voll. di XLVI-796 pp. complessive.

37. *Computational and mathematical linguistics*. Vol. II. A cura di A. Zampolli e N. Calzolari. 1980, 2 voll. di VIII-906 pp. complessive.

38. Semerano, G. *Le origini della cultura europea. Rivelazioni della linguistica storica*. 1984, 2 voll. di LXX-956 pp. complessive. Ristampa 2002.

39. *Fonologia etrusca, fonetica toscana. Il problema del sostrato*. 1983, 204 pp. con 1 tav. f.t.

40. La Stella, T. E. *Dizionario storico di deonomastica*. 1984, 236 pp.

41. Rando, G. *Dizionario degli anglicismi nell'italiano contemporaneo*. 1987, XLII-256 pp.

42. *Lessicografia, filologia e critica*. 1986, 204 pp.

43. Semerano, G. *Le origini della cultura europea*. Vol. II. *Dizionari etimologici. Basi semitiche delle lingue Indeuropee*. I tomo: *Dizionario della lingua greca*. II tomo: *Dizionario della lingua latina*. 1994, 2 voll. di C-726 pp. complessive. III ristampa 2007.

44. Scavuzzo, C. *Studi sulla lingua dei quotidiani messinesi di fine Ottocento*. 1988, 208 pp.

45. Agostiniani, L. - Hjordt-Vetlesen, O. *Lessico etrusco cronologico e topografico dai materiali del «Thesaurus Linguae Etruscae»*. 1988, XXXVI-224 pp.

46. O'Connor, D. *A history of Italian and English bilingual dictionaries*. 1990, 188 pp.

47. Boselli, P. *Dizionario di toponomastica bergamasca e cremonese*. 1990, 346 pp.

48. Delmay, B. *Usi e difese della lingua*. 1990, 154 pp. con 1 tav. f.t.

49. Catenazzi, F. *L'italiano di Svevo. Fra scrittura pubblica e scrittura privata*. 1994, 202 pp.

50. Facchetti, G. M. *Frammenti di diritto privato etrusco*. 2000, 116 pp.

51. *La scrittura professionale: ricerca, prassi, insegnamento.* Atti del I Convegno di studi, Perugia, Università per Stranieri, 23-25 ottobre 2000. A cura di S. Covino. 2001, XXIV-454 pp. con 29 figg. n.t. e 1 pieghevole.

52. Leone, A. *Conversazioni sulla lingua italiana.* 2002, 160 pp.

53. Natella, P. *La parola 'Mafia'.* 2002, 172 pp.

54. Facchetti, G. M. *Appunti di morfologia etrusca. Con un'appendice sulla questione delle identità genetiche dell'etrusco.* 2002, 160 pp.

55. Facchetti, G. M. - Negri, M. *Creta minoica. Sulle tracce delle più antiche scritture d'Europa.* 2003, 200 pp. con 21 figg. n.t. e 2 tavv. f.t.

56. Prandi, M. - Gross, G. - De Santis, C. *La finalità. Strutture concettuali e forme d'espressione in italiano.* 2005, 366 pp.

57. Ferguson, R. *A Linguistic History of Venice.* 2007, 322 pp. con 3 figg. n.t.